PLINY
NATURAL HISTORY

IX

LIBRI XXXIII–XXXV

PLINY

"NATURAL HISTORY,"

WITH AN ENGLISH TRANSLATION
IN TEN VOLUMES

VOLUME IX

LIBRI XXXIII–XXXV

BY

H. RACKHAM, M.A.

FELLOW OF CHRIST'S COLLEGE, CAMBRIDGE

CAMBRIDGE, MASSACHUSETTS
HARVARD UNIVERSITY PRESS
LONDON
WILLIAM HEINEMANN LTD
MCMLII

Printed in Great Britain

INTRODUCTION

Books XXXIII, XXXIV, and XXXV of Pliny's Natural History contain interesting accounts of minerals and mining and of the history of art.

Mr. H. Rackham left when he died a translation in typescript with a few footnotes. The Latin text has been prepared by Prof. E. H. Warmington, who has also added the critical notes on this text, many footnotes on the translation, and marginal helps. Some parts of the translation were completely re-written by him. The sections on Greek art were read and criticised by Prof. T. B. L. Webster, to whom thanks are now duly rendered.

The *codices* cited in the critical notes on the Latin text are as follows: B = Bambergensis; *cd. Leid. Voss.* = V; *cd. Leid. Lips.* = F; *cd. Chiffl(etianus)* = f; *cd. Flor. Ricc.* = R; *cd. Par. Lat.* 6797 = d; *cd. Par.* 6801 = h; *cd. Vind.* CCXXXIV = a; *cd. Tolet.* = T.

CONTENTS

PLINY :
NATURAL HISTORY

BOOK XXXIII

PLINII NATURALIS HISTORIAE

LIBER XXXIII

I. Metalla nunc ipsaeque opes et rerum pretia dicentur, tellurem intus exquirente cura multiplici modo, quippe alibi divitiis foditur [1] quaerente vita aurum, argentum, electrum, aes, alibi deliciis gemmas et parietum lignorumque [2] pigmenta, alibi temeritati ferrum, auro etiam gratius inter bella caedesque. persequimur omnes eius fibras vivimusque super excavatam, mirantes dehiscere aliquando aut intremescere illam, ceu vero non hoc indignatione sacrae 2 parentis exprimi possit. imus in viscera et in sede manium opes quaerimus, tamquam parum benigna fertilique qua calcatur; [3] et inter haec minimum remediorum gratia scrutamur, quoto enim cuique fodiendi causa medicina est? quamquam et hoc summa sui parte tribuit ut fruges, larga facilisque in

---

[1] ante quippe *transferendum aut* fodinis *vel* e fodinis *legendum coni. Mayhoff.*

[2] lignorumque (*vel* signorumque) *Mayhoff*: pictorum *Detlefsen*: digitorumque *cdd.* (*recte?*): deliciis parietum digitorumque gemmas et pigmenta *Bergk.*

[3] *V.ll.* caecatur, cecatur, secatur.

---

[a] *Electrum*, properly amber, was a word applied to an alloy of gold and silver, and also to native argentiferous gold, because of their resemblance in colour.

PLINY: NATURAL HISTORY

BOOK XXXIII

I. Our topic now will be metals, and the actual *Metals.* resources employed to pay for commodities— resources diligently sought for in the bowels of the earth in a variety of ways. For in some places the earth is dug into for riches, when life demands gold, silver, silver-gold *a* and copper, and in other places for luxury, when gems and colours for tinting walls and beams are demanded, and in other places for rash valour, when the demand is for iron, which amid warfare and slaughter is even more prized than gold. We trace out all the fibres of the earth, and live above the hollows we have made in her, marvel- ling that occasionally she gapes open or begins to tremble—as if forsooth it were not possible that this may be an expression of the indignation of our holy parent! We penetrate her inner parts and seek for riches in the abode of the spirits of the departed, as though the part where we tread upon her were not sufficiently bounteous and fertile. And amid all this the smallest object of our searching is for the sake of remedies for illness, for with what fraction of mankind is medicine the object of this delving? Although medicines also earth bestows upon us on her surface, as she bestows corn, bountiful and

3

3 omnibus, quaecumque prosunt. illa nos peremunt,
illa nos ad inferos agunt, quae occultavit atque
demersit, illa, quae non nascuntur repente, ut [1] mens
ad inane evolans reputet, quae deinde futura sit
finis omnibus saeculis exhauriendi eam, quo usque
penetratura avaritia. quam innocens, quam beata,
immo vero etiam delicata esset vita, si nihil aliunde
quam supra terras concupisceret, breviterque, nisi [2]
quod secum est!

4 II. Eruitur aurum et chrysocolla iuxta, ut pretiosior videatur, nomen ex auro custodiens. parum
enim erat unam vitae invenisse pestem, nisi in pretio
esset auri etiam sanies. quaerebat argentum avaritia; boni consuluit interim invenisse minium ruben-
tisque terrae excogitavit usum. heu prodiga ingenia,
quot modis auximus pretia rerum! accessit ars
picturae, et aurum argentumque caelando carius
fecimus. didicit homo naturam provocare. auxere
et artem vitiorum inritamenta; in poculis libidines
5 caelare iuvit ac per obscenitates bibere. abiecta
deinde sunt haec ac [3] sordere coepere, ut [4] auri
argentique nimium fuit. murrina ex eadem tellure
et crystallina effodimus, quibus pretium faceret ipsa
fragilitas. hoc argumentum opum, haec vera luxu-

---

[1] repente ut *Mayhoff*: ut repente *aut* repente.
[2] *V.l.* haberetque non nisi.
[3] ac *Mayhoff*: abs *B*: et *rell.*
[4] ut *Mayhoff*: et.

---

[a] Χρυσοκόλλα, 'gold-solder.' This is malachite, basic copper
carbonate.
[b] See §§ 111 ff.
[c] Or possibly finest agate.

generous as she is in all things for our benefit! The
things that she has concealed and hidden underground, those that do not quickly come to birth,
are the things that destroy us and drive us to the
depths below; so that suddenly the mind soars
aloft into the void and ponders what finally will be
the end of draining her dry in all the ages, what will
be the point to which avarice will penetrate. How
innocent, how blissful, nay even how luxurious life
might be, if it coveted nothing from any source but
the surface of the earth, and, to speak briefly,
nothing but what lies ready to her hand!

II. Gold is dug out of the earth and in proximity to *Gold.*
it gold-solder, which still retains in Greek a name [a]
derived from gold, so as to make it appear more
precious. It was not enough to have discovered one
bane to plague life, without setting value even on
the corrupt humours of gold! Avarice was seeking
for silver, but counted it a gain to have discovered
cinnabar [b] by the way, and devised a use to make of
red earth. Alas for the prodigality of our inventive-
ness! In how many ways have we raised the prices
of objects! The art of painting has come in addition,
and we have made gold and silver dearer by means
of engraving! Man has learnt to challenge nature
in competition! The enticements of the vices have
augmented even art: it has pleased us to engrave
scenes of licence upon our goblets, and to drink
through the midst of obscenities. Afterwards these
were flung aside and began to be held of no account,
when there was an excess of gold and silver. Out
of the same earth we dug supplies of fluor-spar [c]
and crystal, things which their mere fragility
rendered costly. It came to be deemed the proof

5

riae gloria existimata est, habere quod posset statim
perire totum. nec hoc fuit satis. turba gemmarum
potamus et zmaragdis teximus calices, ac temulentiae
causa tenere Indiam iuvat. aurum iam accessio est.
6 III. utinamque posset e vita in totum abdicari
[sacrum fame, ut celeberrimi auctores dixere][1]
proscissum conviciis ab optimis quibusque et ad perniciem vitae repertum, quanto feliciore aevo, cum res
ipsae permutabantur inter sese, sicut et Troianis
temporibus factitatum Homero credi convenit! ita
enim, ut opinor, commercia victus gratia inventa.[2]
7 alios coriis boum, alios ferro captivisque res[3] empti-
tasse tradit. quare,[4] quamquam ipse iam mirator
auri,[5] pecore[6] aestimationes rerum ita fecit, ut c
boum arma aurea permutasse Glaucum diceret cum
Diomedis armis VIIII boum. ex qua consuetudine
multa legum antiquarum pecore constat etiam
Romae.
8 IV. Pessimum vitae scelus fecit qui primus induit
digitis, nec hoc quis fecerit traditur. nam de Pro-
metheo omnia fabulosa arbitror, quamquam illi
quoque ferreum anulum dedit antiquitas vinculumque
id, non gestamen, intellegi voluit. Midae quidem
anulum, quo circumacto habentem nemo cerneret,

---

[1] *Seclusit J. Müller.*
[2] inventa *cd. Par.* 6801 : invecta *rell.*
[3] res *Detlefsen* : merum *coni. Ian* : vinum *Bergk* : rebus
codd. (rerus *B*[1] : rerū *B*[2]).
[4] quare *Mayhoff* : quã *B*[2] : quã *B*[1] : *om. rell.*
[5] *V.l.* miratus auri.
[6] pecore *Mayhoff* : pec : *B*[1] pec *B*[2] : et *aut om. rell.*

---

[a] The MSS. here insert a clause ('accursed by hunger, as
very famous writers have said') adapted from Virgil's famous
phrase in *Aen.* III. 57 : 'auri sacra fames.'

of wealth, the true glory of luxury, to possess something that might be absolutely destroyed in a moment. Nor was this enough: we drink out of a crowd of precious stones, and set our cups with emeralds, we take delight in holding India for the purpose of tippling, and gold is now a mere accessory. III. And would that it [a] could be entirely banished from life, reviled and abused as it is by all the worthiest people, and only discovered for the ruin of human life—how far happier was the period when goods themselves were interchanged by barter, as it is agreed we must take it from Homer [b] to have been the custom even in the days of Troy. That in my view was the way in which trade was discovered, to procure the necessities of life. Homer relates how some people used to make their purchases with ox-hides, others with iron and captives, and consequently, although even Homer himself [c] was already an admirer of gold, he reckoned the value of goods in cattle, saying that Glaucus exchanged gold armour worth 100 beeves with that of Diomede worth 9 beeves. And as a result of this custom even at Rome a fine under the old laws is priced in cattle.

IV. The worst crime against man's life was *Gold rings.* committed by the person who first put gold on his fingers, though it is not recorded who did this, for I deem the whole story of Prometheus mythical, although antiquity assigned to him also an iron ring, and intended this to be understood as a fetter, not an ornament. As for the story of Midas's ring, which when turned round made its wearer invisible,

[b] Homer, *Il.* VII. 472 ff.
[c] *Il.* VI. 234–6.

7

9 quis non etiam fabulosiorem fateatur? manus et prorsus sinistrae maximam auctoritatem conciliavere auro, non quidem Romanae, quarum [1] in more ferrei erant ut [2] virtutis bellicae insigne.

De regibus Romanis non facile dixerim. nullum habet Romuli in Capitolio statua nec praeter Numae Serviique Tullii alia ac ne Lucii quidem Bruti. hoc in Tarquiniis maxime miror, quorum e Graecia fuit origo, unde hic anulorum usus venit, quamquam 10 etiam nunc Lacedaemone ferreo utuntur. sed a Prisco Tarquinio omnium primo filium, cum in praetextae annis occidisset hostem, bulla aurea donatum constat, unde mos bullae duravit, ut eorum, qui equo meruissent, filii insigne id haberent, ceteri lorum; et ideo miror Tarquinii eius statuam sine anulo esse. quamquam et de nomine ipso ambigi video. Graeci a digitis appellavere, apud nos prisci ungulum vocabant, postea et Graeci et nostri 11 symbolum. longo certe tempore ne senatum quidem

---

[1] romani quorum *cd. Par.* 6801.
[2] ut *Hardouin*: et.

---

[a] *Sinistra* suggests 'unlucky,' 'sinister.'
[b] One of the two first consuls (509 B.C.), not a king.
[c] The white toga with a purple border worn by free-born boys at Rome until they were declared to be of age, between 14 and 16, and assumed the *toga pura* or *virilis*, the white woollen cloak of the Roman citizen.
[d] Δακτύλιος, from δάκτυλος.

who would not admit this to be more mythical still? It was the hand and what is more the left [a] hand, that first won for gold such high esteem; not indeed a Roman hand, whose custom it was to wear an iron ring as an emblem of warlike valour.

As to the Roman kings I find it hard to make a statement. The statue of Romulus in the Capitol has nothing, nor has any other king's statue excepting those of Numa and Servius Tullius, and not even that of Lucius Brutus.[b] I am especially surprised at this in the case of the Tarquins, who came originally from Greece, the country from which this fashion in rings came, although an iron ring is worn in Sparta even at the present day. But of all, Tarquinius Priscus, it is well known, first presented his son with a golden amulet when while still of an age to wear the bordered robe [c] he had killed an enemy in battle; and from that time on the custom of the amulet has continued as a distinction to be worn by the sons of those who have served in the cavalry, the sons of all others only wearing a leather strap. Owing to this I am surprised that the statue of that Tarquin has no ring. All the same, I notice that there is a difference of opinion even about the actual word for a ring. The Greek name [d] for it is derived from the word meaning a finger; with ourselves, in early days it was called 'ungulus,' [e] but afterwards both our people and the Greeks give it the name of 'symbolum.' [f] For a long period indeed, it is quite

Rings and brooches in the Roman monarchic period;

---

[e] The word survives in fragments of early poetry.

[f] Greek σύμβολον, originally meaning two parts of a coin or other object broken in half to serve as a means of identification because tallying when put together; and so the word was used to denote any token or symbol.

Romanum habuisse aureos manifestum est, siquidem
iis tantum, qui legati ad exteras gentes ituri essent,
anuli publice dabantur, credo, quoniam ita exterorum
honoratissimi intellegebantur. neque aliis uti mos fuit
quam qui ex ea causa publice accepissent, volgoque sic
triumphabant et, cum corona ex auro Etrusca sus-
tineretur a tergo, anulus tamen in digito ferreus erat
aeque triumphantis et servi prae [1] se coronam sus-
12 tinentis. sic triumphavit de Iugurtha C. Marius
aureumque non ante tertium consulatum sumpsisse
traditur. ii quoque, qui ob legationem acceperant
aureos, in publico tantum utebantur iis, intra domos
vero ferreis, quo argumento etiam nunc sponsae
muneris vice ferreus anulus mittitur, isque sine
gemma. equidem nec Iliacis temporibus ullos fuisse
anulos video. nusquam certe Homerus dicit, cum
et codicillos missitatos epistularum gratia indicet et
conditas arcis vestes ac vasa aurea argenteaque et
eas colligatas [2] nodi, non anuli, nota. sortiri quoque
contra provocationes duces non anulis tradit, fabricae
etiam deum fibulas et alia muliebris cultus, sicut

---

[1] prae se *Mayhoff* (*qui et* forte *coni.*): fortasse *cdd.*: *del.*
edd. vett.

[2] consignatas *coni. Mayhoff.*

---

[a] *I.e.* by the future bridegroom; it was called *anulus
pronubus.*
[b] *Il.* VI. 168-9.
[c] *Od.* VIII. 424, 438-41, 443, 447.
[d] *Il.* VII. 171, 175; κλῆροι 'lots,' were moulded out of clay,
but it is not said that they were marked with the chiefs' signet
rings.
[e] Hephaestus. See *Il.* XVIII. 400 ff.

clear, not even members of the Roman senate had and in the Republican period.
gold rings, inasmuch as rings were bestowed officially
on men about to go as envoys to foreign nations,
and on them only, the reason no doubt being that
the most highly honoured foreigners were recognized
in this way. Nor was it the custom for any others
to wear a gold ring than those on whom one had been
officially bestowed for the reason stated; and
customarily Roman generals went in triumph without one, and although a Tuscan crown of gold was
held over the victor's head from behind, nevertheless
he wore an iron ring on his finger when going in
triumph, just the same as the slave holding the
crown in front of himself. This was the way in
which Gaius Marius celebrated his triumph over
Jugurtha, and it is recorded that he did not assume Jan. 1, 104 B.C. 103 B.C.
a gold ring till his third tenure of the consulship.
Those moreover who had been given gold rings
because they were going on an embassy only wore
them in public, but in their homes wore iron rings;
this is the reason why even now an iron ring and
what is more a ring without any stone in it is sent [a]
as a gift to a woman when betrothed. Indeed I Rings in Homer.
do not find that any rings were worn in the Trojan
period; at all events Homer nowhere mentions
them, although he shows that tablets [b] used to be
sent to and fro in place of letters, and that clothes
and gold and silver vessels were stored away in
chests [c] and were tied up with signet-knots, not
sealed with signet-rings. Also he records the chiefs
as casting lots about meeting a challenge from the
enemy without using signet-rings [d]; and he also says
that the god [e] of handicraft in the original period
frequently made brooches and other articles of

inaures, in primordio factitasse, sine mentione
13 anulorum. et quisquis primus instituit, cunctanter
id fecit: laevis manibus latentibusque induit, cum,
si honos securus fuisset, dextra fuerit ostentandus.
quodsi impedimentum potuit in eo aliquod intellegi,
etiam serioris usus argumentum est:[1] maius in laeva
fuisset, qua scutum capitur. est quidem apud
eundem Homerum virorum crinibus aurum inplexum;
ideo nescio an prior usus feminis coeperit.

14   V. Romae ne fuit quidem aurum nisi admodum
exiguum longo tempore. certe cum a Gallis capta
urbe pax emeretur, non plus quam mille pondo effici
potuere. nec ignoro MM pondo auri perisse Pompeii
III consultu e Capitolini Iovis solio a Camillo ibi
condita, et ideo a plerisque existimari MM pondo
collata. sed quod accessit, ex Gallorum praeda fuit
detractumque ab iis in parte captae urbis delubris—
15 Gallos cum auro pugnare solitos Torquatus indicio
est—; apparet ergo Gallorum templorumque tan-
tundem nec amplius fuisse. quod quidem in

---

[1] est *cdd.* (est et *B*): esset *Mayhoff.*

---

[a] *Il.* XVII. 52.
[b] Victor over the Gauls in 390 B.C.
[c] T. Manlius obtained this surname from the gold necklace,
torques, which he took from a Gaul whom he slew in single
combat in 360 B.C.

feminine finery like earrings—without mentioning finger-rings. And whoever first introduced them did so with hesitation, and put them on the left hand, which is generally hidden by the clothes, whereas it would have been shown off on the right hand if it had been an assured distinction. And if this might possibly have been thought to involve some interference with the use of the right hand, there is the proof of more modern custom; it would have also been more inconvenient to wear it on the left hand, which holds the shield. Indeed it is also stated, by Homer [a] again, that men wore gold plaited in their hair and consequently I cannot say whether the use of gold originated from women.

V. At Rome for a long time gold was actually not to be found at all except in very small amounts. At all events when peace had to be purchased after the capture of the City by the Gauls, not more than a thousand pounds' weight of gold could be produced. I am aware of the fact that in Pompey's third consulship there was lost from the throne of Jupiter of the Capitol two thousand pounds' weight of gold that had been stored there by Camillus,[b] which led to a general belief that 2000 pounds was the amount that had been accumulated. But really the additional sum was part of the booty taken from the Gauls, and it had been stripped by them from the temples in the part of the city which they had captured—the case of Torquatus [c] shows that the Gauls were in the habit of wearing gold ornaments in battle; therefore it appears that the gold belonging to the Gauls and that belonging to the temples did not amount to more than that total; and this in fact was taken to be the meaning contained in the

Roman wealth in gold.

390 B.C.

52 B.C.

13

augurio intellectum est, cum Capitolinus duplum
reddidisset.

Illud quoque obiter indicari convenit—etiam[1] de
anulis sermonem repetivimus—, aedituum custodiae
eius conprehensum fracta in ore anuli gemma statim
16 expirasse et indicium ita extinctum. ergo vel[2] maxime
MM tantum pondo, cum capta est Roma, anno CCCLXIIII
fuere, cum iam capitum liberorum censa essent
C̄LĪI DLXXIII. in eadem post annos CCCVII, quod ex
Capitolinae aedis incendio ceterisque omnibus delubris
C. Marius filius Praeneste detulerat,[3] X̄IIĪI[4] pondo,
quae sub eo titulo in triumpho transtulit Sulla et
argenti V̄I.[5] idem ex reliqua omni victoria pridie
transtulerat auri pondo X̄V, argenti p. C̄XV.

17 VI. Frequentior autem usus anulorum non ante
Cn. Flavium Anni filium deprehenditur. hic namque
publicatis diebus fastis, quos populus a paucis
principum cotidie petebat, tantam gratiam plebei
adeptus est—libertino patre alioqui genitus et ipse
scriba Appi Caeci, cuius hortatu exceperat eos dies
consultando adsidue sagaci ingenio promulgaratque

---

[1] etiam *B*: quoniam *rell.*: et iam *Ian*: etenim *coni.*
Mayhoff.

[2] vel *L. C. Purser*: ut.

[3] detulerat, erant *coni. Mayhoff.*

[4] X̄IIĪI *Ian*: XIIII *B*: XIII *rell.* (tredecim milia *cd. Par*
6801).

[5] V̄I *Ian*: VI.

---

[a] The reference has not been explained.

[b] It contained a poison, cf. § 26.

[c] By the Gauls in 390 B.C.

[d] Appius Claudius, censor in 312 B.C. and builder of the
Appian Way.

augury,[a] when Jupiter the God of the Capitol had repaid twofold.

Also, as we began on this topic from the subject of rings, it is suitable incidentally to point out that the official in charge of the temple of Jupiter of the Capitol when he was arrested broke the stone [b] of his ring between his teeth and at once expired, so putting an end to any possibility of proving the theft. It follows that there was only 2,000 lbs. weight of gold at the outside when Rome was captured in its 364th year,[c] although the census showed there were already 152,573 free citizens. From the same city 307 years later the gold that Gaius Marius 82 B.C. the younger had conveyed to Palestrina from the conflagration of the temple of the Capitol and from all the other shrines amounted to 14,000 lbs., which with a placard above it to that effect was carried along in his triumphal procession by Sulla, as well as 81 B.C. 6,000 lbs. weight of silver. Sulla had likewise on the previous day carried in procession 15,000 lbs. of gold and 115,000 lbs. of silver as the proceeds of all the rest of his victories.

VI. It does not appear that rings were in more *More about* common use before the time of Gnaeus Flavius son *rings in the* of Annius. It was he who first published the dates *period.* for legal proceedings, which it had been customary 305-4 B.C. for the general public to ascertain by daily enquiry from a few of the leading citizens; and this won him such great popularity with the common people—he was also the son of a liberated slave and himself a clerk to Appius Caecus,[d] at whose request he had by dint of natural shrewdness through continual observation picked out those days and published them—that he was appointed a curule

—, ut aedilis curulis crearetur cum Q. Anicio Prae-
nestino, qui paucis ante annis hostis fuisset, prae-
teritis C. Poetilio et Domitio, quorum patres consules
18 fuerant. additum Flavio, ut simul et tribunus plebei
esset, quo facto tanta indignatio exarsit, ut anulos
abiectos in antiquissimis reperiatur annalibus. fallit
plerosque quod tum et equestrem ordinem id fecisse
arbitrantur; etenim adiectum hoc quoque sed et
phaleras positas propterque hoc nomen equitum adiectum est, anulosque depositos a nobilitate in annales
relatum est, non a senatu universo. hoc actum P.
19 Sempronio L. Sulpicio cos. Flavius vovit aedem
Concordiae, si populo reconciliasset ordines, et, cum
ad id pecunia publice non decerneretur, ex multaticia
faeneratoribus condemnatis aediculam aeream fecit
in Graecostasi, quae tunc supra comitium erat,
inciditque in tabella aerea factam eam aedem
20 ccIIII annis post Capitolinam dedicatam. id a.[2]
ccccxxxxvIIII a condita urbe gestum est et primum
anulorum vestigium extat; promiscui autem usus
alterum secundo Punico bello, neque enim aliter

---

[1] propterque hoc *Rackham*: *alii alia*: propterque.
[2] id a. (anno) *C. F. W. Müller*: ita.

---

[a] Probably in the war with the twelve tribes of Etruria,
who were conquered by Fabius at Lake Vadimo, 310 B.C.
[b] A platform Greek and, later, any foreign envoys could
watch proceedings. It was later placed in the Forum.

ædile as a colleague of Quintus Anicius of Palestrina, who a few years previously had been an enemy at war with Rome,[a] while Gaius Poetilius and Domitius, whose fathers had been consuls, were passed over. Flavius had the additional advantage of being tribune of the plebs at the same time. This caused such an outburst of blazing indignation that we find in the oldest annals ' rings were laid aside.' The common belief that the Order of Knighthood also did the same on this occasion is erroneous, inasmuch as the following words were also added : ' but also harness-bosses were put aside as well '; and it is because of this clause that the name of the Knights has been added; and the entry in the annals is that the rings were laid aside by the nobility, not by the entire Senate. This occurrence took place in the consulship of Publius Sempronius 305 B.C. and Lucius Sulpicius. Flavius made a vow to erect a temple to Concord if he succeeded in effecting a reconciliation between the privileged orders and the people; and as money was not allotted for this purpose from public funds, he drew on the fine-money collected from persons convicted of practising usury to erect a small shrine made of bronze on the Graecostasis,[b] which at that date stood above the Assembly-place, and put on it an inscription engraved on a bronze tablet that the shrine had been constructed 204 years after the consecration of the Capitoline temple. This event took place in the 449th year from the foundation of the city, and 305 B.C. is the earliest evidence to be found of the use of rings. There is however a second piece of evidence for their being commonly worn at the time of the Second Punic War, as had this not been the case it

potuisset trimodia anulorum illa Carthaginem ab Hannibale mitti. inter Caepionem quoque et Drusum ex anulo in auctione venali intimicitiae coepere,
21 unde origo socialis belli et exitia rerum. ne tunc quidem omnes senatores habuere. utpote cum memoria avorum multi praetura quoque functi in ferreo consenuerint—sicut Calpurnium et Manilium, qui legatus C. Marii fuerit Iugurthino bello, Fenestella tradit, et multi L. Fufidium illum, ad quem Scaurus de vita sua scripsit—, in Quintiorum vero familia aurum ne feminas quidem habere mos fuerit, nullosque omnino maior pars gentium hominumque, etiam qui sub imperio nostro degunt, hodieque habeat. non signat oriens aut Aegyptus etiam nunc litteris contenta solis.

22 Multis hoc modis, ut cetera omnia, luxuria variavit gemmas addendo exquisiti fulgoris censuque opimo digitos onerando, sicut dicemus in gemmarum volumine, mox et effigies varias caelando, ut alibi ars, alibi materia esset in pretio. alias dein gemmas violari nefas putavit ac, ne quis signandi causam in
23 anulis esse intellegeret, solidas induit. quasdam

---

a This was after the battle of Cannae in 216 B.C. Livy says 3½ pecks, Florus says 2.
b The so-called Social War, 91–88 B.C.
c This statement is untrue.

18

would not have been possible for the three [a] pecks
of rings as recorded to have been sent by Hannibal
to Carthage. Also it was from a ring put up for
sale by auction that the quarrel between Caepio and
Drusus began which was the primary cause of the war
with the allies [b] and the disasters that sprang from it.
Not even at that period did all members of the
senate possess gold rings, seeing that in the memory
of our grandfathers many men who had even held
the office of prætor wore an iron ring to the end of
their lives—for instance, as recorded by Fenestella,
Calpurnius and Manilius, the latter having been
lieutenant-general under Gaius Marius in the war 112–106 B.C.
with Jugurtha, and, according to many authorities,
the Lucius Fufidius to whom Scaurus dedicated his
Autobiography—while another piece of evidence
is that in the family of the Quintii it was not even
customary for the women to have a gold ring, and
that the greater part of the races of mankind, and
even of the people who live under our empire and at
the present day, possess no gold rings at all. The
East and Egypt do not [c] seal documents even now,
but are content with a written signature.

This fashion like everything else luxury has diversi- *Methods of*
fied in numerous ways, by adding to rings gems of *wearing*
exquisite brilliance, and by loading the fingers with *rings.*
a wealthy revenue (as we shall mention in our book XXXVII.
on gems) and then by engraving on them a variety *2 sqq.*
of devices, so that in one case the craftsmanship and
in another the material constitutes the value. Then
again with other gems luxury has deemed it sacrilege
for them to undergo violation, and has caused them
to be worn whole, to prevent anybody's imagin-
ing that people's finger-rings were intended for

19

vero neque ab ea parte, qua digito occultantur,[1]
auro clusit aurumque millis[2] lapillorum vilius fecit.
contra vero multi nullas admittunt gemmas auroque
ipso signant. id Claudii Caesaris principatu repertum. nec non et servitia iam ferrum auro cingunt—
alia per sese mero auro decorant—, cuius licentiae
origo nomine ipso in Samothrace id institutum
declarat.

24 Singulis primo digitis geri mos fuerat, qui sunt
minimis proximi. sic in Numae et Servi Tullii statuis
videmus. postea pollici proximo induere, etiam in[3]
deorum simulacris, dein iuvit et minimo dare.
Galliae Brittanniaeque medio dicuntur usae. hic
nunc solus excipitur, ceteri omnes onerantur, atque
25 etiam privatim articuli minoribus aliis. sunt qui
uni tantum minimo congerant, alii vero et huic
tantum unum, quo signantem signent. conditus
ille, ut res rara et iniuria usus indigna, velut e
sacrario promitur, ut et unum in minimo digito
habuisse pretiosioris in recondito supellectilis osten-
tatio sit. iam alii pondera eorum ostentant. aliis
plures quam unum gestare labor est, alii bratteas

---

[1] *V.l.* quae digito occultatur.
[2] millis *Ian*: micis *Gronov*: vilibus *coni. Urlichs*: millib.
B^1: milibus.
[3] in *add. Mayhoff*.

---

[a] Or possibly 'that finger-rings contained a motive for
sealing documents,' *i.e.*, that people were ready to seal deeds
in order to show off the engraved stones.

[b] Slaves wore iron rings, a symbol of captivity.

[c] *I.e.* they were called Samothracian rings.

sealing documents! [a] Some gems indeed luxury has left showing in the gold even on the side of the ring that is hidden by the finger, and has cheapened the gold with collars of little pebbles. But on the contrary many people do not allow any gems in a signet-ring, and seal with the gold itself; this was a fashion invented when Claudius Cæsar was emperor. A.D. 41-54. Moreover even slaves nowadays encircle the iron of their rings [b] with gold (other articles all over them they decorate with pure gold), an extravagance the origin of which is shown by its actual name [c] to have been instituted in Samothrace.

It had originally been the custom to wear rings on one finger only, the one next the little finger; that is how we see them on the statues of Numa and Servius Tullius. Afterwards people put them on the finger next the thumb, even in the case of statues of the gods, and next it pleased them to give the little finger also a ring. The Gallic Provinces and the British Islands are said to have used the middle finger. At the present day this is the only finger exempted, while all the others bear the burden, and even each finger-joint has another smaller ring of its own. Some people put all their rings on their little finger only, while others wear only one ring even on that finger, and use it to seal up their signet ring, which is kept stored away as a rarity not deserving the insult of common use, and is brought out from its cabinet as from a sanctuary; thus even wearing a single ring on the little finger may advertise the possession of a costlier piece of apparatus put away in store. Some again show off the weight of their rings; others count it hard work to wear more than one; and others consider that filling the gold tinsel

infercire leviore materia propter casum tutius gemmarum sollicitudini putant, alii sub gemmis venena
cludunt, sicut Demosthenes summus Graeciae orator,
26 anulosque mortis gratia habent. denique vel[1] plurima opum scelera anulis fiunt. quae fuit illa vita
priscorum, qualis innocentia, in qua nihil signabatur!
nunc cibi quoque ac potus anulo vindicantur a rapina.
hoc profecere mancipiorum legiones, in domo turba
externa ac iam servorum quoque causa nomenclator
adhibendus. aliter apud antiquos singuli Marcipores
Luciporesve dominorum gentiles omnem victum in
promiscuo habebant, nec ulla domi a domesticis
27 custodia opus erat. nunc rapiendae conparantur
epulae pariterque qui rapiant eas, et claves quoque
ipsas signasse non est satis. gravatis somno aut
morientibus anuli detrahuntur, maiorque vitae ratio
circa hoc instrumentum esse coepit, incertum a quo
tempore. videmur tamen posse in externis auctoritatem eius rei intellegere circa Polycraten Sami
tyrannum, cui dilectus ille anulus in mare abiectus
capto relatus est pisce, ipso circiter ccxxx urbis

[1] vel *Bergk* : ut.

---

[a] Plutarch, *Vit. Demosth.* 29 reports a statement that
Demosthenes always carried a poison in a bracelet on his arm,
and that he killed himself with it to avoid falling into the
hands of Antipater of Macedon, 322 B.C.

[b] *I.e.* documents are forged and sealed with faked signet-
rings.

[c] He was put to death *c.* 515 B.C. by the Persian Oroetes.

of the circle with a lighter material, in case of their
dropping, is a safer precaution for their anxiety
about their gems; others enclose poisons underneath
the stones in their rings, as did Demosthenes,[a] the
greatest orator of Greece, and they wear their rings
as a means of taking their own lives. Finally, a
very great number of the crimes connected with
money are carried out by means of rings.[b] To think
what life was in the days of old, and what innocence
existed when nothing was sealed! Whereas nowadays even articles of food and drink have to be
protected against theft by means of a ring: this is
the progress achieved by our legions of slaves—a
foreign rabble in one's home, so that an attendant
to tell people's names now has to be employed even
in the case of one's slaves! This was not the way
with by-gone generations, when a single servant
for each master, a member of his master's clan,
Marcius's boy or Lucius's boy, took all his meals
with the family in common, nor was there any need
of precautions in the home to keep watch on the
domestics. Nowadays we acquire sumptuous viands
only to be pilfered and at the same time acquire
people to pilfer them, and it is not enough to keep
our keys themselves under seal: while we are fast
asleep or on our death-beds, our rings are slipped off
our fingers; and the prevailing system of our lives
has begun to centre round that portable chattel,
though when this began is doubtful. Still it seems
we can realize the importance this article possesses
abroad in the case of the tyrant of Samos, Polycrates,
who flung his favourite ring into the sea and had it
brought back to him inside a fish which had been
caught: Polycrates himself was put to death [c]

28 nostrae annum interfecto. celebratior quidem usus cum faenore coepisse debet. argumento est consuetudo volgi, ad sponsiones etiamnum [1] anulo exiliente, tracta ab eo tempore, quo nondum erat arra velocior, ut plane adfirmare possimus nummos ante apud nos, mox anulos coepisse. de nummis paulo post dicetur.

29 VII. Anuli distinxere alterum ordinem a plebe, ut semel coeperant esse celebres, sicut tunica ab anulis senatum. quamquam et hoc sero, vulgoque purpura latiore tunicae usos invenimus etiam praecones, sicut patrem L. Aelii Stilonis Praeconini ob id cognominati. sed anuli plane tertium ordinem mediumque plebei et patribus inseruere, ac quod antea militares equi nomen dederant, hoc nunc pecuniae indices tribuunt. nec pridem id factum.

30 divo Augusto decurias ordinante maior pars iudicum in ferreo anulo fuit iique non equites, sed iudices vocabantur. equitum nomen subsistebat in turmis equorum publicorum. iudicum quoque non nisi quattuor decuriae fuere primo, vixque singula milia in decuriis inventa sunt, nondum provinciis ad hoc

----

[1] etiam nunc *coni. Mayhoff.*

----

[a] ' Son of the herald.'
[b] *Eques.*

about the 230th year of the city of Rome. Still the ^{523 B.C.}
employment of a signet-ring must have begun to be
much more frequent with the introduction of usury.
This is proved by the custom of the lower classes,
among whom even at the present day a ring is
whipped out when a contract is being made; the
habit comes down from the time when there was
as yet no speedier method of guaranteeing a bargain,
so we can safely assert that with us money began
first and signet-rings came in afterwards. About
money we shall speak rather later.

VII. As soon as rings began to be commonly *Wearing of*
worn, they distinguished the second order from the *rings by the*
commons, just as a tunic distinguished the senate *Equestrian*
from those who wore the ring, although this distinc- *order.*
tion also was only introduced at a late date, and
we find that a wider purple stripe on the tunic was
commonly worn even by heralds, for instance the
father of Lucius Aelius Stilo Praeconinus, who
received his surname *a* from his father's office. But
wearing rings clearly introduced a third order, intermediate between the commons and the senate, and
the title *b* that had previously been conferred by the
possession of a war-horse is now assigned by money
rates. This however is only a recent introduction:
when his late lamented Majesty Augustus made
regulations for the judicial panels the majority of the
judges belonged to the iron ring class, and these
used to be designated not Knights but Justices;
the title of Knights remained with the cavalry
squadrons mounted at the public charge. Of the
Justices also there were at the first only four panels,
and in each panel scarcely a thousand names were
to be found, as the provinces had not yet been

munus admissis, servatumque in hodiernum est, ne
31 quis e novis civibus in iis iudicaret. decuriae quoque
ipsae pluribus discretae nominibus fuere, tribunorum
aeris et selectorum et iudicum. praeter hos etiamnum nongenti vocabantur ex omnibus electi ad
custodiendas suffragiorum cistas in comitiis. et
divisus hic quoque ordo erat superba usurpatione
nominum, cum alius se nongentum, alius selectum,
alius tribunum appellaret.

32 VIII. Tiberii demum principatu [1] nono anno in
unitatem venit equester ordo, anulorumque aucto-
ritati forma constituta est C. Asinio Pollione C.
Antistio Vetere cos. anno urbis conditae DCCLXXV,
quod miremur, futtili paene de causa, cum C. Sulpicius Galba, iuvenalem famam apud principem popina-
rum poenis aucupatus, questus esset in senatu, volgo
institores eius culpae defendi anulis. hac de causa
constitutum, ne cui ius esset nisi qui ingenuus ipse,
⟨ingenuo⟩ [2] patre, avo paterno HS $\overline{CCCC}$ census fuisset
et lege Iulia theatrali in quattuordecim ordinibus
33 sedisset. postea gregatim insigne id adpeti coeptum.

---

[1] principatu *B* : principatus *rell.*
[2] ingenuus ipse ingenuo *Detlefsen* : qui ingenuus ipse *aut* cui ingenuo ipsi.

---

[a] Originally it seems officials (tribuni aerarii) who collected the property-tax from Roman citizens (until 167 B.C.), and paid the soldiers out of a special fund. But in the first century B.C. they appear as an *ordo* in the state next below the *equites.*
[b] Tiberius.
[c] *I.e.* the gold ring of the Order of Knighthood, whose members often practised banking, tax-farming and other businesses.
[d] The financial standing of an *eques.*

admitted to this duty; and the regulation has survived to the present day that nobody newly admitted to citizenship shall serve as a justice on one of the panels. The panels themselves also were distinguished by various designations, as consisting of Tribunes of the Money,[a] Selected Members and Justices. Moreover beside these there were those styled the Nine Hundred, selected from the whole body as keepers of the ballot-boxes at elections. And the proud adoption of titles had made divisions in this order also, one person styling himself a member of the Nine Hundred, another one of the Select, another a Tribune.

VIII. Finally in the ninth year in office of the Emperor Tiberius the Order of Knights was united A.D. 14-37. into a single body; and in the Consulship of Gaius A.D. 22. Asinius Pollio and Gaius Antistius Vetus, in the 775th year since the foundation of Rome, a regulation was established authorizing who should wear rings; the motive for this, a thing that may surprise us, was virtually the futile reason that Gaius Sulpicius Galba had made a youthful effort to curry favour with the emperor [b] by enacting penalties for keeping eating-houses and had made a complaint in the senate that peddling tradesmen when charged with that offence commonly protected themselves by means of their rings.[c] Consequently a rule was made that nobody should have this right except one who was himself a free-born man whose father and father's father had been free-born also, and who had been rated as the owner of 400,000 sesterces [d] and had been entitled under the Julian law as to the theatre to sit in the fourteen front rows of seats. Sub-sequently people began to apply in crowds for this

27

propter haec discrimina C. princeps decuriam quintam
adiecit, tantumque enatum est fastus, ut, quae sub
divo Augusto impleri non potuerant, decuriae non
capiant eum ordinem, passimque ad ornamenta ea
etiam servitute liberati transiliant, quod antea numquam erat factum, quoniam [1] ferreo anulo et equites
iudicesque intellegebantur. adeoque id promiscuum
esse coepit, ut apud Claudium Caesarem in censura
eius unus ex equitibus Flavius Proculus cccc ex ea
causa reos postularet. ita dum separatur ordo ab
34 ingenuis, communicatus est cum servitiis. iudicum
autem appellatione separare eum ordinem primi
omnium instituere Gracchi discordi popularitate in
contumeliam senatus, mox debellata auctoritas
nominis vario seditionum eventu circa publicanos
substitit et aliquamdiu tertiae sortis viri publicani
fuere. M. Cicero demum stabilivit equestre nomen
in consulatu suo Catilinianis rebus, ex eo ordine pro-
fectum se celebrans eiusque vires peculiari popularitate quaerens. ab illo tempore plane hoc tertium
corpus in re p. factum est, coepitque adici senatui

---

[1] quoniam *Mayhoff*: q̅m̅ in.

---

[a] In fact C. Gracchus, tribune 123–2 B.C.

mark of rank; and in consequence of the disputes thus occasioned the Emperor Gaius Caligula added A.D. 37-41. a fifth panel, and so much conceit has this occasioned that the panels which under his late lamented Majesty Augustus it had not been possible to fill will not hold that order, and there are frequent cases of men who are actually liberated slaves making a leap over to these distinctions, a thing that previously never occurred, since the iron ring was the distinguishing mark even of knights and judges. And the thing began to be so common that during the censorship of the Emperor Claudius a member A.D. 48. of the Order of Knighthood named Flavius Proculus laid before him information against 400 persons on this ground, so that an order intended to distinguish the holder from other men of free birth has been shared with slaves. It was the Gracchi[a] who first instituted the name of Justices or Judges as the distinguishing name of that order of knights— seditiously currying favour with the people in order to humiliate the senate; but subsequently the importance of the title of Knight was swamped by the shifting currents of faction, and came down to be attached to the farmers of public revenues, and for some time these revenue officers constituted the third rank in the state. Finally Marcus Cicero, thanks to the Catilinarian affair, during his consulship 63 B.C. put the title of knighthood on a firm footing, boasting that he himself sprang from that order, and winning its powerful support by methods of securing popu- larity that were entirely his own. From that time onward the Knighthood definitely became a third element in the state, and the name of the Equestrian Order came to be added to the formula 'The

29

populoque Romano et equester ordo. qua de causa et nunc post populum scribitur, quia novissime coeptus est adici.

35 IX. Equitum quidem etiam nomen ipsum saepe variatum est, in iis quoque, qui id ab equitatu trahebant. celeres sub Romulo regibusque sunt appellati, deinde flexuntes, postea trossuli, cum oppidum in Tuscis citra Volsinios p. $\overline{\text{VIIII}}$ sine ullo peditum adiumento cepissent eius vocabuli, idque duravit

36 ultra C. Gracchum. Iunius certe, qui ab amicitia eius Gracchanus appellatus est, scriptum reliquit his verbis: Quod ad equestrem ordinem attinet, antea trossulos vocabant, nunc equites vocant ideo, quia non intellegunt trossulos nomen quid valeat, multosque pudet eo nomine appellari. et causam, quae supra indicata est, exponit invitosque etiamnum [1] tamen trossulos vocari.

37 X. Sunt adhuc aliquae non omittendae in auro differentiae. auxilia quippe et externos torquibus aureis donavere, at cives non nisi argenteis, praeterque armillas civibus dedere, quas non dabant externis.

38 XI. Iidem, quo magis miremur, coronas ex auro dedere et civibus. quis primus donatus sit ea, non

[1] etiamtum *coni. Mayhoff.*

---

[a] But in fact the regular order of words was senate, equites, Roman people.

[b] Trossum or Trossulum; there are still remains of a town at Trosso, two miles from Monte Fiascone in Tuscany.

Senate and People of Rome.' This is the reason why it is even now written after [a] ' People,' because it was the latest addition introduced.

IX. Indeed the very name of the Knights has itself frequently been altered, even in the case of those who derived the title from the fact of their serving as cavalry. Under Romulus and the Kings they were called the Celeres, then the Flexuntes and afterwards the Trossuli, because of their having without any assistance from infantry captured a town of that name [b] in Tuscany nine miles this side of Volsinii; and the name survived till after the time of Gaius Gracchus. At all events in the writings left by Junius, who owing to his friendship with Gaius Gracchus was called Gracchanus, these words occur: ' So far as concerns the Equestrian Order they were previously called the Trossuli, but are now simply designated the Cavalry, because people do not know what the word Trossuli means and many of them are ashamed of being called by that name.' He goes on to explain the reason above indicated, and says that they were even in his time still called Trossuli, though they did not wish to be.

X. There are some additional particulars in regard to gold which must not be omitted. For instance our authorities actually bestowed gold necklaces on foreign soldiers, but only awarded silver ones to Roman citizens, and what is more they gave bracelets to citizens, which it was not their custom to give to foreigners. *Necklaces.*

XI. But at the same time, as is even more surprising, they gave crowns of gold even to citizens. Who was the first person to receive one I have not *Crowns of gold.*

inveni equidem; quis primus donaverit, a L. Pisone traditur: A. Postumius dictator apud lacum Regillum castris Latinorum expugnatis eum, cuius maxime opera capta essent. hanc coronam ex praeda is dedit II l.,[1] item L. Lentulus consul Servio Cornelio Merendae Samnitum oppido capto, sed hic quinque librarum; trium[2] Piso Frugi filium ex privata pecunia donavit eamque coronam testamento ei praelegavit.

39 XII. Deorum honoris causa in sacris nihil aliud excogitatum est quam ut auratis cornibus hostiae, maiores dumtaxat, immolarentur. sed in militia quoque in tantum adolevit haec luxuria, ut M. Bruti e Philippicis campis epistulae reperiantur frementis fibulas tribunicias ex auro geri. ita, Hercules? idem enim tu, Brute, mulierum pedibus aurum gestatum[3] tacuisti. et nos sceleris arguimus illum, qui primus auro dignitatem per anulos fecit! habeant in lacertis iam quidem et viri, quod ex Dardanis venit—itaque et Dardanium vocabatur; 40 viriolae Celtice dicuntur, viriae Celtiberice—; habeant feminae in armillis digitisque totis, collo, auribus, spiris; discurrant catenae circa latera et in secreto

---

[1] dedit II l. (*i.e.* librarum) *Mayhoff*: dedit *cdd.* (dedit l. *B*[1]).
[2] trium *add. Hardouin coll. Val. Max.* IV.3.10.
[3] gestatum *B, cd. Colb.*: gestari *rell.*

myself been able to ascertain, but Lucius Piso records who was the first person to bestow one, namely the dictator Aulus Postumius, who when the camp of the Latins at Lake Regillus had been taken by storm awarded a gold crown to the soldier who had been chiefly responsible for taking the place. In this case the crown which he bestowed was made of gold taken from the booty captured, and weighed two pounds. Also Lucius Lentulus as consul awarded a gold crown to Servius Cornelius Merenda after the taking of a town belonging to the Samnites, but Servius's crown weighed five pounds; while Piso Frugi bestowed on his son one weighing three pounds out of his personal resources, leaving it to him by will as a specific legacy. 497 B.C.

275 B.C.

XII. As a mark of honour to the gods at sacrifices no other means has been devised but to gild the horns of the victims to be immolated, at all events of full-grown animals. But in military service also this form of luxury has grown to such dimensions that we find a letter of Marcus Brutus sent from the Plains of Philippi expressing his indignation at the brooches made of gold that were worn by the tribunes. Really I must protest! Why, even you, Brutus, did not mention the gold worn on their feet by women, and we accuse of crime the man who first conferred dignity on gold by using gold rings! Let even men nowadays wear gold bracelets—called 'Dardania' because the fashion came from the Dardani—the Celtic name for them is 'viriolae' and the Celtiberian 'viriae'; let women have gold in their bracelets and covering their fingers and on their neck, ears and tresses, let gold chains run at random round their waists; and let little bags of

Further remarks about gold.

42 B.C.

33

margaritarum sacculi e collo dominarum auro pen-
deant, ut in somno quoque unionum conscientia
adsit: etiamne pedibus induetur atque inter stolam
plebemque hunc medium feminarum equestrem
ordinem faciet? honestius viri paedagogis id damus,

41 balineasque dives puerorum forma convertit.[1] iam
vero et Harpocraten statuasque Aegyptiorum numinum in digitis viri quoque portare incipiunt. fuit et
alia Claudii principatu differentia insolens iis, quibus
admissiones[2] liberae[3] ius[4] dedissent[5] imaginem
principis ex auro in anulo gerendi, magna criminum
occasione, quae omnia salutaris exortus Vespasiani
imperatoris abolevit aequaliter publicando principem.
de anulis aureis usuque eorum hactenus sit dictum.

42 XIII. Proximum scelus fuit eius, qui primus ex
auro denarium signavit, quod et ipsum latet auctore
incerto. populus Romanus ne argento quidem signato ante Pyrrhum regem devictum usus est.
libralis—unde etiam nunc libella dicitur et dupondius
—adpendebatur assis; quare aeris gravis poena
dicta, et adhuc expensa in rationibus dicuntur, item

---

[1] converrit *J. Müller.*
[2] admissiones *Mommsen*: admissionem *cdd. pler.*: admis-
sion//i//s *B²*.
[3] liberae *B*: liberti *rell.*
[4] ius *Lips*: eius.
[5] dedisset *coni. Ian.*

---

[a] *I.e.* gold ornaments on the sandal-straps.
[b] Said to have been the Egyptian god of silence.
[c] *I.e.* committed against the welfare of mankind. The
worst crime was the introduction of gold rings, § 8.
[d] Equal in value to 25 silver denarii.
[e] The *as* was reduced in weight in the 1st Punic War or
soon after.
[f] A piece worth two *asses.*

pearls hang invisible suspended by gold chains from their lady owners' neck, so that even in their sleep they may retain the consciousness of possessing gems: but are even their feet to be shod with gold,[a] and shall gold create this female Order of Knighthood, intermediate between the matron's robe and the common people? Much more becomingly do we men bestow this on our page-boys, and the wealthy show these lads make has quite transformed the public baths! But nowadays even men are beginning to wear on their fingers a representation of Harpocrates [b] and figures of Egyptian deities. In the time of the Emperor Claudius there was also another unusual distinction, belonging to those whose rights of free access to the presence had given them the privilege of wearing a gold likeness of the emperor on a ring, this affording a great opportunity for informations; but all of this was however entirely abolished by the opportune rise to power of the Emperor Vespasian, by making the emperor equally accessible to all. Let this suffice for a discussion of the subject of gold rings and their employment. {.margin A.D. 41-54. ... A.D. 69-79.}

XIII. Next in degree was the crime committed by the person who first coined a gold denarius,[d] a crime which itself also is hidden and its author unknown. The Roman nation did not even use a stamped silver coinage before the conquest of King Pyrrhus. The *as* weighed one pound—hence the term still in use, ' little pound ' [e] and ' two pounder ' [f]; this is the reason why a fine is specified in ' heavy bronze,' [g] and why in book-keeping outlay is still designated as ' sums weighed out,' and likewise {.margin c Roman coinage in three metals. ... 275 B.C.}

[g] On *aes,* see XXXIV. 1, note.

43 inpendia et dependere, quin et militum stipendia,
hoc est stipis pondera, dispensatores, libripendes, qua
consuetudine in iis emptionibus, quae mancipi sunt,
etiam nunc libra interponitur. Servius rex primus
signavit aes. antea rudi usos Romae Timaeus tradit.
signatum est nota pecudum, unde et pecunia appel-
lata. maximus census $\overline{\text{cxx}}$ assium fuit illo rege, et
ideo haec prima classis.

44 Argentum signatum anno urbis ccccLxxxv,[1] Q.
Ogulnio C. Fabio cos., quinque annis ante primum
Punicum bellum. et placuit denarium pro x libris
aeris valere, quinarium pro v, sestertium pro dupon-
dio ac semisse. librale autem pondus aeris inminutum
est bello Punico primo, cum inpensis res p. non
sufficeret, constitutumque ut asses sextantario pondere
ferirentur. ita quinque partes lucri factae, dissolu-
45 tumque aes alienum. nota aeris eius fuit ex altera
parte Ianus geminus, ex altera rostrum navis, in
triente vero et quadrante rates. quadrans antea
teruncius vocatus a tribus unciis. postea Hannibale

---

[1] ccccLxxxv *Cellarius*: ccccLxxxxv *B*: dLxxxv *rell.*

---

[a] *I.e.* an *as.*

interest as 'weighed on account' and paying as
'weighing down,' and moreover it explains the
terms 'soldiers' stipend,' which means 'weights of
heaped money,' and the words for accountants and
paymasters that mean 'weighers' and 'pound-
weighers,' and owing to this custom in purchases that
deal with all larger personal property, even at the
present day, an actual pair of 'pound'-scales is
introduced. King Servius was the first to stamp a *Traditional dates 578–534 B.C.*
design on bronze; previously, according to Timaeus,
at Rome they used raw metal. The design stamped
on the metal was an ox or a sheep, *pecus*, which is the
origin of the term 'pecunia.' The highest assess-
ment of one man's property in the reign of Servius was
120,000 *as-pieces*, and consequently that amount of
property was the standard of the first class of citizens.

Silver was first coined in the 485th year of the city, 269–8 B.C.
in the consulship of Quintus Ogulnius and Gaius
Fabius, five years before the first Punic War. It
was decided that the value of a denarius should be
ten pounds of bronze, that of a half-denarius five
pounds, that of a sesterce two pounds and a half.
The weight of a standard pound [a] of bronze was
however reduced during the first Punic War, when
the state could not meet its expenditure, and it was
enacted that the *as* should be struck weighing two
ounces. This effected a saving of five sixths, and the
national debt was liquidated. The design of this
bronze coin was on one side a Janus facing both
ways and on the other the ram of a battleship; the
third of an *as* and the quarter *as* had a ship. The
latter had previously been called a *teruncius*, as
weighing three ounces. Subsequently when the
presence of Hannibal was being felt, in the dictator- 217 B.C.

urguente [1] Q. Fabio Maximo dictatore asses unciales
facti, placuitque denarium XVI assibus permutari,
quinarium octonis, sestertium quaternis. ita res p.
dimidium lucrata est, in militari tamen stipendio
46 semper denarius pro X assibus datus est. notae
argenti fuere bigae atque quadrigae; inde bigati
quadrigatique dicti.

Mox lege Papiria semunciarii asses facti. Livius
Drusus in tribunatu plebei octavam partem aeris
argento miscuit. is, qui nunc victoriatus appellatur,
lege Clodia percussus est; antea enim hic nummus ex
Illyrico advectus mercis loco habebatur. est autem
signatus Victoria, et inde nomen.

47 Aureus nummus post annos LI [2] percussus est
quam argenteus ita, ut scripulum valeret sestertios
vicenos,[3] quod effectum [4] in librali [5] ratione sestertii,[6]
qui tunc erat, CCCC.[7] postea placuit X XXXX signari
ex auri libris, paulatimque principes inminuere pondus, et novisissime Nero ad XXXXV.

48 XIV. Sed a nummo prima origo avaritiae faenore
excogitato quaestuosaque segnitia; nec paulatim
exarsit rabie quadam non iam avaritia, sed fames auri,

---

[1] urguente *B, cd. Par.* 6801: urguente Marcum *rell.*:
urguente Marcum Minucium *Brotier.*

[2] LI *B*: LXII *rell.*

[3] sestertios vicenos *Brotier*: sestertius vicenus *aut* sestertiis
vicenis (sestertio *B*: viciens *B¹*: vincens *B²*).

[4] effectum *K. C. Bailey*: efficit *B, cd. Par.* 6801: effecit
rell.

[5] librali *B*: libras *rell.*: libram *Mayhoff.*

[6] sestertium *Urlichs*: sestertiorum *Caesarius.*

[7] erat *cd. Par.* 6801: erant *rell.*: cccc *B*: D nongenti *rell.*
(sestertios DCCCC *cd. Par.* 6801): *varia editores.*

ship of Quintus Fabius Maximus, *asses* of one ounce weight were coined, and it was enacted that the exchange-value of the denarius should be sixteen *asses*, of the half-denarius eight and of the quarter-denarius four; by this measure the state made a clear gain of one half. But nevertheless in the pay of soldiers one denarius has always been given for ten asses. The designs on silver were a two-horse and a four-horse chariot, and consequently the coins were called a *pair of horses* and a *four-in-hand*.

Next according to a law of Papirius *asses* 89 B.C. weighing half an ounce were struck. Livius Drusus [a] when holding the office of tribune of the plebs alloyed the silver with one-eighth part of bronze. The coin now named the victory coin was struck under the law of Clodius; previously a coin c. 104 B.C. of this name was imported from Illyria and was looked on as an article of trade. The design on it was a figure of Victory, which gives it its name.

The first gold coin was struck 51 years later than 217 B.C. the silver coinage, a scruple of gold having the value of twenty sesterces; this was done at 400 to the pound of silver, at the then rating of the sesterce. It was afterwards decided to coin denarii at the rate 49 B.C. of 40 from a pound of gold, and the emperors gradually reduced the weight of the gold denarius, and most recently Nero brought it down to 45 denarii to A.D. 54–68. the pound.

XIV. But from the invention of money came the *Examples of misuse of gold.* original source of avarice when usury was devised, and a profitable life of idleness; by rapid stages what was no longer mere avarice but a positive

[a] Probably the tribune of 123 B.C., not his son who was tribune in 91 B.C.

utpote cum Septumuleius, C. Gracchi familiaris, auro
rependendum caput eius abscisum ad Opimium
tulerit plumboque in os addito parricido suo rem p.
etiam circumscripserit; nec iam Quiritium aliquis,
sed universo nomine Romano infami rex Mithridates
Aquilio duci capto aurum in os infudit. haec parit

49 habendi cupido! pudet intuentem nomina ista, quae
subinde nova Graeco sermone excogitantur insperso
argenteis vasis auro et incluso, quibus deliciis pluris
veneunt inaurata quam aurea, cum sciamus inter-
dixisse castris suis Spartacum, ne quis aurum haberet
aut argentum. tanto plus fuit animi fugitivis nos-

50 tris! Messalla orator prodidit Antonium triumvirum
aureis usum vasis in omnibus obscenis desideriis,
pudendo crimine etiam Cleopatrae. summa apud
exteros licentiae fuerat Philippum regem poculo
aureo pulvinis subdito dormire solitum, Hagnonem
Teium, Alexandri Magni praefectum, aureis clavis
suffigere crepidas: Antonius solus contumelia naturae vilitatem auro fecit. o dignum proscriptione,
sed Spartaci!

51 XV. Equidem miror populum Romanum victis
gentibus in tributo semper argentum imperasse, non

---

a Consul in 121 B.C.
b After the battle of Protomachium in Asia Minor, 88 B.C.
c Leader of a great slave-rising in Italy, 73–71 B.C.
d *I.e.* by a slave, not by a fellow freeman. Antony was
infamous for the proscription which he inflicted in 43 B.C.

40

hunger for gold flared up with a sort of frenzy, inasmuch as the friend of Gaius Gracchus, Septumuleius, a price having been set on Gracchus's head to the amount of its weight in gold, when Gracchus's head had been cut off, brought it to Opimius,[a] after adding to his unnatural murder by putting lead in the mouth of the corpse, and so cheated the state in addition. Nor was it now some Roman citizen, but King Mithridates who disgraced the whole name of Roman when he poured molten gold into the mouth of the General Aquilius whom he had taken prisoner.[b] These are the things that the lust for possessions engenders! One is ashamed to see the new-fangled names that are invented every now and then from the Greek to denote silver vessels filigreed or inlaid with gold, niceties which make gilded plate fetch a higher price than gold plate, when we know that Spartacus[c] issued an order to his camp forbidding anybody to possess gold or silver: so much more spirit was there then in our run-away slaves! The orator Messala has told us that the triumvir Antony used vessels of gold in satisfying all the indecent necessities, an enormity that even Cleopatra would have been ashamed of. Till then the record in extravagance had lain with foreigners—King Philip sleeping with a gold goblet under his pillows and Alexander the Great's prefect Hagnon of Teos having his sandals soled with gold nails; but Antony alone cheapened gold by this contumely of nature. How he deserved to be proscribed! but proscribed by Spartacus![d]

XV. It does indeed surprise me that the Roman nation always imposed a tribute of silver, not of gold, on races that it conquered, for instance on Carthage

121 B.C.

c. 83–30 B.C.

69-8-30 B.C.

Ruled 359-336 B.C.

Ruled 336-323 B.C.

Examples of luxury and wealth in precious metals.

41

aurum, sicut Carthagini cum Hannibale victae octin-
genta milia, $\overline{XVI}$[1] pondo annua in quinquaginta
annos, nihil auri. nec potest videri paenuria mundi
id evenisse. iam Midas et Croesus infinitum pos-
sederant, iam Cyrus devicta Asia pondo $\overline{XXIIII}$
invenerat praeter vasa aurumque factum et in eo
solium,[2] platanum, vitem. qua victoria argenti $\overline{D}$[3]
talentorum reportavit et craterem Semiramidis,
52 cuius pondus xv talentorum colligebat. talentum
Aegyptium pondo LXXX patere M.[4] Varro tradit.
iam regnaverat in Colchis Saulaces Aeetae suboles,
qui terram virginem nactus plurimum auri argentique
eruisse dicitur in Suanorum gente, et alioqui velleribus aureis incluto regno. et illius aureae camarae,
argenteae trabes et columnae atque parastaticae
narrantur victo[5] Sesostri, Aegypti rege tam superbo,
ut prodatur annis quibusque sorte reges singulos e
subiectis iungere ad currum solitus atque ita trium-
phare.
53 XVI. Et nos fecimus quae posteri fabulosa arbi-
trentur. Caesar, qui postea dictator fuit, primus in
aedilitate munere patris funebri omni apparatu

[1] $\overline{XVI}$ *Ian*: AVT *cdd.* (*om. B*: argenti *cd. Par.* 6801).
[2] soliũ (*i.e.* solium) *Mayhoff*: solia ac *Pintianus*: foliatam *Ian*: folia *B*: folia ac *rell.*
[3] D *coni. Warmington.*
[4] patere M. *Detlefsen*: capere *Gelen*: pendere *aut* habere *coni. Mayhoff*: patere *cdd.* (paterem *cd. Leid. Voss.*).
[5] victae *cd. Par.* 6801.

[a] Probably the right reading is D = 500.
[b] The legend was that Phrixus flew there on a ram with a
fleece of gold to escape from his stepmother, and married the

when conquered together with Hannibal, 800,000 202 B.C.
pounds weight of silver in yearly instalments of
16,000 pounds spread over 50 years, but no gold. Nor
can it be considered that this was due to the world's
poverty. Midas and Croesus had already possessed
wealth without limit, and Cyrus had already on conquering Asia Minor found booty consisting of 24,000 546–5 B.C.
pounds weight of gold, besides vessels and articles
made of gold, including a throne, a plane-tree and a
vine. And by this victory he carried off 500,000 [a]
talents of silver and the wine-bowl of Semiramis the
weight of which came to 15 talents. The Aegyptian
talent according to Marcus Varro amounts to 80
pounds of gold. Saulaces the descendant of Aeetes
had already reigned in Colchis, who is said to have
come on a tract of virgin soil in the country of the
Suani and elsewhere and to have dug up from it a great
quantity of gold and silver, his realm being moreover
famous for golden fleeces.[b] We are also told of his
gold-vaulted ceilings and silver beams and columns
and pilasters, belonging to Sesostris King of Egypt
whom Saulaces conquered, so proud a monarch that
he is reported to have been in the habit every year
of harnessing to his chariot individual kings selected
by lot from among his vassals and so going in
triumphal procession.

XVI. We too have done things to be deemed
mythical by those who come after us. Caesar, the
future dictator, was the first person in the office of
aedile to use nothing but silver for the appointments 65 B.C.
of the arena—it was at the funeral games presented
in honour of his father; and this was the first

daughter of King Aeetes. The fleece was later carried away
by Jason and the Argonauts.

43

harenae argenteo usus est, ferasque etiam argenteis vasis incessivere tum primum noxii, quod iam etiam [1] in municipiis aemulantur. C. Antonius ludos scaena argentea fecit, item L. Murena; Gaius princeps in circo pegma duxit, in quo fuere argenti pondo

54 $\overline{\text{CXXIIII}}$.[2] Claudius successor eius, cum de Brittannia triumpharet, inter coronas aureas $\overline{\text{VII}}$ [3] pondo habere quam contulisset Hispania citerior, $\overline{\text{VIIII}}$ [4] quam Gallia comata, titulis indicavit. huius deinde succesor Nero theatrum Pompei operuit auro in unum diem, quo Tiridati Armeniae regi ostenderet. et quota pars ea fuit aureae domus ambientis urbem!

55 XVII. Auri in aerario populi R. fuere Sex. Iulio L. Aurelio cos., septem annis ante bellum Punicum tertium, pondo $\overline{\text{XVII}}$ [5] CCCCX, argenti $\overline{\text{XXII}}$ LXX, et in numerato $\overline{|\text{LXI}|}$ $\overline{\text{XXXV}}$ CCCC, Sexto Iulio L. Marcio cos., hoc est belli socialis initio, auri . . .[6] $\overline{|\text{XVI}|}$ $\overline{\text{XX}}$ DCCCXXXI.

56 C. Caesar primo introitu urbis civili bello suo ex aerario protulit laterum aureorum $\overline{\text{XV}}$, argenteorum $\overline{\text{XXX}}$, et in numerato [7] $\overline{|\text{CCC}|}$. nec fuit aliis temporibus

---

[1] iam etiam *B*: etiam *rell*.: iam et *coni. Mayhoff.*
[2] $\overline{\text{CXXIIII}}$ *B*: CXXIIII *aut* CXXXIIII *rell.*
[3] $\overline{\text{VII}}$ *B*: VII *rell.*
[4] $\overline{\text{VIIII}}$ *B*: VIIII *rell.*
[5] *In §§ 55–56 numeri varie traduntur.*
[6] *lac. Detlefsen, Mommsen.*
[7] *V.ll.* nummo, numero (*add.* HS *cdd. nonnulli*: pondo *cd. Par.* 6801).

---

[a] A wooden edifice on wheels in two or more stages, which were raised and lowered, opened and closed, by machinery; on them performances were given.
[b] So *cd. B.* The number 124 or 134 of the other *cdd.* is of course much too small.
[c] So *cd. B.* The other MSS. give 7 and 9. The higher number is so absurd that perhaps we should omit *inter* and

occasion on which criminals made to fight with wild animals had all their equipment made of silver, a practice nowadays rivalled even in our municipal towns. Gaius Antonius gave plays on a silver stage, and so did Lucius Murena; and the emperor Gaius Caligula brought on a scaffolding [a] in the circus which had on it 124,000 [b] pounds weight of silver. A.D. 37-41. His successor Claudius when celebrating a triumph after the conquest of Britain, advertised by placards A.D. 43. that among the gold coronets there was one having a weight of 7000 [c] pounds contributed by Hither Spain and one of 9000 [c] from Gallia Comata. His immediate successor Nero covered the theatre of A.D. 54-68. Pompey with gold for one day's purpose, when he was to display it to Tiridates King of Armenia. Yet how small was the theatre in comparison with Nero's Golden Palace which goes all round the city!

XVII. The gold contained in the national treasury of Rome in the consulship of Sextus Julius 156 B.C. and Lucius Aurelius, seven years before the third Punic War, amounted to 17,410 lbs., the silver to 22,070 lbs., and in specie there was 6,135,400 sesterces; in the consulship of Sextus Julius and 91 B.C. Lucius Marcius, that is to say, at the beginning of the war with the allies,[d] there was . . . lbs. of gold and 1,620,831 lbs. of silver. Gaius Julius Caesar, on first entering Rome during the civil war that bears 49 B.C. his name, drew from the treasury 15,000 gold ingots, 30,000 silver ingots, and 30,000,000 sesterces in coin; at no other periods was the state more wealthy.

translate: ' that there were crowns weighing in all 7000 pounds contributed by Hither Spain and 9000 pounds from Gallia Comata.'

[d] See n. on § 20.

res p. locupletior. intulit et Aemilius Paulus Perseo
rege victo e Macedonica praeda $\overline{\text{MMM}}$, a quo tempore
populus Romanus tributum pendere desiit.

57   XVIII. Laquearia, quae nunc et in privatis domibus auro teguntur, post Carthaginem eversam primo
in Capitolio inaurata sunt censura L. Mummi. inde
transiere in camaras quoque et parietes, qui iam et
ipsi tamquam vasa inaurantur, cum varie sua aetas
de Catulo existimaverit, quod tegulas aereas Capitoli
inaurasset.

58   XIX. Primos inventores auri, sicut metallorum
fere omnium, septimo volumine diximus. praecipuam
gratiam huic materiae fuisse arbitror non colore, qui
clarior in argento est magisque diei similis, ideo
militaribus signis familiarior, quoniam [1] longius fulget,
manifesto errore eorum, qui colorem siderum placuisse in auro arbitrantur, cum in gemma aliisque
59 rebus non sit praecipuus. nec pondere aut facilitate
materiae praelatum est ceteris metallis, cum cedat
per utrumque plumbo,[2] sed quia rerum uni nihil igne
deperit, tuto [3] etiam in incendiis rogisque. quin
immo quo saepius arsit, proficit ad bonitatem,

---

[1] *V.l.* quo nimis : quoniam in iis *coni. Mayhoff.*
[2] plumbo *cdd.* : plumbum *coni. K. C. Bailey.*
[3] *V.ll.* tota, toto.

---

[a] King of Macedonia, defeated at Pydna, 168 B.C.
[b] It was not levied after 167 B.C.
[c] *I.e.* he was by no means universally approved.
[d] This is not true.

Aemilius Paulus also after the defeat of King Perseus [a] paid in to the treasury from the booty won in Macedonia 300 million sesterces; and from that date onward the Roman nation left off paying the citizens' property-tax.[b]

XVIII. At the present day we see ceilings covered with gold even in private houses, but they were first gilded in the Capitol during the censorship of Lucius Mummius after the fall of Carthage. 146 B.C. From ceilings the use of gilding passed over also to vaulted roofs and walls, these too being now gilded like pieces of plate, whereas a variety of judgements were passed [c] on Catulus by his contemporaries for *Between 79 and 60 B.C.* having gilded the brass tilings of the Capitol.

XIX. We have already said in Book VII who *VII. 97.* were the people who first discovered gold, and *Popularity of gold.* almost all of the metals likewise. I think that the chief popularity of this substance has been won not by its colour, that of silver being brighter and more like daylight, which is the reason why it is in more common use for military ensigns because its brilliance is visible at a greater distance; those persons who think that it is the colour of starlight in gold that has won it favour being clearly mistaken because in the case of gems and other things with the same tint it does not hold an outstanding place. Nor is it its weight or its malleability that has led to its being preferred to all the rest of the metals, since in both qualities it yields [d] the first place to lead, but because gold is the only thing that loses no substance by the *Special qualities of gold.* action of fire, but even in conflagrations and on funeral pyres receives no damage. Indeed as a matter of fact it improves in quality the more often it is fired, and fire serves as a test of its goodness,

aurique experimentum ignis est, ut simili colore
rubeat ignescatque et ipsum; obrussam vocant.
60 primum autem bonitatis argumentum quam difficillime accendi. praeterea mirum, prunae [1] violentis-
simi ligni indomitum palea citissime ardescere atque,
ut purgetur, cum plumbo coqui.

Altera causa pretii maior, quod minimum usus
deterit, cum argento, aere, plumbo lineae praedu-
61 cantur manusque sordescant decidua materia. nec
aliud laxius dilatatur aut numerosius dividitur, utpote
cuius unciae in septingenas quinquagenas pluresque
bratteas quaternum utroque digitorum spargantur.
crassissimae ex iis Praenestinae vocantur, etiamnum
retinentes [2] nomen Fortunae inaurato fidelissime ibi
62 simulacro. proxima brattea quaestoria appellatur.
Hispania strigiles [3] vocat auri parvolas massas. super
omnia solum in massa aut ramento capitur. cum
cetera in metallis reperta igni perficiantur, hoc
statim aurum est consummatamque materiam suam
protinus habet, cum ita invenitur. haec enim inventio eius naturalis est; alia, quam dicemus, coacta.
super cetera non robigo ulla, non aerugo, non aliud
ex ipso, quod consumat bonitatem minuatve pondus.
iam contra salis et aceti sucos, domitores rerum,
constantia [4] superat omnia, superque [5] netur ac

---

[1] pruna *cd. deperd. recte*?
[2] retinente *B.*
[3] *V.ll.* strigile, striges (*B*).
[4] constantiam *B.*
[5] superat omnia superque *Mayhoff*: superque superat
omnia *B*: superque omnia *rell.*

---

[a] Cf. Schol. ad Thuc. II. 13 ὄβρυζον χρυσίον.
[b] A variant reading 'striges' gives 'grooves.'

making it assume a similar red hue and itself becomes the colour of fire; this process is called assaying.[a] The first proof of quality in gold is however its being affected by fire with extreme difficulty; beside that, it is remarkable that though invincible to live coal made of the hardest wood it is very quickly made red hot by a fire of chaff, and that for the purpose of purifying it it is roasted with lead.

Another more important reason for its value is that it gets extremely little worn by use; whereas, with silver, copper and lead, lines may be drawn, and stuff that comes off them dirties the hand. Nor is any other material more malleable or able to be divided into more portions, seeing that an ounce of gold can be beaten out into 750 or more leaves 4 inches square. The thickest kind of gold leaf is called Palestrina leaf, still bearing the name taken from the faithfully gilded statue of Fortune in that place. The foil next in thickness is styled Quaestorian leaf. In Spain tiny pieces of gold are called scrapers.[b] Gold more than all other metals is found unalloyed in nuggets or in the form of detritus. Whereas all other metals when found in the mines are brought into a finished condition by means of fire, gold is gold straight away and has its substance in a perfect state at once, when it is obtained by mining. This is the natural way of getting it, while another which we shall describe is artificial. More §§ 68 sqq. than any other substance gold is immune from rust or verdigris or anything else emanating from it that wastes its goodness or reduces its weight. Moreover in steady resistance to the overpowering effect of the juices of salt and vinegar it surpasses all things, and over and above that it can be spun

63 texitur lanae modo vel sine lana. tunica aurea triumphasse Tarquinium Priscum Verrius docet; nos
vidimus Agrippinam Claudi principis, edente eo
navalis proelii spectaculum, adsidentem et indutam
paludamento aureo textili sine alia materia. Atta-
licis vero iam pridem intexitur, invento regum Asiae.

64 XX. Marmori et iis, quae candefieri non possunt,
ovi candido inlinitur, ligno glutini ratione conposita,[1]
leucophorum vocant. quid sit hoc aut quemadmo-
dum fiat, suo loco docebimus. aes inaurari argento
vivo aut certe hydrargyro legitimum erat, de quis, ut [2]
dicemus illorum naturam reddentes, excogitata fraus

65 est. namque aes cruciatur in primis [3] accensumque
restinguitur sale, aceto, alumine, postea examinatur,[4]
an satis recoctum sit, splendore deprehendente,
iterumque exhalatur [5] igni, ut possit, edomitum mixtis pumice et [6] alumine, argento vivo inductas accipere
bratteas. alumen et in purgando vim habet qualem
esse diximus plumbo.

66 XXI. Aurum invenitur in nostro orbe, ut omitta-
mus Indicum a formicis aut apud Scythas grypis

[1] composito *vel* re composita *coni. Mayhoff.*
[2] ut *cd. Par.* 6801 : *om. rell.*
[3] primis *cdd.* : prunis *coni. D'Arcy Thompson.*
[4] examinatur *K. C. Bailey* : exharenatur.
[5] exhalatur *cdd.* : exharenatur *Detlefsen* : excitatur *coni. Mayhoff.*
[6] et *add. K. C. Bailey.*

[a] Probably Attalus I of Pergamum, 241–197 B.C.
[b] Literally ' fluid silver.'
[c] See XXXV. 183 ff.
[d] *I.e.* alum purifies copper as lead purifies gold.

into thread and woven into a fabric like wool, even without an addition of wool. Verrius informs us that Tarquinius Priscus celebrated a triumph wearing a golden tunic. We have in our own times seen the Emperor Claudius's wife Agrippina, at a show at which he was exhibiting a naval battle, seated at his side wearing a military cloak made entirely of cloth of gold. For a long period gold has been woven into the fabric called cloth of Attalus,[a] an invention of Kings of Asia.

Traditional dates 616–578 B.C.

XX. On marble and other materials incapable of being raised to a white heat gold is laid with white of egg; on wood it is laid with glue according to a formula; it is called leucophorum, white-bearing; what this is and how it is made we will explain in its proper place. The regular way to gild copper would be to use natural or at all events artificial quicksilver,[b] concerning which a method of adulteration has been devised, as we shall relate in describing the nature of those substances. The copper is first subjected to the violence of fire; then, when it is red hot, it is quenched with a mixture of brine, vinegar, and alum,[c] and afterwards put to a test, its brilliance of colour showing whether it has been sufficiently heated; then it is again dried in the fire, so that, after a thorough polishing with a mixture of pumice and alum, it is able to take the gold-leaf laid on with quicksilver. Alum has the same cleansing property here that we said is found in lead.[d]

XXXV. 36.

§§ 100. 125.

§ 60.

XXI. Gold in our part of the world—not to speak of the Indian gold obtained from ants or the gold dug up by griffins in Scythia [e]—is obtained in three

Methods for discovering gold.

---

[e] This Indian and Scythian gold was perhaps got from Tibet. The stories about it go back to Herodotus.

erutum, tribus [1] modis: fluminum ramentis, ut in
Tago Hispaniae, Pado Italiae, Hebro Thraciae,
Pactolo Asiae, Gange Indiae, nec ullum absolutius
aurum est, ut cursu ipso attrituque perpolitum. alio
modo puteorum scrobibus effoditur aut in ruina
montium quaeritur [2]; utraque ratio dicatur.

67 Aurum qui quaerunt, ante omnia segullum [3] tol-
lunt; ita vocatur indicium. alveus hic est harenae,
quae lavatur, atque ex eo, quod resedit, coniectura
capitur. invenitur aliquando in summa tellure
protinus rara felicitate, ut nuper in Delmatia principatu Neronis singulis diebus etiam quinquagenas
libras fundens. cum ita inventum est in summo
caespite, talutium [4] vocant, si et aurosa tellus subest.
cetero montes Hispaniarum, aridi sterilesque et in
quibus nihil aliud gignatur, huic [5] bono fertiles esse
coguntur.

68 Quod puteis foditur, canalicium vocant, alii cana-
liense, marmoris glareae inhaerens, non illo modo,
quo in oriente [6] sappiro atque Thebaico aliisque in
gemmis scintillat, sed micans [7] amplexu [8] marmoris.
vagantur hi venarum canales per latera puteorum et
huc illuc, inde nomine invento, tellusque ligneis
69 columnis suspenditur. quod effossum est, tunditur,

[1] tribus *Bergk*: apud nos tribus.
[2] quaeritur *B*: quare *rell.*
[3] *V.l.* segutilum.
[4] talutium *B*: talutatium *rell.*: alutatium *Hardouin*:
alutiatum *Gronov*: *an* alutium (*cf.* XXXIV, 157)?
[5] hoc *coni. Warmington.*
[6] orientis *coni. Mayhoff.*
[7] micans *B²*: micas.
[8] amplexu *Salmasius*: amplexum.

ways: in the detritus of rivers, for instance in the Tagus in Spain, the Po in Italy, the Maritza in Thrace, the Sarabat in Asia Minor and the Ganges in India; and there is no gold that is in a more perfect state, as it is thoroughly polished by the mere friction of the current. Another method is by sinking shafts; or it is sought for in the fallen debris of mountains. Each of these methods must be described.

People seeking for gold begin by getting up *segullum* [a]—that is the name for earth that indicates the presence of gold. This is a pocket of sand, which is washed, and from the sediment left an estimate of the vein is made. Sometimes by a rare piece of luck a pocket is found immediately, on the surface of the earth, as occurred recently in Dalmatia when Nero was emperor, one yielding fifty pounds A.D. 54-68. weight of gold a day. Gold found in this way in the surface crust is called *talutium* if there is also auriferous earth underneath. The otherwise dry, barren mountains of the Spanish provinces which produce nothing else whatever are forced into fertility in regard to this commodity.

Gold dug up from shafts is called ' channelled ' or *Gold-mining.* ' trenched ' gold; it is found sticking to the grit of marble, not in the way in which it gleams in the lapis lazuli of the East and the stone [b] of Thebes and in other precious stones, but sparkling in the folds of the marble. These channels of veins wander to and fro along the sides of the shafts, which gives the gold its name; and the earth is held up by wooden props. The substance dug out is crushed, washed,

[a] *Segullo* is still the miners' name for surface earth in auriferous deposits in Castile.
[b] Apparently some micaceous granite.

lavatur, uritur, molitur in farinam; farinam[1] a pila
scudem[2] vocant; argentum, quod exit a fornace,
sudorem. quae e camino iactatur spurcitia in omni
metallo scoria appellatur. haec in auro tunditur
iterumque coquitur. catini fiunt ex tasconio; hoc
est terra alba similis argillae, neque enim alia flatum
ignemque et ardentem materiam tolerat.

70 Tertia ratio opera vicerit Gigantum. cuniculis
per magna spatia actis cavantur montes lucernarum
ad lumina; eadem mensura vigiliarum est, multisque
mensibus non cernitur dies.

Arrugias id genus vocant. siduntque rimae subito
et opprimunt operatos,[3] ut iam minus temerarium
videatur e profundo maris petere margaritas atque
purpuras. tanto nocentiores fecimus terras! relin-
quuntur itaque fornices crebri montibus sustinendis.

71 occursant in utroque genere silices; hos igne et
aceto rumpunt, saepius vero, quoniam id cuniculos[4]
vapore et fumo strangulat, caedunt fractariis CL libras
ferri habentibus egeruntque umeris noctibus ac
diebus per tenebras proximis tradentes; lucem
novissimi cernunt. si longior videtur silex, latus

---

[1] molitur (*aut* mollitur) in farinam; farinam *Warmington*:
mollitur (*B, cd. Par.* 6801: molitur *rell.*) farinam (in farinam
cd. Flor. Ricc. et ut videtur cd. Par. 6801).
[2] a pila scudem *Madvig*: a.p. cudem *Detlefsen*: apitascudem
B: *varia rell. cdd. et edd.*
[3] *V.ll.* operantes, operarios.
[4] id cuniculos *B*: in cuniculis *rell.*

---

[a] A given amount of oil is known to last a specified time.
[b] *Arrugia* is said to be the term for a deep mine in Spain
to-day. The word is probably connected ὀρύσσω, dig.

fired and ground to a soft powder. The powder from the mortar is called the ' *scudes* ' and the silver that comes out from the furnace the ' sweat '; the dirt thrown out of the smelting-furnace in the case of every metal is called ' *scoria*,' slag. In the case of gold the scoria is pounded and fired a second time; the crucibles for this are made of tasconium, which is a white earth resembling clay. No other earth can stand the blast of air, the fire, or the intensely hot material.

The third method will have outdone the achievements of the Giants. By means of galleries driven for long distances the mountains are mined by the light of lamps—the spells of work are also measured by lamps,[a] and the miners do not see daylight for many months.

The name for this class of mines is *arrugiae*;[b] also cracks give way suddenly and crush the men who have been at work, so that it actually seems less venturesome to try to get pearls and purple-fishes out of the depth of the sea: so much more dangerous have we made the earth! Consequently arches are left at frequent intervals to support the weight of the mountain above. In both kinds of mining masses of flint are encountered, which are burst asunder by means of fire and vinegar, though more often, as this method makes the tunnels suffocating through heat and smoke, they are broken to pieces with crushing-machines carrying 150 lbs. of iron, and the men carry the stuff out on their shoulders, working night and day, each man passing them on to the next man in the dark, while only those at the end of the line see daylight. If the bed of flint seems too long, the miner follows along the side of it

sequitur fossor ambitque. et tamen in silice facilior
72 existimatur opera; est namque terra ex quodam
argillae genere glarea mixta—gangadiam vocant—
prope inexpugnabilis. cuneis eam ferreis adgredi-
untur et isdem malleis nihilque durius putant, nisi
quod inter omnia auri fames durissima est. peracto
opere cervices fornicum ab ultimo caedunt.[1] dat
signum rima, eamque[2] solus intellegit in cacumine
73 eius montis vigil. hic voce, nutu[3] evocari iubet
operas pariterque ipse devolat. mons fractus cadit
ab sese longe fragore qui concipi humana mente
non possit, aeque et flatu incredibili. spectant victores ruinam naturae. nec tamen adhuc aurum est
nec sciere esse, cum foderent, tantaque ad pericula
et inpendia satis causae fuit sperare quod cuperent.

74 Alius par labor ac vel maioris inpendii: flumina ad
lavandam hanc ruinam iugis[4] montium obiter duxere
a centesimo plerumque lapide; corrugos vocant, a
conrivatione credo. mille et hic labores: praeceps

[1] cadunt *B.*
[2] rima eamque *cd. Par.* 6801: ruina eamque *B et al.*:
ruinamque *rell.*: ruinae eamque *Gelen*: ruinae rima eamque
Detlefsen.
[3] voce nutu *B*: voce ictuve *cd. Tolet.*: vocent utve *rell.*:
voce in tutum *Detlefsen.*
[4] *fortasse* ⟨a⟩ *vel* ⟨ab⟩ iugis.

and goes round it. And yet flint is considered to involve comparatively easy work, as there is a kind of earth consisting of a sort of potter's clay mixed with gravel, called *gangadia*, which it is almost impossible to overcome. They attack it with iron wedges and the hammer-machines mentioned above; and it is thought to be the hardest thing that exists, except greed for gold, which is the most stubborn of all things. When the work is completely finished, beginning with the last, they cut through, at the tops, the supports of the arched roofs. A crack gives warning of a crash, and the only person who notices it is the sentinel on a pinnacle of the mountain. He by shout and gesture gives the order for the workmen to be called out and himself at the same moment flies down from his pinnacle. The fractured mountain falls asunder in a wide gap, with a crash which it is impossible for human imagination to conceive, and likewise with an incredibly violent blast of air. The miners gaze as conquerors upon the collapse of Nature. And nevertheless even now there is no gold so far, nor did they positively know there was any when they began to dig; the mere hope of obtaining their coveted object was a sufficient inducement for encountering such great dangers and expenses.

Another equally laborious task involving even greater expense is the incidental operation of previously bringing streams along mountain-heights frequently a distance of 100 miles for the purpose of washing away the débris of this collapse; the channels made for this purpose are called *corrugi*, a term derived I believe from *conrivatio*, a uniting of streams of water. This also involves a thousand

esse libramentum oportet, ut ruat verius quam fluat;
itaque altissimis partibus ducitur. convalles et in-
tervalla substructis canalibus iunguntur. alibi rupes
inviae caeduntur sedemque trabibus cavatis praebere
75 coguntur. qui caedit, funibus pendet, ut procul
intuenti species ne ferarum quidem, sed alitum fiat.
pendentes maiore ex parte librant et lineas itineri
praeducunt, quaque insistentis vestigiis hominis
locus non est, amnes trahuntur ab homine.[1] vitium
lavandi est, si fluens amnis lutum inportet; id genus
terrae urium vocant. ergo per silices calculosve
ducunt et urium evitant. ad capita deiectus in
superciliis montium piscinae cavantur ducenos pedes
in quasque partes et in altitudinem denos. emissaria
in iis quina pedum quadratorum ternum fere relin-
quuntur, ut repleto stagno excussis opturamentis
76 erumpat torrens tanta vi ut saxa provolvat. alius
etiamnum in plano labor. fossae, per quas profluat,
cavantur—agogas vocant—; hae sternuntur grada-
tim ulice. frutex est roris marini similis, asper
aurumque retinens. latera cluduntur tabulis, ac per

---

[1] trahuntur ab homine *B*: trahuntur ad homines *rell.*:
trahunt. omne *Hardouin.*

---

[a] *I.e.* the gold-bearing débris.
[b] The identification is doubtful in view of the alleged
resemblance to rosemary. Rosemary may be called ' rough,'
but it is not prickly like gorse.

tasks; the dip of the fall must be steep, to cause a rush rather than a flow of water, and consequently it is brought from very high altitudes. Gorges and crevasses are bridged by aqueducts carried on masonry; at other places impassable rocks are hewn away and compelled to provide a position for hollowed troughs of timber. The workman hewing the rock hangs suspended with ropes, so that spectators viewing the operations from a distance seem to see not so much a swarm of strange animals as a flight of birds. In the majority of cases they hang suspended in this way while taking the levels and marking out the lines for the route, and rivers are led by man's agency to run where there is no place for a man to plant his footsteps. It spoils the operation of washing if the current of the stream carries mud along with it: an earthy sediment of this kind is called *urium*. Consequently they guide the flow over flint stones and pebbles, and avoid *urium*. At the head of the waterfall on the brow of the mountains reservoirs are excavated measuring 200 ft. each way and 10 ft. deep. In these there are left five sluices with apertures measuring about a yard each way, in order that when the reservoir is full the stopping-barriers may be struck away and the torrent may burst out with such violence as to sweep forward the broken rock.[a] There is also yet another task to perform on the level ground. Trenches are excavated for the water to flow through —the Greek name for them means ' leads '; and these, which descend by steps, are floored with gorse [b]—this is a plant resembling rosemary, which is rough and holds back the gold. The sides are closed in with planks, and the channels are carried

59

praerupta suspenduntur canales. ita profluens terra in mare labitur ruptusque mons diluitur, ac longe terras in mare his de causis iam promovit Hispania.
77 in priore genere quae exhauriuntur inmenso labore, ne occupent puteos, in hoc rigantur. aurum arrugia quaesitum non coquitur, sed statim suum [1] est. inveniuntur ita massae, nec non in puteis, et denas excedentes libras; palagas,[2] alii palacurnas,[3] iidem quod minutum est balucem vocant. ulex siccatur, uritur, et cinis eius lavatur substrato caespite her-
78 boso, ut sidat aurum. vicena milia pondo ad hunc modum annis singulis Asturiam atque Callaeciam et Lusitaniam praestare quidam prodiderunt, ita ut plurimum Asturia gignat. neque in alia terrarum parte tot saeculis perseverat haec fertilitas. Italiae parci vetere interdicto patrum diximus; alioqui nulla fecundior metallorum quoque erat tellus. extat lex censoria Victumularum [4] aurifodinae in Vercellensi agro, qua cavebatur, ne plus quinque milia hominum in opere publicani haberent.
79   XXII. Aurum faciendi est etiamnum una ratio ex auripigmento, quod in Syria foditur pictoribus in summa tellure, auri colore, sed fragile lapidum specularium modo. invitaveratque spes Gaium prin-

[1] sudum *coni. Hermolaus Barbarus.*
[2] palagas *B*: palacas *rell.*: palacras *ed. Basil.*
[3] *V.ll.* psalacurnas, palacranas.
[4] Victumularum *B*: *V.ll.* victim-, vittim-: (vici) Ictimulorum *Hermolaus Barbarus coll. Strab.*

---

[a] See § 70.   [b] Yellow sulphide of arsenic.

on arches over steep pitches. Thus the earth carried along in the stream slides down into the sea and the shattered mountain is washed away; and by this time the land of Spain owing to these causes has encroached a long way into the sea. The material drawn out at such enormous labour in the former kind of mining is in this latter process washed out, §67. so as not to fill up the shafts. The gold obtained by means of an *arrugia*ᵃ does not have to be melted, but is pure gold straight away. In this process nuggets are found and also in the shafts, even weighing more than ten pounds. They are called palagae or else palacurnae, and also the gold in very small grains baluce. The gorse is dried and burnt and its ash is washed on a bed of grassy turf so that the gold is deposited on it. According to some accounts Asturia and Callaecia and Lusitania produce in this way 20,000 lbs. weight of gold a year, Asturia supplying the largest amount. Nor has there been in any other part of the world such a continuous production of gold for so many centuries. We have stated that by an old prohibiting decree of the senate Italy is protected from exploitation; otherwise no country would have been more productive in metals, as well as in crops. There is extant a ruling of the censors relating to the gold mines of Victumulae in the territory of Vercellae which prohibited the farmers of public revenues from having more than 5000 men engaged in the work.

XXII. There is moreover one method of making gold out of orpiment ᵇ which is dug up in Syria for use by painters; it is found on the surface of the earth, and is of a gold colour, but is easily broken, like looking-glass stone. Hopes inspired by it had

cipem avidissimum auri; quam ob rem iussit excoqui
magnum pondus et plane fecit aurum excellens, sed
ita parvi ponderis, ut detrimentum sentiret propter
avaritiam expertus, quamquam auripigmenti librae
X IIII permutarentur. nec postea temptatum ab
ullo est.

80   XXIII. Omni auro inest argentum vario pondere,
aliubi decuma parte,[1] aliubi octava. in uno tantum
Callaeciae metallo, quod vocant Albucrarense, tri-
censima sexta portio invenitur; ideo ceteris praestat.
ubicumque quinta argenti portio est, electrum
vocatur; scobes hae reperiuntur in canaliensi. fit et
cura electrum argento addito. quod si quintam portio-
81 nem excessit, incudibus non resistit. vestusta et
electro auctoritas Homero teste, qui Menelai regiam
auro, electro, argento, ebore fulgere tradit. Minervae templum habet Lindos insulae [2] Rhodiorum, in
quo Helena sacravit calicem ex electro; adicit
historia, mammae suae mensura. electri natura est
ad lucernarum lumina clarius argento splendere.
quod est nativum, et venena deprehendit. namque
discurrunt in calicibus arcus caelestibus similes cum
igneo stridore et gemina ratione praedicunt.

[1] decuma parte B : non rell. (nona cd. Par. Lat. 6797 :
dena alibi nona cd. Par. 6801).
[2] insula B : in insula coni. Mayhoff.

---

[a] Properly the word means ' amber.' See § 1, note.
[b] Od. IV. 71 ff.

attracted the Emperor Gaius Caligula, who was _{A.D. 37–41.} extremely covetous for gold, and who consequently gave orders for a great weight of it to be smelted; and as a matter of fact it did produce excellent gold, but so small a weight of it that he found himself a loser by his experiment that was prompted by avarice, although orpiment sold for 4 denarii a pound; and no one afterwards has repeated the experiment.

XXIII. All gold contains silver in various pro- '*Electrum*.' portions, a tenth part in some cases, an eighth in others. In one mine only, that of Callaecia called the Albucrara mine, the proportion of silver found is one thirty-sixth, and consequently this one is more valuable than all the others. Wherever the proportion of silver is one-fifth, the ore is called electrum[a]; grains of this are found in ' channelled ' _{Cf. § 68.} gold. An artificial electrum is also made by adding silver to gold. If the proportion of silver exceeds one-fifth, the metal produced offers no resistance on the anvil. Electrum also held a high position in old times, as is evidenced by Homer[b] who represents the palace of Menelaus as resplendent with gold, electrum, silver and ivory. There is a temple of Athena at Lindus of the island of Rhodes in which there is a goblet made of electrum, dedicated by Helen; history further relates that it has the same measurement as her breast. A quality of electrum is that it shines more brightly than silver in lamp-light. Natural electrum also has the property of detecting poisons; for semicircles resembling rainbows run over the surface in poisoned goblets and emit a crackling noise like fire, and so advertise the presence of poison in a twofold manner.

82 XXIV. Aurea statua prima omnium nulla inani-
tate et antequam ex aere aliqua modo fieret, quam
vocant holosphyraton, in templo Anaitidis posita dicitur quo situ terrarum nomen hoc signavimus,
83 numine gentibus illis sacratissimo. direpta ea est
Antonii Parthicis rebus, scitumque narratur vetera-
norum unius Bononiae hospitali divi Augusti cena,
cum interrogatus [1] esset, sciretne [2] eum, qui primus
violasset id numen, oculis membrisque captum ex-
spirasse; respondit enim cum maxime Augustum e
crure [3] eius cenare seque illum esse totumque sibi
censum ex ea rapina. hominum primus et auream
statuam et solidam LXX [4] circiter olympiade Gorgias
Leontinus Delphis in templo posuit sibi. tantus erat
docendae artis oratoriae quaestus.

84 XXV. Aurum pluribus modis pollet in remediis
volneratisque et infantibus adplicatur, ut minus
noceant quae inferantur veneficia. est et ipsi
superlato [5] vis malefica, gallinarum quoque et
pecuariorum [6] feturis. remedium abluere inlitum [7]
et spargere eos, quibus mederi velis. torretur et cum

[1] interrogatus *B* : interrogaretur *rell.*
[2] esset sciretne *B²* : esset *B¹* : essetne (*aut* essene) verum
rell. : esset verumne esset *Ian.*
[3] cruore *cd. Par.* 6801.
[4] LXXXX *Bergk.*
[5] superlito *Gronov.*
[6] pecuariorum *cd. Flor. Ricc.* : pecorum *rell.*
[7] inlitum *Gronov.* : inlatum.

[a] V. 83, where Anaitica is said to be a region divided from
Cappadocia by the upper Euphrates.
[b] 500–497 B.C. But Gorgias the 'sophist' visited Athens
in 427 B.C. and professed rhetoric and philosophy there in
subsequent years. Probably the right date is the 90th
Olympiad (420–417 B.C.).

XXIV. The first gold statue of all that was made *Golden* of solid metal and even before any was made of *statues.* bronze, of the kind called 'made of solid beaten metal,' is said to have been erected in the temple of Anaitis, in the region of the earth where we have designated this name,[a] that goddess' deity being held in the highest reverence by those races. This statue was taken as booty during the campaigns of *c. 36 B.C.* Antonius in Parthia, and a story is told of a witty saying of one of the veterans of our army who was being entertained as a guest at dinner by his late lamented Majesty Augustus at Bologna. He was asked whether it was true that the man who was the first to commit this sacrilege against that deity was struck blind and paralysed and so expired. His answer was that the emperor was at that very moment eating his dinner off one of the goddess's legs, and that he himself was the perpetrator of the sacrilege and owed his entire fortune to that piece of plunder. The first solid gold statue of a human being was one of himself set up by Gorgias of Leontini in the temple at Delphi about the 70th Olympiad.[b] So great were the profits to be made by teaching the art of oratory!

XXV. Gold is efficacious as a remedy in a variety *Medicinal* of ways, and is used as an amulet for wounded *uses of* people and for infants to render less harmful *gold.* poisonous charms that may be directed against them. Gold has itself however a maleficent effect if carried over the head, in the case of chickens and the young of cattle as well as human beings. As a remedy it is smeared on, then washed off and sprinkled on the persons you wish to cure. Gold is also heated with twice its weight of salt and

65

salis gemino pondere, triplici misyis ac rursus cum II
salis portionibus et una lapidis, quem schiston vocant.
ita virus trahit rebus una crematis in fictili vase,
85 ipsum purum et incorruptum. reliquus cinis ser-
vatus in fictili olla, ex qua [1] inlitas [2] lichenas in facie
lomento eo convenit ablui. fistulas etiam sanat et
quae vocantur haemorroides. quodsi tritus pumex
adiciatur, putria ulcera et taetri odoris emendat, ex
melle vero decoctum cum melanthio inlitum umbilico
leniter solvit alvum. auro verrucas curari M. Varro
auctor est.
86 XXVI. Chrysocolla umor est in puteis, quos
diximus, per venam auri defluens crassescente limo
rigoribus hibernis usque in duritiam pumicis. lauda-
tiorem eandem in aerariis metallis et proximam in
aegentariis fieri conpertum est. invenitur et in
plumbariis vilior etiam [3] auraria. in omnibus autem
his metallis fit et cura multum infra naturalem illam
inmissis in venam aquis leniter hieme tota usque in
Iunium mensem, dein siccatis Iunio et Iulio, ut plane
intellegatur nihil aliud chrysocolla quam vena putris.
87 nativa duritia maxime distat; uvam vocant. et
tamen illa quoque herba, quam lutum appellant,
tinguitur. natura est, quae lino lanaeve, ad sucum

---

[1] qua *B* : aqua *rell.*   [2] inlitas *B²* : inlitus.
[3] *fortasse* tamen.

---

[a] *Lomentum* is properly barley-meal mixed with rice.
[b] See § 4, note.
[c] Or : ' Another sort is found in lead mines, but it is inferior
to the true " gold "-kind.'

three times its weight of copper pyrites, and again with two portions of salt and one of the stone called 'splittable.' Treated in this way it draws poison out, when the other substances have been burnt up with it in an earthenware crucible while it remains pure and uncorrupted itself. The ash remaining is kept in an earthenware jar, and eruptions on the face may well be cleansed away by being smeared with this lotion [a] from the jar. It also cures fistulas and what are called hæmorrhoids. With the addition of ground pumicestone it relieves putrid and foul-smelling ulcers, while boiled down in honey and git, and applied as a liniment to the navel it acts as a gentle aperient. According to Marcus Varro gold is a cure for warts.

XXVI. Gold-solder [b] is a liquid found in the shafts we spoke of, flowing down along a vein of gold, with a slime that is solidified by the cold of winter even to the hardness of pumicestone. A more highly spoken of variety of the same metal has been ascertained to be formed in copper mines, and the next best in silver-mines. A less valuable sort also with an element of gold is also found in lead mines. [c] In all these mines however an artificial variety is produced that is much inferior to the natural kind referred to; the method is to introduce a gentle flow of water into the vein all winter and go on till the beginning of June and then to dry it off in June and July, clearly showing that gold-solder is nothing else than the putrefaction of a vein of metal. Natural gold-solder, known as 'grape,' differs very greatly from the artificial in hardness, and neverthe-less it also takes a dye from the plant called yellow-weed. It is of a substance that absorbs moisture,

Gold-solder. §§ 67 sqq.

bibendum. tunditur in pila, dein tenui cribro cernitur, postea molitur ac deinde tenuius cribratur.
quidquid non transmeat, repetitur in pila, dein
88 molitur. pulvis semper in catinos digeritur et ex
aceto maceratur, ut omnis duritia solvatur, ac rursus
tunditur, dein lavatur, in conchis siccatur, tum
tinguitur alumine schisto et herba supra dicta pin-
giturque, antequam pingat. refert quam bibula
docilisque sit. nam nisi rapuit colorem, adduntur et
scytanum atque turbistum; ita vocant medicamenta
sorbere cogentia.

89   XXVII. Cum tinxere pictores, orobitin vocant
eiusque duo genera faciunt: elutam,[1] quae servatur
in lomentum, et liquidam globulis sudore resolutis.
haec utraque genera in Cypro fiunt. laudatissima
autem est in Armenia, secunda in Macedonia, largis-
sima in Hispania; summa[2] commendationis, ut
colorem in herba segetis laete virentis quam simil-
90 lime reddat. visumque iam est Neronis principis
spectaculis harenam circi chrysocolla sterni, cum ipse
concolori panno aurigaturus esset. indocta opificum
turba tribus eam generibus distinguit: asperam, quae

---

[1] luteam *Hermolaus Barbarus fortasse recte* (cf. § 91).
[2] *V.l.* summae: summa est *Ian.*

---

[a] Cf. XXXV. 186.
[b] These two substances have not been identified.
[c] Perhaps we should adopt the reading *luteam*.

like flax or wool. It is pounded in a mortar and then passed through a fine sieve, and afterwards milled and then sifted again with a finer sieve, everything that does not pass through the sieve being again treated in the mortar and then milled again. The powder is all along separated off into bowls and steeped in vinegar so as to dissolve all hardness, and then is pounded again and then rinsed in shells and left to dry. Then it is dyed by means of 'splittable' alum [a] and the plant above mentioned and so given a colour before it serves as a colour itself. It is important how absorbent it is and ready to take the dye; for if it does not at once catch the colour, scytanum and turbistum [b] must be added as well— those being the names of two drugs producing absorption.

XXVII. When painters have dyed gold-solder, they call it orobitis, vetch-like, and distinguish two kinds, the purified,[c] which is kept for a cosmetic, and the liquid, in which the little balls are made into a paste with a liquid. Both of these kinds are made in Cyprus, but the most highly valued is in Armenia and the second best in Macedonia, while the greatest quantity is produced in Spain, the highest recommendation in the latter being the quality of reproducing as closely as possible the colour in a bright green blade of corn. We have before now seen at the shows given by the emperor Nero the sand of the circus sprinkled with gold- A.D. 54- 68. solder when the emperor in person was going to give an exhibition of chariot-driving wearing a coat of that colour. The unlearned multitude of artisans distinguish three varieties of the substance, the rough, which is valued at 7 denarii a pound, the

taxatur[1] in libras X vii, mediam quae X v, attritam,
quam et herbaceam vocant, X iii. sublinunt autem
harenosam, priusquam inducant, atramento et
91 Paraetonio. haec sunt tenacia eius, colore blanda.
Paraetonium, quoniam est natura pinguissimum et
propter levorem tenacissimum, atramento aspergitur,
ne Paraetonii candor pallorem chrysocollae adferat.
luteam putant a luto herba dictam, quam ipsam
caeruleo subtritam pro chrysocolla inducunt, vilis-
simo genere atque fallacissimo.

92   XXVIII. Usus chrysocollae et in medicina est ad
purganda volnera cum cera atque oleo. eadem per
se arida siccat et[2] contrahit. datur et in angina
orthopnoeave lingenda cum melle. concitat vomitiones, miscetur et collyriis ad cicatrices oculorum ac
viridibus emplastris ad dolores mitigandos, cicatrices
trahendas. hanc chrysocollam medici acesim appellant, quae non est orobitis.

93   XXIX. Chrysocollam et aurifices sibi vindicant
adglutinando auro, et inde omnes appellatas similiter
virentes dicunt. temperatur autem Cypria aerugine
et pueri inpubis urina addito nitro teriturque Cyprio

---

[1] mataxatur *B*: iam taxatur *coni. Mayhoff.*
[2] siccat et *B, cd. Par.* 6801 : et sicca *rell.*

---

[a] *Paraetonium*; see XXXV, 30, 36.
[b] ἄκεσις, a remedy, healing.
[c] Or, child (of either sex).

middling, which is 5 denarii, and the crushed, also called the grass-green kind, 3 denarii. Before applying the sandy variety they put on a preliminary coating of black dye and pure white chalk [a] : these serve to hold the gold-solder and give a softness of colour. As the pure chalk is of a very unctuous consistency and extremely tenacious owing to its smoothness, it is sprinkled with a coat of black, to prevent the extreme whiteness of the chalk from imparting a pale hue to the gold-solder. The yellow gold-solder is thought to derive its name from the plant yellow-weed, which is itself often pounded up with steel-blue and applied for painting instead of gold-solder, making a very inferior and counterfeit kind of colour.

XXVIII. Gold-solder is also used in medicine, mixed with wax and olive oil, for cleansing wounds; likewise applied dry by itself it dries wounds and draws them together. It is also given in cases of quinsy or asthma, to be taken as an electuary with honey. It acts as an emetic, and also is used as an ingredient in salves for sores in the eyes and in green plasters for relieving pains, and drawing together scars. This kind of gold-solder is called by medical men ' remedial solder,' [b] and is not the same as orobitis.

XXIX. The goldsmiths also use a special gold-solder of their own for soldering gold, and according to them it is from this that all the other substances with a similar green colour take the name. The mixture is made with Cyprian copper verdigris and the urine of a boy [c] who has not reached puberty with the addition of soda [d] ; this is ground with a pestle

[d] Sodium carbonate.

aere in cypriis mortariis; santernam vocant nostri.
ita feruminatur aurum, quod argentosum vocant.
signum est, si addita santerna nitescit. e diverso
aerosum contrahit se hebetaturque et difficulter
feruminatur. ad id glutinum fit auro et septima
argenti parte ad supra dicta additis unaque tritis.

94   XXX. Contexique par est reliqua circa hoc, ut
universa naturae contingat admiratio. auro glutinum est tale, argilla ferro, cadmea aeris massis,
alumen lamnis, resina plumbo et marmori, at plumbum nigrum albo iungitur ipsumque album sibi oleo,
item stagnum aeramentis, stagno argentum. pineis
optume lignis aes ferrumque funditur, sed et Aegyptio
papyro, paleis aurum. calx aqua accenditur et
Thracius lapis, idem oleo restinguitur, ignis autem
aceto maxime et visco et ovo. terra minime flagrat,
carboni vis maior exusto iterumque flagranti.

95   XXXI. Ab his argenti metalla dicantur, quae
sequens insania est. non nisi in puteis reperitur
nullaque spe sui nascitur, nullis, ut in auro, lucentibus

---

^a Here zinc oxide. See also XXXIV. 100.
^b Tin.
^c Or *stannum*, an alloy of silver and lead.
^d Perhaps a kind of asphalt.

made of Cyprian copper in mortars of the same metal, and the Latin name for the mixture is santerna. It is in this way used in soldering the gold called silvery-gold; a sign of its having been so treated is if the application of borax gives it brilliance. On the other hand 'coppery' gold shrinks in size and becomes dull, and is difficult to solder; for this purpose a solder is made by adding some gold and one seventh as much silver to the materials above specified, and grinding them up together.

XXX. While speaking of this it will be well to annex the remaining particulars, so as to occasion all-round admiration for Nature. The proper solder for gold is the one described; for iron, potter's clay; for copper in masses, *cadmea* [a]; for copper in sheets, alum; for lead and marble, resin. Black lead however is joined by means of white lead,[b] and white lead to white lead by using oil; stagnum [c] likewise with copper filings, and silver with stagnum. For smelting copper and iron pine-wood makes the best fuel, though Egyptian papyrus can also be used; gold is best smelted with a fire made of chaff. Water sets fire to quicklime and Thracian stone,[d] and olive-oil puts it out; fire however is most readily quenched by vinegar, mistletoe and eggs. Earth it is quite impossible to ignite, but charcoal gives a more powerful heat if it is burned till it goes out and then catches fire again.

XXXI. After these details let us speak about the *Silver.* varieties of silver ore, the next madness of mankind. Silver is only found in deep shafts, and raises no hopes of its existence by any signs, giving off no shining sparkles such as are seen in the case of gold.

scintillis. terra est alias rubra, alias[1] cineracea.
excoqui non potest, nisi cum plumbo nigro aut cum
vena plumbi—galenam vocant—, quae iuxta argenti
venas plerumque reperitur. et eodem opere ignium
discedit pars in plumbum, argentum autem innatat[2]
superne, ut oleum aquis.

96 Reperitur in omnibus paene provinciis, sed in
Hispania pulcherrimum, id quoque in sterili solo atque
etiam montibus, et ubicumque una inventa vena est,
non procul invenitur alia. hoc quidem et in omni
fere materia, unde metalla Graeci videntur dixisse.
mirum, adhuc per Hispanias ab Hannibale inchoatos
durare puteos. sua nomina ab inventoribus habent,
97 ex quis Baebelo appellatur hodie, qui CCC pondo
Hannibali subministravit in dies, ad MD passus iam
cavato monte, per quod spatium aquatini[3] stantes
noctibus diebusque egerunt aquas lucernarum men-
98 sura amnemque faciunt. argenti vena in summo
reperta crudaria appellatur. finis antiquis fodiendi
solebat esse alumen inventum; ultra nihil quaere-
batur. nuper inventa aeris vena infra alumen nullam
finem spei fecit. odor ex argenti fodinis inimicus
omnibus animalibus, sed maxime canibus. aurum
argentumque quo mollius, eo pulchrius. lineas ex
argento nigras praeduci plerique mirantur.

99 XXXII. Est et lapis in iis venis, cuius vomica

---

[1] *V.l.* rufa alia.   [2] natat *B*.
[3] aquatini *coni. Sillig*: Accitani *coni. Hardouin*: Iacetani
Pintianus: aquitani.

---

[a] Still so called. It is lead sulphide, the most useful lead
ore. For *galena* in a different sense, see XXXIV. 159.
[b] Taking μέταλλα as (ἄλλα) μετ' ἄλλα 'one after another'.
[c] Possibly carbon dioxide, which, since it lies low, would
affect dogs before men.

The ore is sometimes red, sometimes ash-coloured. It cannot be smelted except when combined with lead or with the vein of lead, called galena,[a] lead ore, which is usually found running near veins of silver ore. Also when submitted to the same process of firing, part of the ore precipitates as lead while the silver floats on the surface, like oil on water.

Silver is found in almost all the provinces, but the finest is in Spain, where it, as well as gold, occurs in sterile ground and even in the mountains; and wherever one vein is found another is afterwards found not far away. This indeed also occurs in the case of almost every metal, and accounts it seems for the word ' metals ' used by the Greeks.[b] It is a remarkable fact that the shafts initiated by Hannibal 221-219 B.C. all over the Spanish provinces are still in existence; they are named from the persons who discovered them; one of these mines, now called after Baebelo, furnished Hannibal with 300 pounds weight of silver a day, the tunnelling having been carried a mile and a half into the mountain. Along the whole of this distance watermen are posted who all night and day in spells measured by lanterns bale out the water and make a stream. The vein of silver nearest the surface is called ' the raw.' In early days the excavations used to stop when they found alum, and no further search was made; but recently the discovery of a vein of copper under the alum has removed all limit to men's hopes. The exhalations [c] from silver mines are dangerous to all animals, but specially to dogs. Gold and silver are more beautiful the softer they are. It surprises most people that silver traces black lines.

XXXII. There is also a mineral found in these Quicksilver.

liquoris aeterni argentum vivum appellatur. venenum rerum omnium est perrumpitque vasa perma-
nans tabe dira. omnia ei innatant praeter aurum ;
id unum ad se trahit. ideo et optime purgat, ceteras
eius sordes expuens crebro iactatu fictilibus in vasis.
ita vitiis [1] eiectis [2] ut et ipsum ab auro discedat, in
pelles subactas effunditur, per quas sudoris vice
100 defluens purum relinquit aurum. ergo et cum aera
inaurentur, sublitum bratteis pertinacissime retinet,
verum pallore detegit simplices aut praetenues bratteas. quapropter id furtum quaerentes ovi liquore
candido usum eum adulteravere, mox et hydrargyrum,[3] de quo dicemus suo loco. et alias argentum
vivum non largum inventu est.

101 XXXIII. In isdem argenti metallis invenitur, ut
proprie dicatur,[4] spumae lapis candidae nitentisque,
non tamen tralucentis ; stimi appellant, alii stibi,
alii alabastrum, aliqui larbasim.[5] duo eius genera,
mas ac femina. magis probant feminam, horridior

---

[1] V.ll. ita vitis, ita ut iis (hiis), avitis: ita autem iis *Sillig*:
alutis *Brotier*: vestibus *Hardouin*.
[2] eiectis *Ian*: tectis *B*: V.ll. abiectis, iniectis, invectis.
[3] hydrargyrum *L. Poinsinet de Sivry*: hydrargyro.
[4] dicatur *Mayhoff*: dicatus *B¹*: dictus *B²*: dicemus *rell.*
(dicamus *cd. Par.* 6801).
[5] larbasim *B*: turbasim *rell.*: larbason *Hermolaus Barbarus*
coll. Diosc. V. 99.

veins of silver which contains a humour, in round drops, that is always liquid, and is called quicksilver. It acts as a poison on everything, and breaks vessels by penetrating them with malignant corruption. All substances float on its surface except gold, which is the only thing that it attracts to itself; consequently it is also excellent for refining gold, as if it is briskly shaken in earthen vessels it rejects all the impurities contained in it. When these blemishes have been thus expelled, to separate the quicksilver itself from the gold it is poured out on to hides that have been well dressed, and exudes through them like a kind of perspiration and leaves the gold behind in a pure state. Consequently when also things made of copper are gilded, a coat of quicksilver is applied underneath the gold leaf and keeps it in its place with the greatest tenacity: but if the gold-leaf is put on in one layer or is very thin it reveals the quicksilver by its pale colour. Consequently persons intending this fraud adulterated the quick-silver used for this purpose with white of egg; and later they falsified also hydrargyrum or artificial quicksilver, which we shall speak about in its proper § 123. place. Otherwise quicksilver is not to be found in any large quantity.

XXXIII. In the same mines as silver there is found *Antimony.* what is properly to be described as a stone, made of white and shiny but not transparent froth; several names are used for it, stimi, stibi, alabastrum and sometimes larbasis. It is of two kinds, male and female.[a] The female variety is preferred, the male

---

[a] Probably stibnite (sulphide of antimony), and native metallic antimony respectively (K. C. Bailey, *The Elder Pliny's Chapters on Chemical Subjects*, I, p. 213).

est mas scabriorque et minus ponderosus minusque
radians et harenosior, femina contra nitet, friabilis
fissurisque, non globis, dehiscens.

102   XXXIV. Vis eius adstringere ac refrigerare,
principalis autem circa oculos, namque ideo etiam
plerique platyophthalmon id appellavere, quoniam in
calliblepharis mulierum dilatet oculos, et fluctiones
inhibet oculorum exulcerationesque farina eius ac
turis cummi admixto. sistit et sanguinem e cerebro
profluentem, efficacissime[1] et contra recentia volnera
et contra veteres canum morsus inspersa farina et
contra ambusta igni cum adipe ac spuma argenti
103 cerussaque et cera. uritur autem offis bubuli fimi
circumlitum in clibanis, dein restinguitur mulierum
lacte teriturque in mortariis admixta aqua pluvia;
ac subinde turbidum transfunditur in aereum vas
emundatum nitro. faex eius intellegitur plumbosis-
sima, quae subsedit in mortario, abiciturque.[2] dein
vas, in quod turbida transfusa sint, opertum linteo
per noctem relinquitur et postero die quidquid
104 innatet effunditur spongeave tollitur. quod ibi subsedit, flos intellegitur ac linteo interposito in sole
siccatur, non ut perarescat, iterumque in mortario
teritur et in pastillos dividitur. ante omnia autem

___

[1] efficacissime *B* : efficaci *rell.* (efficacior *cd. Par.* 6801).
[2] abiciturque *Gelen* : abigiturque *aut* abicitur.

___

[a] See XXXIV. 175.

being more uneven and rougher to the touch, as well as lighter in weight, not so brilliant, and more gritty; the female on the contrary is bright and friable and splits in thin layers and not in globules.

XXXIV. Antimony has astringent and cooling properties, but it is chiefly used for the eyes, since this is why even a majority of people have given it a Greek name meaning 'wide-eye,' because in beauty-washes for women's eyebrows it has the property of magnifying the eyes. Made into a powder with powdered frankincense and an admixture of gum it checks fluxes and ulcerations of the eyes. It also arrests discharge of blood from the brain, and is also extremely effective with a sprinkling of its powder against new wounds and old dog-bites and against burns if mixed with fat and litharge of silver, or lead acetate[a] and wax. It is prepared by being smeared round with lumps of ox dung and burnt in ovens, and then cooled down with women's milk and mixed with rain water and pounded in mortars. And next the turbid part is poured off into a copper vessel after being purified with soda. The lees are recognized by being full of lead, and they settle to the bottom of the mortars and are thrown away. Then the vessel into which the turbid part was poured off is covered with a cloth and left for a night, and the next day anything floating on the surface is poured off or removed with a sponge. The sediment on the bottom is considered the choicest part and is covered with a linen cloth and put to dry in the sun but not allowed to become very dry, and is ground up a second time in the mortar and divided into small tablets. But it is above all essential to limit the amount of heat

urendi modus necessarius est, ne plumbum fiat.
quidam non fimo utuntur coquentes, sed adipe.
alii tritum in aqua triplici linteo saccant [1] faecemque
abiciunt idque, quod defluxit,[2] transfundunt, quidquid subsidat colligentes. emplastris quoque et
collyriis miscent.

105 XXXV. Scoriam in argento Graeci vocant hel-
cysma. vis eius adstringere et refrigerare corpora,
ac remedio est [3] addita [4] emplastris ut molybdaena, de
qua dicemus in plumbo, cicatricibus maxime glutin-
andis, et contra tenesmos dysenteriasque infusa
clysteribus cum myrteo oleo. addunt et in medicamenta, quae vocant liparas, ad excrescentia ulcerum
aut ex attritu facta aut in capite manantia.

106 Fit in isdem metallis et quae vocatur spuma argenti. genera eius tria: optima quam chrysitim vocant,
sequens quam argyritim, tertia quam molybditim.
et plerumque omnes hi colores in isdem tubulis inveniuntur. probatissima est Attica, proxima His-
paniensis. chrysitis ex vena ipsa fit, argyritis ex
argento, molybditis e plumbi ipsius [5] fusura—quae
107 fit Puteolis—et inde habet nomen. omnis autem fit
excocta sua materia ex superiore catino defluens in
inferiorem et ex eo sublata vericulis ferreis atque in

[1] *V.l.* siccant.
[2] effluxit *B.*
[3] ac remedio est *Mayhoff qui et* acribus *aut* viridibus *aut*
a Graecis *coni.*: hac de re *Detlefsen*: quare his *J. Müller*:
acre dies *B*[1]: hac re *B*[2]: *om. rell.*
[4] addita *Mayhoff*: additur.
[5] potius *coni. Mayhoff.*

---

[a] Really into metallic antimony, mistaken for lead.
[b] *I.e.* débris scraped off.

applied to it, so that it may not be turned into lead.[a]
Some people do not employ dung in boiling it but
fat. Others pound it in water and strain it through
three thicknesses of linen cloth and throw away the
dregs, and pour off the liquor that comes through,
collecting all the deposit at the bottom, and this
they use as an ingredient in plasters and eye-washes.

XXXV. The slag in silver is called by the Greeks *Slag of*
the 'draw-off.'[b] It has an astringent and cooling *silver.*
effect on the body, and like sulphuret of lead, of XXXIV.
which we shall speak in dealing with lead, it has 173 *sqq.*
healing properties as an ingredient in plasters, being
extremely effective in causing wounds to close-up,
and when injected by means of syringes, together
with myrtle-oil, as a remedy for straining of the
bowels and dysentery. It is also used as an ingredient in the remedies called emollient plasters used
for proud flesh of gathering sores, or sores caused
by chafing or running ulcers on the head.

The same mines also produce the mineral called *Litharge.*
scum[c] of silver. Of this there are three kinds, with
Greek names meaning respectively golden, silvery
and leaden; and for the most part all these colours
are found in the same ingots. The Attic kind is the
most approved, next the Spanish. The golden scum
is obtained from the actual vein, the silvery from
silver, and the leaden from smelting the actual lead,
which is done at Pozzuoli, from which place it takes
its name.[d] Each kind however is made by heating
its raw material till it melts, when it flows down from
an upper vessel into a lower one and is lifted out of
that with small iron spits and then twisted round on

[c] Litharge, lead monoxide.
[d] Argyritis Puteolana.

ipsa flamma convoluta vericulo, ut sit modici ponderis. est autem, ut ex nomine intellegi potest, fer-
vescentis et futurae [1] materiae spuma. distat a
scoria quo potest spuma a faece distare: alterum,
purgantis se materiae, alterum purgatae vitium est.
108 quidam duo genera faciunt spumae, quae vocant
scirerytida [2] et peumenen,[3] tertium molybdaenam in
plumbo dicendam. spuma, ut sit utilis, iterum
coquitur confractis tubulis ad magnitudinem anulorum.[4] ita accensa follibus ad separandos carbones
cineremque abluitur aceto aut vino simulque restinguitur. quodsi sit argyritis, ut candor ei detur,
magnitudine fabae confracta in fictili coqui iubetur ex
aqua addito in linteolis tritico et hordeo novis, donec
109 ea purgentur. postea vi diebus terunt in mortariis,
ter die abluentes aqua frigida et, cum desinant,[5]
calida, addito sale fossili in libram spumae obolo.
novissimo die dein condunt in plumbeo vase. alii
cum faba candida et tisana cocunt siccantque sole,
alii in lana candida cum faba, donec lanam non
denigret. tunc salem fossilem adiciunt subinde
aqua mutata siccantque diebus xl calidissimis aestatis
nec non in ventre suillo in aqua coquunt exemptamque

[1] et foturae B^2: e fusura coni. Mayhoff: del. Hardouin.
[2] scirerytida B: varia rell.: lythrida Brotier: sclererytida Detlefsen.
[3] reumenen Detlefsen.
[4] avellanarum Caesarius coll. Diosc. V. 102: nucularum coni. Ian.
[5] desinant cd. Par. 6801: desinat rell.: dies desinat Mayhoff: denigrare desinat C. F. W. Müller.

---

[a] Native lead sulphide.

a spit in the actual flame, in order to make it of moderate weight. Really, as may be inferred from its name, it is the scum of a substance in a state of fusion and in process of production. It differs from dross in the way in which the scum of a liquid may differ from the lees, one being a blemish excreted by the material when purifying itself and the other a blemish in the metal when purified. Some people make two classes of scum of silver which they call 'scirerytis' and 'peumene,' and a third, leaden scum,[a] which we shall speak of under the head of lead. To make the scum available for use it is boiled a second time after the ingots have been broken up into pieces the size of finger-rings. Thus after being heated up with the bellows to separate the cinders and ashes from it it is washed with vinegar or wine, and cooled down in the process. In the case of the silvery kind, in order to give it brilliance the instructions are to break it into pieces the size of a bean and boil it in water in an earthen-ware pot with the addition of wheat and barley wrapped in new linen cloths, until the silvery scum is cleaned of impurities. Afterwards they grind it in mortars for six days, three times daily washing it with cold water and, when they have ceased opera-tions, with hot, and adding salt from a salt-mine, an obol weight to a pound of scum. Then on the last day they store it in a lead vessel. Some boil it with white beans and pearl-barley and dry it in the sun, and others boil it with beans in a white woollen cloth till it ceases to discolour the wool; and then they add salt from a salt-mine, changing the water from time to time, and put it out to dry on the 40 hottest days of summer. They also boil it in a sow's

XXXIV. 173.
Lead
sulphide.

nitro fricant et ut supra terunt in mortariis cum sale.
sunt qui non coquant, sed cum sale terant et adiecta
110 aqua abluant. usus eius ad collyria et cuti [1] mulierum cicatricum foeditates tollendas maculasque,
abluendum [2] capillum. vis autem siccare, mollire,
refrigerare, temperate purgare, explere ulcera,
tumores lenire; talibusque [3] emplastris additur et
liparis supra dictis. ignes etiam sacros tollit cum ruta
myrtisque et aceto, item perniones cum myrtis et
cera.[4]

111 XXXVI. In argentariis metallis invenitur minium
quoque, et nunc inter pigmenta magnae auctoritatis
et quondam apud Romanos non solum maximae, sed
etiam sacrae. enumerat auctores Verrius, quibus
credere necesse sit Iovis ipsius simulacri faciem diebus
festis minio inlini solitam triumphantiumque corpora;
112 sic Camillum triumphasse; hac religione etiamnum
addi in unguenta cenae triumphalis et a censoribus
in primis Iovem miniandum locari. cuius rei causam
equidem miror, quamquam et hodie id expeti constat
Aethiopum populis totosque eo tingui proceres, hunc

---

[1] cuti *Mayhoff* (*qui et* cutem *coni.*): litum B^1: situm B^2:
varia rell.

[2] *V.l.* et (*aut* et ad) abluendum: ad alendum *Ian.*

[3] albisque *Fröhner.*

[4] cera *edd. vett.*: cetera.

---

[a] Sulphide of mercury (' cinnabar ') is meant here. True
red lead was properly called *minium secundarium*. See § 119.

paunch in water, and when they take it out rub it with soda, and grind it in mortars with salt as above. In some cases people do not boil it but grind it up with salt and then add water and rinse it. It is used to make an eye-wash and for women's skins to remove ugly scars and spots and as a hair-wash. Its effect is to dry, to soften, to cool, to act as a gentle purge, to fill up cavities caused by ulcers, and to soften tumours; it is used as an ingredient in plasters serving these purposes, and for the emollient plasters mentioned above. Mixed with § 105. rue and myrtle and vinegar, it also removes erysipelas, and likewise chilblains if mixed with myrtle and wax.

XXXVI. *Minium* or cinnabar [a] also is found in *Cinnabar.* silver mines; it is of great importance among pigments at the present day, and also in old times it not only had the highest importance but even sacred associations among the Romans. Verrius gives a list of writers of unquestionable authority who say that on holidays it was the custom for the face of the statue of Jupiter himself to be coloured with cinnabar, as well as the bodies of persons going in a triumphal procession, and that Camillus was so coloured in his triumph, and that under the same ritual it was usual even in their day for cinnabar to be added to the unguents used at a banquet in honour of a triumph, and that one of the first duties of the Censors was to place a contract for painting Jupiter with cinnabar. For my own part I am quite at a loss to explain the origin of this custom, although at the present day the pigment in question is known to be in demand among the nations of Ethiopia whose chiefs colour themselves all over with it, and

ibi deorum simulacris colorem esse. quapropter
diligentius persequemur omnia de eo.

113 XXXVII. Theophrastus lxxxx annis ante Praxi-
bulum Atheniensium magistratum—quod tempus
exit in urbis nostrae cccxlviiii [1] annum—tradit inventum minium a Callia Atheniense initio sperante
aurum excoqui posse harenae rubenti in metallis
argenti; hanc fuisse originem eius, reperiri autem
114 iam tum [2] in Hispania, sed durum et harenosum, item
apud Colchos in rupe quadam inaccessa, ex qua
iaculantes decuterent; id esse adulterum, optimum
vero supra Ephesum Cilbianis agris harena cocci
colorem habente, hanc teri, dein lavari farinam et
quod subsidat iterum lavari; differentiam artis esse,
quod alii minium faciant prima lotura, apud alios id
esse dilutius, sequentis autem loturae optimum.

115 XXXVIII. Auctoritatem colori fuisse non miror.
iam enim Troianis temporibus rubrica in honore erat
Homero teste, qui naves ea commendat, alias circa
pigmenta picturasque rarus. milton vocant Graeci
116 miniumque cinnabarim. unde natus error Indicae [3]
cinnabaris nomine.[4] sic enim appellant illi saniem
draconis elisi elephantorum morientium pondere

[1] cccxlviiii *Hermolaus Barbarus*: ccccxxxviiii *Casaubon*:
ccxlviiii.

[2] nativum *coni. Hardouin.*

[3] Indicae *K. C. Bailey*: indicio *B*: indico *rell.*

[4] cinnabaris nomine *cd. Par.* 6801: nominum (*om.* cinnabaris) *B*: nomine (*om.* cinnabaris) *rell.*: inscitia nominum
Mayhoff.

---

[a] *De Lap.* 59, 58.
[b] 315 b.c.
[c] This was really an exudation (still called ' dragon's
blood ') from species of the oriental plant *Dracaena* or *Pterocarpus.*

with whom the statues of the gods are of that colour.
On that account we will investigate all the facts
concerning it more carefully.

XXXVII. Theophrastus [a] states that cinnabar
was discovered by an Athenian named Callias, 90
years before the archonship [b] of Praxibulus at 405 B.C.
Athens—this date works out at the 349th year of
our city, and that Callias was hoping that gold
could by firing be extracted from the red sand
found in silver mines; and that this was the origin
of cinnabar, although cinnabar was being found
even at that time in Spain, but a hard and sandy
kind, and likewise in the country of the Colchi on a
certain inaccessible rock from which the natives
dislodged it by shooting javelins, but that this is
cinnabar of an impure quality whereas the best is
found in the Cilbian territory beyond Ephesus, where
the sand is of the scarlet colour of the kermes-insect;
and that this is ground up and then the powder is
washed and the sediment that sinks to the bottom
is washed again; and that there is a difference of
skill, some people producing cinnabar at the first
washing while with others this is rather weak and
the product of the second washing is the best.

XXXVIII. I am not surprised that the colour had
an important rank, for as far back as Trojan times *Il.* II. 637.
red ochre was highly valued, as evidenced by Homer,
who speaks of it as a distinguished colour for ships,
although otherwise he rarely alludes to colours and
paintings. The Greek name for it is ' miltos,' and
they call *minium* ' cinnabar.' This gave rise to a
mistake owing to the name ' Indian cinnabar,' for
that is the name the Greeks give to the gore [c] of a
snake crushed by the weight of dying elephants,

87

permixto utriusque animalis sanguine, ut diximus,
neque est alius colos, qui in pictura proprie sanguinem reddat. illa cinnabaris antidotis medicamen-
tisque utilissima est. at, Hercules, medici, quia
cinnabarim vocant, utuntur hoc minio, quod[1]
venenum esse paulo mox docebimus.

117 XXXIX. Cinnabari veteres quae etiam nunc
vocant monochromata pingebant. pinxerunt et
Ephesio minio, quod derelictum est, quia curatio
magni operis erat. praeterea utrumque nimis acre
existimabatur. ideo transiere ad rubricam et Sino-
pidem, de quibus suis locis dicam. cinnabaris
adulteratur sanguine caprina aut sorbis tritis. pretium sincerae nummi L.

118 XL. Iuba minium nasci et in Carmania tradit,
Timagenes[2] et in Aethiopia, sed neutro ex loco
invehitur ad nos nec fere aliunde quam ex Hispania,
celeberrimo Sisaponensi regione in Baetica miniario
metallo vectigalibus populi Romani, nullius rei dili-
gentiore custodia. non licet ibi perficere id excoquique[3]; Romam adfertur[4] vena signata, ad bina
milia fere pondo annua, Romae autem lavatur, in
vendendo pretio statuta[5] lege, ne modum excederet
HS LXX in libras. sed adulteratur multis modis, unde
119 praeda societati. namque est alterum genus omni-

---

[1] cinnabarim minium v., u.h. quod *coni. Warmington.*
[2] Timaeus *coni. Pintianus.*
[3] excoquique *cdd.* (excoqui quae *B* : excoctique *cd. Leid.
Voss.*) : excoctaque *coni. Mayhoff.*
[4] *V.ll.* refertur, defertur, deferuntur : perferuntur *edd. vett.*
[5] *V.l.* statuto.

when the blood of each animal gets mixed together, as we have said; and there is no other colour that properly represents blood in a picture. That kind of cinnabar is extremely useful for antidotes and medicaments. But our doctors, I swear, because they give the name of cinnabar to *minium* also, employ this *minium*, which as we shall soon show is a poison. §124

XXXIX. In old times 'dragon's-blood' cinnabar was used for painting the pictures that are still called monochromes, 'in one colour.' Cinnabar from Ephesus was also used for painting, but this has been given up because pictures in that colour were a great amount of trouble to preserve. Moreover both colours were thought excessively harsh; consequently painters have gone over to red-ochre and Sinopic ochre, pigments about which I shall XXXV. speak in the proper places. Cinnabar is adulterated 30 *sqq.* with goat's blood or with crushed service-berries. The price of genuine cinnabar is 50 sesterces a pound.

XL. Juba reports that cinnabar is also produced in Carmania, and Timagenes says it is found in Ethiopia as well, but from neither place is it exported to us, and from hardly any other either except from Spain, the most famous cinnabar mine for the revenues of the Roman nation being that of Almaden in the Baetic region, no item being more carefully safeguarded: it is not allowed to smelt and refine the ore upon the spot, but as much as about 2000 lbs. per annum is delivered to Rome in the crude state under seal, and is purified at Rome, the price in selling it being fixed by law established at 70 sesterces a pound, to prevent its going beyond limit. But it is adulterated in many ways, which is a source *Red lead and* of plunder for the company. For there is in fact *cinnabar.*

bus fere argentariis itemque plumbariis metallis,
quod fit exusto lapide venis permixto, non ex illo,
cuius vomicam argentum vivum appellavimus—is
enim et ipse in argentum ⟨vivum⟩ [1] excoquitur—, sed
ex aliis simul repertis. steriles etiam plumbi deprehenduntur [2] solo colore nec nisi in fornacibus
rubescentes exustique tunduntur in farinam. hoc
est secundarium minium perquam paucis notum,
120 multum infra naturales illas harenas. hoc ergo
adulteratur minium in officinis sociorum, et vilius [3]
Syrico. quonam modo Syricum fiat suo loco docebimus; sublini autem Syrico minium compendi ratio
demonstrat. et alio modo pingentium furto oppor-
tunum est, plenos subinde abluentium penicillos.
121 sidit autem in aqua constatque furantibus. sincero
cocci nitor esse debet, secundarii autem splendor in
parietibus sentit [4] robiginem,[5] quamquam hoc robigo quaedam metalli est. Sisaponensibus autem
miniariis sua vena harenae sine argento. excoquitur
auri modo; probatur auro candente, fucatum enim

[1] vivum *add. K. C. Bailey.*
[2] de micae prehenduntur *B*[1]: micae. deprehenduntur *B*[2]:
deprehenduntur.
[3] et vilius *Mayhoff*: et ubivis *Ian*: et vivis *B*: item *rell.*
[4] *V.l.* sentire.
[5] robiginem *K. C. Bailey*: plumbaginem *Mayhoff*: uliginem
Caesarius: imaginem.

---

[a] Probably the true red-lead (prepared from cerusite,
natural lead carbonate).
[b] Of sulphide of mercury. See § 111, note; § 118.
[c] This is not true.

another kind *a* of *minium*, found in almost all silver-mines, and likewise lead-mines, which is made by smelting a stone that has veins of metal running through it, and not obtained from the stone the round drops of which we have designated quick-silver—for that stone also if fired yields quicksilver—but from other stones found at the same time. These have no quicksilver and are detected only by their leaden colour, and only when they turn red in the furnaces, and after being thoroughly smelted they are pulverized by hammering. This gives a *minium* of second rate quality, which is known to very few people, and is much inferior to the natural sands we have mentioned. It is this then that is used for adulterating real *minium* in the factories of the company, but a cheaper kind is adulterated with Syrian : the preparation of the latter will be des-cribed in the proper place ; but the process of giving cinnabar and red-lead a treatment of Syrian is detected by calculation when the one is weighed against the other. Cinnabar also, with red-lead, affords an opportunity for pilfering by painters in another way, if they wash out their brushes imme-diately when full of paint; the cinnabar or the red-lead settles at the bottom of the water and stays there for the pilferers. Pure cinnabar ought to have the brilliant colour of the scarlet kermes-insect, while the shine of that of the second quality when used on wall-paintings is affected by rust, although this is itself a sort of metallic rust. In the cinnabar mines *b* of Almaden the vein of sand is pure, without silver. It is melted like gold *c*; it is assayed by means of gold made red hot, as if it has been adulterated it turns black, but if genuine

§ 99.

XXXV. 40.

nigrescit, sincerum retinet colorem. invenio et calce
adulterari, ac simili ratione ferri candentis lamna, si
122 non sit aurum, deprehendi. inlito solis atque lunae
contactus inimicus. remedium, ut pariete siccato
cera Punica cum oleo liquefacta candens saetis in-
ducatur iterumque admotis gallae carbonibus inura-
tur ad sudorem usque, postea candelis subigatur ac
deinde linteis puris, sicut et marmora nitescunt. qui
minium in officinis poliunt, faciem laxis vesicis inli-
gant, ne in respirando pernicialem pulverem trahant
et tamen super [1] illas spectent. minium in voluminum quoque scriptura usurpatur clarioresque
litteras vel in muro [2] vel in marmore, etiam in sepulchris, facit.

123 XLI. Ex secundario invenit vita et hydrargyrum
in vicem argenti vivi, paulo ante dilatum. fit autem
duobus modis: aereis mortariis pistillisque trito minio
ex aceto aut patinis fictilibus impositum ferrea con-
cha, calice coopertum, argilla superinlita, dein sub
patinis accenso [3] follibus continuis igni atque ita
calici [4] sudore deterso, qui fit argenti colore et aquae
liquore. idem guttis dividi facilis et lubrico umore

---

[1] super *Mayhoff coll. Diosc.* V. 109 : ut per *cdd.*
[2] muro *Detlefsen* : aere *Hübner* : auro *cdd.*
[3] accenso *quidam ap. Dalecamp* : accensum.
[4] *V.l.* calicis.

---

[a] This seems to be the meaning here; *secundario* would
not refer to the minium of the second quality (see above,
§ 111, note; § 119), for hydrargyrum was made from the
sulphide of mercury of § 111.

it keeps its colour. I find that it is also adulterated with lime, and this can be detected in a similar way with a sheet of red-hot iron if there is no gold available. A surface painted with cinnabar is damaged by the action of sunlight and moonlight. The way to prevent this is to let the wall dry and then to coat it with Punic wax melted with olive oil and applied by means of brushes of bristles while it is still hot, and then this wax coating must be again heated by bringing near to it burning charcoal made of plant-galls, till it exudes drops of perspiration, and afterwards smoothed down with waxed rollers and then with clean linen cloths, in the way in which marble is given a shine. Persons polishing cinnabar in workshops tie on their face loose masks of bladder-skin, to prevent their inhaling the dust in breathing, which is very pernicious, and nevertheless to allow them to see over the bladders. Cinnabar is also used in writing books, and it makes a brighter lettering for inscriptions on a wall or on marble even in tombs.

XLI. Of secondary importance [a] is the fact that experience has also discovered a way of getting hydrargyrum or artificial quicksilver as a substitute for real quicksilver; we postponed the description of this a little previously. It is made in two ways, §§ 64, 100. by pounding red-lead in vinegar with a copper pestle in a copper mortar, or it is put in an iron shell in flat earthenware pans, and covered with a convex lid smeared on with clay, and then a fire is lit under the pans and kept constantly burning by means of bellows, and so the surface moisture (with the colour of silver and the fluidity of water) which forms on the lid is wiped off it. This moisture is also easily divided into drops and rains down freely with slippery

124 compluere.[1] quod cum venenum esse conveniat,
omnia, quae de minio in medicinae usu traduntur,
temeraria arbitror, praeterquam fortassis inlito
capiti ventrive sanguinem sisti, dum ne qua penetret
in viscera ac volnus attingat. aliter utendum non
equidem censeam.

125 XLII. Hydrargyro argentum inauratur solum nunc
prope, cum et in aerea simili modo duci debeat.
sed eadem fraus, quae in omni parte vitae ingenio-
sissima est, viliorem excogitavit materiam, ut docuimus.

126 XLIII. Auri argentique mentionem comitatur
lapis, quem coticulam appellant, quondam non solitus
inveniri nisi in flumine Tmolo, ut auctor est Theophrastus, nunc vero passim. alii Heraclium, alii
Lydium vocant. sunt autem modici, quaternas
uncias longitudinis binasque latitudinis non excedentes. quod a sole fuit in iis, melius quam quod
a terra. his coticulis periti cum e vena ut lima
rapuerunt experimento ramentum,[2] protinus dicunt
quantum auri sit in ea, quantum argenti vel aeris,
scripulari differentia, mirabili ratione non fallente.

127 XLIV. Argenti duae differentiae. vatillis ferreis
candentibus ramento inposito, quod candidum per-

---

[1] compluere B : confluere rell.
[2] experimento ramentum L. C. Purser : experimentum.

---

[a] Both kinds of minium—the sulphide of mercury and the
lead carbonate—are poisonous.
[b] De Lap. 47, 46.

fluidity. And as cinnabar and red-lead [a] are admitted to be poisons, all the current instructions on the subject of its employment for medicinal purposes are in my opinion decidedly risky, except perhaps that its application to the head or stomach arrests hæmorrhage, provided that it does not find access to the vital organs or come in contact with a lesion. In any other way for my own part I would not recommend its employment.

XLII. At the present time silver is almost the only substance that is gilded with artificial quicksilver, though really a similar method ought to be used in coating copper. But the same fraudulence which is so extremely ingenious in every department of life has devised an inferior material, as we have § 100. shown.

XLIII. With the mention of gold and silver *Touchstone.* goes a description of the stone called the touch stone, formerly according to Theophrastus [b] not usually found anywhere but in the river Tmolus, but now found in various places. Some people call it Heraclian stone and others Lydian. The pieces are of a moderate size, not exceeding four inches in length and two in breadth. The part of these pieces that has been exposed to the sun is better than the part on the ground. When experts using this touchstone, like a file, have taken with it a scraping from an ore, they can say at once how much gold it contains and how much silver or copper, to a difference of a scruple, their marvellous calculation not leading them astray.

XLIV. There are two points in which silver shows a variation. A shaving that remains perfectly white when placed on white-hot iron shovels is passed

95

maneat, probatur.   proxima bonitas rufo, nulla nigro.
sed experimento quoque fraus intervenit. servatis
in urina virorum vatillis inficitur ita ramentum obiter
dum uritur candoremque mentitur. est aliquod
experimentum politi et in halitu hominis, si sudet
protinus nubemque discutiat.

128   XLV. Lamnas duci,[1] speciem[2] fieri[3] non nisi ex
optimo posse creditum.[4] fuerat id integrum, sed id
quoque iam fraude corrumpitur.

Est[5] natura mira imagines reddendi, quod reper-
cusso aëre atque in oculos regesto fieri convenit.
eadem vi sic[6] in speculi usu polita crassitudine pau-
lumque propulsa dilatatur in inmensum magnitudo
imaginum. tantum interest, repercussum illum ex-
129 cipiat an respuat. quin etiam pocula ita figurantur
expulsis[7] intus crebris ceu speculis, ut vel uno intuente
totidem populus imaginum fiat. excogitantur et
monstrifica, ut in templo Zymrnae dicata. id evenit
figura materiae. plurimum refert concava sint et
poculi modo an parmae Threcidicae, media depressa
an elata, transversa an obliqua, supina an infesta,

---

[1] duci et *cd. Par. Lat.* 6801: duci *rell.*: duci in *Mayhoff.*
[2] speciem *B*: specula *rell.*
[3] fieri *cdd.*: vitri *Mayhoff.*
[4] credimus *B.*
[5] est *Mayhoff*: sed.
[6] vi sic *Mayhoff*: vis *Sillig*: visi *B*: vi *cd. Par.* 6801:
nisi *rell.*
[7] *V.l.* exsculptis.

---

[a] When it is concave.
[b] When it is convex or plane.
[c] *I.e.* the major axis of an oval mirror, or of a convex or
concave oval centre of a special kind of mirror.

as good, while if it turns red it is of the next quality, and if black it has no value at all. But fraud has found its way even into this test; if the shovels are kept in men's urine the silver shaving is stained by it during the process of being burnt, and counterfeits whiteness. There is also one way of testing polished silver in a man's breath—if it at once forms surface moisture and dissipates the vapour.

XLV. It has been believed that only the best silver is capable of being beaten out into plates and producing an image. This was formerly a sound test, but nowadays this too is spoiled by fraud.

Reflecting qualities and uses of silver. Mirrors.

Still, the property of reflecting images is marvellous; it is generally agreed that it takes place owing to the repercussion of the air which is thrown back into the eyes. In a similar way, owing to the same force, in employing a mirror if the thickness of the metal has been polished and beaten out into a slightly concave shape the size of the objects reflected is enormously magnified: such a difference does it make whether the surface welcomes[a] the air in question or flings it back.[b] Moreover bowls can be made of such a shape, with a number of looking-glasses so to speak beaten outward inside them, that if only a single person is looking into them a crowd of images is formed of the same number as the facets in question. Ingenuity even devises vessels that do conjuring tricks, for instance those deposited as votive offerings in the temple at Smyrna: this is brought about by the shape of the material, and it makes a very great difference whether the vessels are concave and shaped like a bowl or convex like a Thracian shield, whether their centre is recessed or projecting, whether the oval[c] is horizontal or oblique,

97

qualitate excipientis figurae torquente venientes um-
130 bras; neque enim est aliud illa imago quam digesta
claritate materiae accipientis umbra.[1] atque ut
omnia de speculis peragantur in hoc loco, optima aput
maiores fuerant Brundisina, stagno et aere mixtis.
praelata sunt argentea; primus fecit Pasiteles Magni
Pompei aetate. nuper credi coeptum certiorem
imaginem reddi auro opposito vitris.[2]

131   XLVI. Tinguit Aegyptus argentum, ut in vasis
Anubim suum spectet, pingitque, non caelat, argentum. unde transiit materia et ad triumphales
statuas; mirumque, crescit pretium fulgoris excae-
cati. id autem fit hoc modo; miscentur argento
tertiae aeris Cyprii tenuissimi, quod coronarium
vocant, et sulpuris vivi quantum argenti; conflantur
ita in fictili circumlito argilla; modus coquendi,
donec se ipsa opercula aperiant. nigrescit et ovi
indurati luteo, ut tamen aceto et creta deteratur.

132   Miscuit denario triumvir Antonius ferrum, miscent
aera falsae monetae,[3] alii et [4] ponderi [5] subtrahunt,
cum sit iustum LXXXIIII e libris signari. igitur ars

[1] umbra *Ian*: umbram.
[2] vitris *K. C. Bailey*: vitris aversis *coni. D'Arcy Thompson*: aversis.
[3] *V.l.* falsa moneta.
[4] alii et *Mayhoff*: alii e *cd. Par.* 6801: alia *aut* aliae *aut* alii (*om.* et) *rell.*
[5] ponderi *Urlichs*: pondere *aut* ponderae *aut* pondera. alii de pondere *coni. Mayhoff.*

---

[a] Cf. § 94 and note.
[b] *vitris* is K. C. Bailey's conjecture. The sentence cannot refer to *silver* mirrors. Roman glass mirrors, backed usually with lead, have been found, but seem to belong to a later time than Pliny.
[c] This was employed to make imitation gold crowns for use on the stage.

laid flat or placed upright, as the quality of the shape
receiving the shadows twists them as they come:
for in fact the image in a mirror is merely the shadow
arranged by the brilliance of the material receiving
it. And in order to complete the whole subject of
mirrors in this place, the best of those known in old
days were those made at Brindisi of a mixture of
stagnum *a* and copper. Silver mirrors have come to
be preferred; they were first made by Pasiteles in
the period of Pompey the Great. But it has *c.* 106-48.
recently come to be believed that a more reliable ᴮ·ᶜ·
reflection is given by applying a layer of gold to the
back of glass.*b*

XLVI. The people of Egypt stain their silver so
as to see portraits of their god Anubis in their
vessels; and they do not engrave but paint their
silver. The use of that material thence passed
over even to our triumphal statues, and, wonderful
to relate, its price rises with the dimming of its
brilliance. The method adopted is as follows: with
the silver is mixed one third its amount of the very
fine Cyprus copper called chaplet-copper *c* and the
same amount of live sulphur as of silver, and then
they are melted in an earthenware vessel smeared
round with potter's clay; the heating goes on till the
lids of the vessels open of their own accord. Silver
is also turned black by means of the yolk of a hard-
boiled egg, although the black can be rubbed off
with vinegar and chalk.

The triumvir Antony alloyed the silver denarius *Debased and*
with iron, and forgers put an alloy of copper in silver *forged silver*
coins, while others also reduce the weight, the *coinage.*
proper coinage being 84 denarii from a pound of
silver. Consequently a method was devised of

facta denarios probare, tam iucunda plebei lege, ut
Mario Gratidiano vicatim tota [1] statuas dicaverit.
mirumque, in hac artium sola vitia discuntur et falsi
denarii spectatur exemplar pluribusque veris denariis
adulterinus emitur.

133   XLVII. Non erat apud antiquos numerus ultra
centum milia; itaque et hodie multiplicantur haec,
ut decies centena aut saepius dicantur. faenus hoc
fecit nummusque percussus, et sic quoque aes alienum
etiamnum appellatur. postea Divites cognominati,
dummodo notum sit eum, qui primus hoc cognomen
134 acceperit, decoxisse creditoribus suis. ex eadem
gente M. Crassus negabat locupletem esse nisi qui
reditu annuo legionem tueri posset. in agris ʜs
⎡ᴍᴍ⎤ possedit Quiritium post Sullam divitissimus, nec
fuit satis nisi totum Parthorum usurpasset aurum;
atque ut memoriam quidem opum occupaverit—
iuvat enim insectari inexplebilem istam habendi
cupidinem—: multos postea cognovimus servitute
liberatos opulentiores, pariterque tres Claudii principatu paulo ante Callistum, Pallantem, Narcissum.
135 atque ut hi omittantur, tamquam adhuc rerum

---

[1] totas B, cd. Par. 6801.

---

[a] Crassus the so-called 'triumvir' was defeated by the
Parthians at Carrhae (Haran) in 53 ʙ.ᴄ., and assassinated
when treating for peace. His head was cut off and sent to
the Parthian king, who caused melted gold to be poured into
its mouth, saying 'Sate thyself now with the metal for which
when alive thou wert so greedy' (Dio Cassius XL. 27).

assaying the denarius, under a law that was so popular that the common people unanimously district by district voted statues to Marius *Before* 82 Gratidianus. And it is a remarkable thing that in B.C. this alone among arts spurious methods are objects of study, and a sample of a forged denarius is carefully examined and the adulterated coin is bought for more than genuine ones.

XLVII. In old days there was no number standing *Examples of* for more than 100,000, and accordingly even to-day *private and* we reckon by multiples of that number, using the *public* expression times ' ten times one hundred thousand ' *wealth.* or larger multiples. This was due to usury and to the introduction of coined money, and also on the same lines we still speak of money owed as ' somebody else's copper.' Afterwards ' Dives,' ' Rich,' became a family surname, though it must be stated that the man who first received this name ran through his creditors' money and went bankrupt. Afterwards Marcus Crassus, who was a member of the Rich family, used to say that nobody was a wealthy man except one who could maintain a legion of troops on his yearly income. He owned landed property worth two hundred million sesterces, being the richest Roman citizen after Sulla. Nor was he satisfied without getting possession of the whole of the Parthians' gold [a] as well; and although it is true he was the first to win lasting reputation for wealth—it is a pleasant task to stigmatize insatiable covetousness of that sort—we have known subsequently of many liberated slaves who have been wealthier, and three at the same time not long before our own days in the period of the emperor Claudius, namely Callistus, A.D. 41-54. Pallas and Narcissus. And to omit these persons,

potiantur, C. Asinio Gallo C. Marcio Censorino cos.
a. d. vi Kal. Febr. C. Caecilius C. l. Isidorus testamento suo edixit, quamvis multa bello civili perdidis-
set, tamen relinquere servorum [1] $\overline{\text{IIII}}$ cxvi, iuga boum
$\overline{\text{III}}$ DC, reliqui pecoris $\overline{\text{CCLVII}}$, in numerato HS $\overline{|DC|}$,

136 funerari se iussit HS $\overline{|XI|}$.[2] congerant excedentes
numerum opes, quota tamen portio erunt Ptolemaei,
quem Varro tradit Pompeio res gerente circa Iudae-
am octona milia equitum sua pecunia toleravisse,
mille convivas totidem aureis potoriis, mutantem
ea vasa cum ferculis, saginasse! quota vero ille ipse

137 —neque enim de regibus loquor—portio fuerit
Pythis Bithyni, qui platanum auream vitemque
nobiles illas Dario regi donavit, Xerxis copias, hoc est
$|\overline{\text{VII}}|$ $\overline{\text{LXXXVIII}}$[3] hominum, excepit epulo, stipendium
quinque mensum frumentumque pollicitus, ut e
quinque liberis in dilectu senectuti suae unus saltem
concederetur! hunc quoque ipsum aliquis comparet
Croeso regi! quae, malum, amentia est id in vita
cupere, quod aut et servis contigerit aut ne in regibus
quidem invenerit finem!

138   XLVIII. Populus R. stipem spargere coepit Sp.
Postumio Q. Marcio cos.; tanta abundantia pecuniae
erat, ut eam conferret L. Scipioni, ex qua is ludos

---

[1] se servorum *coni. Mayhoff.*
[2] $|\overline{\text{XII}}|$ *Ian*; $|\overline{\text{X}}|$ *Detlefsen*: xi milibus *cd. Par.* 6801: xi
rell. (ixi *B*).
[3] $|\overline{\text{VII}}|$ *Ian*: $\overline{\text{LXXXVIII}}$ *Sillig*: *varia cdd.*

---

[a] *I.e.* still alive and ruling the Empire, so that it would be
dangerous to speak of them.
[b] Probably Auletes, King of Egypt 80–51 B.C.
[c] See Herodotus VII. 27, 38.

as if they were still in sovereign power,[a] there is
Gaius Caecilius Isidorus, the freedman of Gaius
Caecilius who in the consulship of Gaius Asinius 8 B.C.
Gallus and Gaius Marcius Censorinus executed a
will dated January 27 in which he declared that in
spite of heavy losses in the civil war he nevertheless
left 4116 slaves, 3600 pairs of oxen, 257,000 head of
other cattle, and 60 million sesterces in cash, and he
gave instructions for 1,100,000 to be spent on his
funeral.   But let them amass uncountable riches,
yet what fraction will they be of the riches of the
Ptolemy [b] who is recorded by Varro, at the time
when Pompey was campaigning in the regions 63 B.C.
adjoining Judaea, to have maintained 8000 horse at
his own charges, to have given a lavish feast to a
thousand guests, with 1,000 gold goblets, which were
changed at every course; and then what fraction
would his own estate have been (for I am not speaking
about kings) of that of the Bithynian Pythes,[c] who presented the famous gold plane tree and vine to King
Darius, and gave a banquet to the forces of Xerxes, 480 B.C.
that is 788,000 men, with a promise of five months'
pay and corn on condition that one at least of his
five children when drawn for service should be left to
cheer his old age?   Also let anyone compare even
Pythes himself with King Croesus!   What madness
it is (damn it all!), to covet a thing in our lifetime that
has either fallen to the lot even of slaves or has
reached no limit even in the desires of Kings!

XLVIII. The   Roman   nation   began   lavishing
donations in the consulship of Spurius Postumius 186 B.C.
and Quintus Marcius: so abundant was money at
that date that they contributed funds for Lucius
Scipio to defray the cost of games which he cele-

fecit. nam quod Agrippae Menenio sextantes aeris
in funus contulit, honoris id necessitatisque propter
paupertatem Agrippae, non largitionis esse duxerim.[1]

139 XLIX. Vasa ex argento mire inconstantia humani
ingenii variat nullum genus officinae diu probando.
nunc Furniana,[2] nunc Clodiana, nunc Gratiana—
etenim tabernas mensis adoptamus—, nunc ana-
glypta asperitatemque exciso [3] circa liniarum picturas
140 quaerimus, iam vero et mensas repositoriis inponi-
mus ad sustinenda opsonia, interradimus alia, ut
quam plurimum lima perdiderit. vasa cocinaria ex
argento fieri Calvus orator quiritat; at nos carrucas
argento caelare invenimus, nostraque aetate Poppaea
coniunx Neronis principis soleas delicatioribus iu-
mentis suis ex auro quoque induere iussit.

141 L. Triginta duo libras argenti Africanus sequens
heredi reliquit idemque, cum de Poenis triumpharet,
IIII CCCLXX pondo transtulit. hoc argenti tota Car-
thago habuit illa terrarum aemula, quot mensarum
postea apparatu victa! Numantia quidem deleta
idem Africanus in triumpho militibus ✕ VII dedit. o

---

[1] *V.l.* dixerim.   [2] *V.l.* Firmiana.
[3] *V.l.* excisa.

---

[a] In performance of a vow that he made in the war with
Antiochus III, King of Syria, victoriously concluded in
190 B.C.
[b] These various kinds of plate are named after the silversmiths who introduced them. For the last cf. Martial IV. 39:
Argenti genus omne comparasti . . . Nec desunt tibi vera
Gratiana.
[c] Scipio Aemilianus.

brated.[a] As for the national contribution of one-sixth of an *as* per head for the funeral of Menenius Agrippa, I should consider this as a mark of respect and also a measure rendered necessary by Agrippa's poverty, and not a matter of lavish generosity.

XLIX. Fashions in silver plate undergo marvellous variations owing to the vagaries of human taste, no kind of workmanship remaining long in favour. At one time Furnian plate is in demand, at another Clodian, at another Gratian [b]—for we make even the factories feel at home at our tables—at another time the demand is for embossed plate and rough surfaces, where the metal has been cut out along the painted lines of the designs, while now we even fit removable shelves on our sideboards to carry the viands, and other pieces of plate we decorate with filigree, so that the file may have wasted as much silver as possible. The orator Calvus complainingly cries that cooking-pots are made of silver; but it is we who invented decorating carriages with chased silver, and it was in our day that the emperor Nero's wife Poppaea had the idea of even having her favourite mules shod with gold.

L. The younger Africanus [c] left his heir thirty-two pounds weight of silver, and the same person paraded 4370 pounds of silver in his triumphal procession after the conquest of Carthage. This was the amount of silver owned by the whole of Carthage, Rome's rival for the empire of the world, yet subsequently beaten in the show of plate on how many dinner-tables! Indeed after totally destroying Numantia the same Africanus at his triumph gave a largess of seven denarii a head to his troops—warriors not unworthy of such a general who were

Silver plate etc.

491 B.C.

82–c. 47 B.C.

129 B.C.

146 B.C.

133–2 B.C.

viros illo imperatore dignos, quibus hoc satis fuit!
frater eius Allobrogicus primus omnium pondo mille
habuit, at Drusus Livius in tribunatu plebei $\overline{\text{x}}$.[1]
142 nam propter x [2] pondo notatum a censoribus trium-
phalem senem fabulosum iam videtur, item Catum
Aelium, cum legati Aetolorum in consulatu pranden-
tem in fictilibus adissent, missa ab iis vasa argentea
non accepisse neque aliud habuisse argenti ad supre-
mum vitae diem quam duo pocula, quae L. Paulus
socer ei ob virtutem devicto Perseo rege donavisset.
143 invenimus legatos Carthaginiensium dixisse nullos
hominum inter sese benignius vivere quam Romanos.
eodem enim argento apud omnes cenitavisse ipsos.
at, Hercules, Pompeium Paulinum, Arelatensis
equitis Romani filium paternaque gente pellitum,
$\overline{\text{XII}}$ pondo argenti habuisse apud exercitum ferocissi-
144 mis gentibus oppositum scimus; (LI.) lectos vero iam
pridem mulierum totos operiri argento, pridem [3] et
triclinia. quibus argentum addidisse primus traditur
Carvilius Pollio eques Romanus, non ut operiret aut
Deliaca specie faceret, sed Punicana; eadem et

---

[1] $\overline{\text{x}}$ coni. Mayhoff: XI milia cd. Par. 6801: x rell. ($\times$ B).
[2] x Freinshem coll. Liv. Ep. XIV, Val. Max. II.9.4, etc.:
quinque.
[3] pridem coni. Mayhoff: quaedam cdd. (quidem cd. Par.
Lat. 6797).

---

[a] Q. Fabius Maximus Allobrogicus, consul in 121, and in
reality a nephew of Aemilianus.
[b] P. Cornelius Rufinus.
[c] C. Fabricius and Q. Aemilius. Cf. § 153.

satisfied with that amount! His brother
Allobrogicus [a] was the first person who ever owned 121 B.C.
1000 lbs. weight of silver, whereas Livius Drusus
when tribune of the people had 10,000 lbs. For 91 B.C.
that an old warrior,[b] honoured with a triumphal
procession, incurred the notice of the censors [c] for 275 B.C.
possessing ten pounds weight of silver—that nowadays
seems legendary, and the same as to Catus Aelius's
not accepting the silver plate presented to him by
the envoys from Aetolia who during his consulship 198 B.C.
had found him eating his lunch off earthenware, and
as to his never till the last day of his life having
owned any other silver but the two bowls given to
him by his wife's father Lucius Paulus in recognition
of his valour at the time when King Perseus was 168 B.C.
conquered. We read that the Carthaginian ambas-
sadors declared that no race of mankind lived on
more amicable terms with one another than the
Romans, inasmuch as in a round of banquets they had
found the same service of plate in use at every
house! But, good heavens, Pompeius Paulinus the
son of a Knight of Rome at Arles and descended on
his father's side from a tribe that went about clad
in skins, to our knowledge had 12,000 lbs. weight of
silver plate with him when on service with an army
confronted by tribes of the greatest ferocity; (LI.)
while we know that ladies' bedsteads have for a long
time now been entirely covered with silver plating,
and so for long have banqueting-couches also. It is
recorded that Carvilius Pollio, Knight of Rome, was
the first person who had silver put on these latter,
though not so as to plate them all over or make
them to the Delos pattern, but in the Carthaginian
style. In this latter style he also had bedsteads

aureos fecit, nec multo post argentei Deliacos imitati
sunt.   quae omnia expiavit bellum civile Sullanum.
145   LII. Paulo enim ante haec factae sunt lances e
centenis libris argenti, quas tunc cuper CL numero
fuisse Romae constat multosque ob eas proscriptos
dolo concupiscentium.   erubescant annales, qui bellum civile illud talibus vitiis inputavere; nostra aetas
fortior fuit.   Claudii principatu servus eius Drusillanus [1] nomine Rotundus, dispensator Hispaniae citeri-
oris, quingenariam lancem habuit, cui fabricandae [2]
officina prius exaedificata fuerat, et comites eius octo
ad CCL libras, quaeso, ut quam multi eas conservi eius
146 inferrent, aut quibus cenantibus? Cornelius Nepos
tradit ante Sullae victoriam duo tantum triclinia
Romae fuisse argentea, repositoriis argentum addi
sua memoria coeptum.   Fenestella, qui obiit novissimo Tiberii Caesaris principatu, ait et testudinea
tum in usum venisse, ante se autem paulo lignea,
rotunda, solida nec multo maiora quam mensas fuisse,
se quidem puero quadrata et conpacta aut acere
operta aut citro coepisse, mox additum argentum in

---

[1] *V.l.* Drusilianus.
[2] *V.l.* cum fabricando : quam fabricando *Detlefsen.*

---

[a] By Sulla in 82 B.C.
[b] Fenestella died in A.D. 21, Tiberius in A.D. 37.

made of gold, and not long afterwards silver bedsteads were made, in imitation of those of Delos.
All this extravagance however was expiated by the 83-2 B.C.
civil war of Sulla.

LII. In fact it was shortly before this period that *Other silver*
silver dishes were made weighing a hundred pounds, *furniture.*
and it is well-known that there were at that date over
150 of those at Rome, and that many people were
sentenced to outlawry [a] because of them, by the
intrigues of people who coveted them. History
which has held vices such as these to be responsible
for that civil war may blush with shame, but our
generation has gone one better. Under the
Emperor Claudius his slave Drusillanus, who bore A.D. 41-54.
the name of Rotundus, the Emperor's steward of
Nearer Spain, possessed a silver dish weighing 500
lbs., for the manufacture of which a workshop had
first been specially built, and eight others of 250 lbs.
went with it as side-dishes, so that how many of his
fellow-slaves, I ask, were to bring them in or who
were to dine off them? Cornelius Nepos records
that before the victory won by Sulla there were 82 B.C.
only two silver dinner-couches at Rome, and that
silver began to be used for decorating sideboards
within his own recollection. And Fenestella who
died towards the end of the principate of Tiberius [b]
says that tortoiseshell sideboards also came into
fashion at that time, but a little before his day they
had been solid round structures of wood, and not
much larger than tables; but that even in his boyhood they began to be made square and of planks
morticed together and veneered either with maple
or citrus wood, while later silver was laid on at the
corners and along the lines marking the joins, and

angulos lineasque per commissuras, tympana vero
se iuvene appellata, tum a stateris et lances, quas
antiqui magides vocaverant.

147   LIII. Nec copia argenti tantum furit vita, sed
valdius paene manipretiis, idque iam pridem, ut
ignoscamus nobis. delphinos quinis milibus sestertium [1] in libras emptos C. Gracchus habuit, L. vero
Crassus orator duos scyphos Mentoris artificis manu
caelatos HS c̄,[2] confessus tamen est numquam iis uti
propter verecundiam ausum. scimus [3] eundem HS
148 VI in singulas libras vasa empta habuisse. Asia
primum devicta luxuriam misit in Italiam, siquidem
L. Scipio in triumpho transtulit argenti caelati pondo
mille et [4] CCCC [5] et vasorum aureorum pondo MD anno
conditae urbis DLXV. at eadem Asia donata [6] multo
etiam gravius adflixit mores, inutiliorque victoria illa
149 hereditas Attalo rege mortuo fuit. tum enim haec
emendi Romae in auctionibus regiis verecundia ex-
empta est urbis anno DCXXII, mediis LVII annis erudita
civitate amare etiam, non solum admirari, opulentiam
externam, inmenso et Achaicae victoriae momento ad
inpellendos mores, quae et ipsa in hoc intervallo anno

----

[1] sestertium *Hardouin*: sestertiis.
[2] c̄ *Urlichs*: c.
[3] scimus *Mayhoff*: scitum *coni. Ian*: constat *ed. Basil.*:
sicut.
[4] mille et *Mayhoff*: milia *aut* M.
[5] CCCC *B*: CCCL *rell.*
[6] domita *cd. Par.* 6801: domata *Gelen.*

----

    [a] *I.e.* Asia Minor and Syria, peace having been concluded
with King Antiochus in 189 B.C.
    [b] Attalus III, King of Pergamum 138–133 B.C., bequeathed
his kingdom to Rome.   Part of it became the province Asia.
    [c] *I.e.* the destruction of Corinth by L. Mummius, 146 B.C.

when he was a young man they were called ' drums,'
and then also the dishes for which the old name had
been magides came to be called basins from their
resemblance to the scales of a balance.

LIII. Yet it is not only for quantities of silver
that there is such a rage among mankind but there
is an almost more violent passion for works of fine
handicraft; and this goes back a long time, so that
we of to-day may excuse ourselves from blame.
Gaius Gracchus had some figures of dolphins for 153–121 B.C.
which he paid 5000 sesterces per pound, while the
orator Lucius Crassus had a pair of chased goblets, 140–91 B.C.
the work of the artist Mentor, that cost 100,000;
yet admittedly he was too ashamed ever to use them.
It is known to us that he likewise owned some vessels
that he bought for 6000 sesterces per pound. It
was the conquest of Asia *a* that first introduced
luxury into Italy, inasmuch as Lucius Scipio carried
in procession at his triumph 1400 lbs. of chased
silverware and vessels of gold weighing 1500 lbs.:
this was in the 565th year from the foundation of
the city of Rome. But receiving Asia also as a gift 189 B.C.
dealt a much more serious blow to our morals, and
the bequest of it that came to us on the death of
King Attalus *b* was more disadvantageous than the
victory of Scipio. For on that occasion all scruples
entirely disappeared in regard to buying these
articles at the auctions of the king's effects at Rome—
the date was the 622nd year of the city, and in the 132 B.C.
interval of 57 years our community had learnt not
merely to admire but also to covet foreign opulence;
an impetus having also been given to manners by
the enormous shock of the conquest of Achaia,*c*
that victory itself also having during this interval

150 urbis DCVIII parta signa et tabulas pictas invexit. ne
quid deesset, pariter quoque [1] luxuria nata est et
Cathago sublata, ita congruentibus fatis, ut et
liberet amplecti vitia et liceret. petiere et digna-
tionem hinc aliqui veterum. C. Marius post victoriam Cimbricam cantharis potasse Liberi patris exemplo traditur, ille arator Arpinas et manipularis
imperator.

151   LIV. Argenti usum in statuas primum divi Augusti
temporum adulatione transisse falso existimatur.
iam enim triumpho Magni Pompei reperimus trans-
latam Pharnacis, qui primus regnavit in Ponto,
argenteam statuam, item Mithridatis Eupatoris et
152 currus aureos argenteosque. argentum succedit
aliquando et auro luxu feminarum plebis compedes
sibi facientium, quas induere aureas mos tritior vetet.
vidimus et ipsi Arellium Fuscum motum equestri
ordine ob insignem calumniam, cum celebritatem [2]
adsectaretur[3] adulescentium scholae, argenteos anulos
habentem. et quid haec attinet colligere, cum
capuli militum ebore etiam fastidito caelentur argento, vaginae catellis, baltea lamnis crepitent, iam vero

---

[1] quoque *Mayhoff* : que.
[2] celebritatem *B* : celebritate *rell.*
[3] adsectaretur *Warmington* : adsectarentur *cdd.* : expec-
tarentur *edd. vett.* : assectationem *Hermolaus Barbarus.*

---

[a] He became king *c.* 190 B.C.
[b] Mithridates VI, King of Pontus, finally quelled in Pompey's
campaigns.

of time introduced the statues and pictures won in
the 608th year of the city. That nothing might be 146 B.C.
lacking, luxury came into being simultaneously,
with the downfall of Carthage, a fatal coincidence
that gave us at one and the same time a taste for the
vices and an opportunity for indulging in them.
Some of the older generation also sought to gain
esteem from these sources. It is recorded that
Gaius Marius after his victory over the Cimbrians 101 B.C.
drank from Bacchic tankards, in imitation of Father
Liber—he, the ploughman of Arpino who rose to
the position of general from the ranks!

LIV. The view is held that the extension of *Silver*
the use of silver to statues was made in the case of *statues,*
statues of his late lamented Majesty Augustus, *rings, etc.*
owing to the sycophancy of the period, but this is
erroneous. We find that previously a silver statue
of Pharnaces the First,[a] King of Pontus, was carried
in the triumphal procession of Pompey the Great, 61 B.C.
as well as one of Mithridates Eupator,[b] and also
chariots of gold and silver were used. Likewise
silver has at some periods even supplanted gold,
female luxury among the plebeians having its shoe
buckles made of silver, as wearing gold buckles
would be prohibited by the more common fashion.
We have ourselves seen Arellius Fuscus (who was
expelled from the Equestrian order on a singularly
grave charge) wearing silver rings when he sought to
acquire celebrity for his school for youths. But
what is the point of collecting these instances, when
our soldiers' sword hilts are made of chased silver,
even ivory not being thought good enough; and
when their scabbards jingle with little silver chains
and their belts with silver tabs, nay now-a-days our

paedagogia in transitu virilitatis custodiantur argento, feminae laventur et nisi argentea solia fastidiant, eademque materia et cibis et probris serviat?

153 videret haec Fabricius et stratas argento mulierum balineas ita, ut vestigio locus non sit, cum viris lavantium! Fabricius, qui bellicos imperatores plus quam pateram et salinum habere ex argento vetabat, videret hinc dona fortium fieri aut in haec frangi! heu mores, Fabrici nos pudet!

154 LV. Mirum auro caelando neminem inclaruisse, argento multos. maxime tamen laudatus est Mentor, de quo supra diximus. quattuor paria ab eo omnino [1] facta sunt, ac iam nullum extare dicitur Ephesiae Dianae templi ac [2] Capitolini incendiis.

155 Varro se et aereum signum [3] eius habuisse scribit. proximi ab eo in admiratione Acragas et Boëthus et Mys fuere. exstant omnium opera hodie in insula Rhodiorum, Boëthi apud Lindiam Minervam, Acragantis in templo Liberi patris in ipsa Rhodo Centauros Bacchasque caelati scyphi, Myos in eadem aede Silenos et Cupidines. Acragantis et venatio in

156 scyphis magnam famam habuit. post hos celebratus est Calamis, et Antipatro qui [4] Satyrum in phiala

---

[1] vasorum *Thiersch.*
[2] ac *Warmington*: aut (iacet *cd. Par.* 6801).
[3] sinum *Havet.*
[4] Antipatro qui *Mayhoff*: Antipater quoque qui *Urlichs*: Antipater quoq *B*: A. quinque *rell.* (quique *cd. Par.* 6801).

---

[a] C. Fabricius Luscinus, a man who held high offices *c.* 285–275, but died a poor man.
[b] See § 142.
[c] In 356 B.C.
[d] In 83 B.C.

schools for pages just at the point of adolescence wear silver badges as a safeguard, and women use silver to wash in and scorn sitting-baths not made of silver, and the same substance does service both for our viands and for our baser needs? If only Fabricius [a] could see these displays of luxury— women's bathrooms with floors of silver, leaving nowhere to set your feet—and the women bathing in company with men—if only Fabricius, who forbade [b] gallant generals to possess more than a dish and a saltcellar of silver, could see how nowadays the rewards of valour are made from the utensils of luxury, or else are broken up to make them! Alas for our present manners—Fabricius makes us blush!

LV. It is a remarkable fact that the art of chasing gold has not brought celebrity to anyone, whereas persons celebrated for chasing silver are numerous. The most famous however is Mentor of whom we spoke above. Four pairs of goblets were all that he ever made, but it is said that none of them now survive, owing to the burning of the Temple of Artemis of Ephesus [c] and of the Capitol.[d] Varro says in his writings that he also possessed a bronze statue by this sculptor. Next to Mentor the artists most admired were Acragas, Boethus and Mys. Works by all of these exist at the present day in the island of Rhodes—one by Boethus in the temple of Athena at Lindus, some goblets engraved with Centaurs and Bacchants by Acragas in the temple of Father Liber or Dionysus in Rhodes itself, goblets with Sileni and Cupids by Mys in the same temple. Hunting scenes by Acragas on goblets also had a great reputation. After these in celebrity is Calamis, and Diodorus who was said to have placed

Famous examples of artistic work in silver.

§ 147 *and* VII. 127.

gravatum somno conlocavisse verius quam caelasse
dictus est Diodorus,[1] Stratonicus mox Cyzicenus,
Tauriscus, item Ariston et Eunicus Mitylenaei laudantur et Hecataeus et circa Pompei Magni aetatem
Pasiteles, Posidonius Ephesius, Hedys, Thracides,[2]
qui proelia armatosque caelavit, Zopyrus, qui Areo-
pagitas et iudicium Orestis in duobus scyphis HS XII[3]
aestimatis. fuit et Pytheas, cuius II unciae X x̄
venierunt : Ulixes et Diomedes erant in phialae em-
157 blemate Palladium subripientes. fecit idem et cocos
magiriscia appellatos parvolis potoriis et e quibus ne
exemplaria quidem liceret exprimere ; tam oppor-
tuna iniuriae subtilitas erat. habuit et Teucer
crustarius famam, subitoque ars haec ita exolevit, ut
sola iam vestustate censeatur usuque attritis cae-
laturis si nec[4] figura discerni possit auctoritas constet.
158 Argentum medicatis aquis inficitur atque adflatu
salso, sic et[5] in mediterraneis Hispaniae.

LVI. In argenti et auri metallis nascuntur etiamnum pigmenta sil et caeruleum. sil proprie limus
est. optimum ex eo quod Atticum vocatur, pretium
in pondo libras X II ; proximum marmorosum di-

---

[1] *lac. Mayhoff* qui Diodorus *suppl.* : *coni. et* Antipatro
poetae Satyrum . . . caelasse dictus ⟨Diodorus⟩, Stratonicus.
[2] Hedys, Thracides *Furtwängler* : hedystrachides *B* : haedi-
stadices, iedisthracides *aut alia rell.* : Hedystratides *Sillig* :
Telesarchides *coni. Dilthey.*
[3] XII *Gelen* : |XII| *B* : XII *rell.*
[4] si nec *Urlichs* : si ne *B* : ne *rell.*
[5] sic et *L. Poinsinet de Sivry* : sicut.

---

[a] Who probably wrote an epigram (*Anth. Plan.* 248)
stating that Diodorus ' put to sleep ' the satyr. The MSS.
of Pliny make Antipater the engraver, while in the Anthology
the epigram is attributed to Plato the younger.
[b] At Athens for the murder of his mother, according to the
story.

in a condition of heavy sleep rather than engraved
on a bowl a Slumbering Satyr for Antipater.[a]
Next praise is awarded to Stratonicus of Cyzicus,
Tauriscus, also Ariston and Eunicus of Mitylene,
and Hecataeus, and, around the period of Pompey
the Great, Pasiteles, Posidonius of Ephesus, Hedys, _{B.C.}
Thracides who engraved battle scenes and men in
armour, and Zopyrus who engraved the Athenian
Council of Areopagus and the Trial of Orestes [b] on
two goblets valued at 12,000 sesterces. There was
also Pytheas, one of whose works sold at the price of
10,000 denarii for two ounces: it consisted of an
embossed base of a bowl representing Odysseus
and Diomede in the act of stealing the Palladium.
The same artist also carved some very small drinking
cups in the shape of cooks known as 'The Chefs in
Miniature,' which it was not allowed even to
reproduce by casts, so liable to damage was the
fineness of the work. Also Teucer the artist in
embossed work attained celebrity, and all of a sudden
this art so declined that it is now only valued in old
specimens, and authority attaches to engravings
worn with use even if the very design is invisible.

Silver becomes tarnished by contact with water
from springs containing minerals and by the salt
breezes, as happens also even in the interior regions
of Spain.[c]

LVI. In gold and silver mines also are formed *Oxides and*
the pigments yellow ochre and blue. Yellow ochre [d] *hydroxides*
is strictly speaking a slime. The best kind comes *of iron.*
from what is called Attic slime; its price is two
denarii a pound. The next best is marbled ochre,

c The sentence is probably misplaced.
d Various oxides and hydroxides of iron.

midio Attici pretio. tertium genus est pressum,
159 quod alii Scyricum vocant, ex insula Scyro, iam et ex
Achaia, quo utuntur ad picturae umbras, pretium in
libras HS bini; dupondiis vero detractis quod lucidum
vocant, e Gallia veniens. hoc autem et Attico ad
lumina utuntur, ad abacos non nisi marmoroso,
quoniam marmor in eo resistit amaritudini calcis.
effoditur et ad XX ab urbe lapidem in montibus;
postea uritur pressum appellantibus qui adulterant.
sed esse falsum exustumque, amaritudine apparet et
quoniam resolutum in pulverem est.

160   Sile pingere instituere primi Polygnotus et Micon,
Attico dumtaxat. secuta aetas hoc ad lumina usa
est, ad umbras autem Scyrico et Lydio. Lydium
Sardibus emebatur, quod nunc obmutuit.[1]

161   LVII. Caeruleum harena est, huius genera tria
fuere antiquitus: Aegyptium maxime[2] probatur,[3]
Scythicum mox[4] diluitur facile et, cum teritur, in
quattuor colores mutatur, candidiorem nigrioremve
et crassiorem tenuioremve; praefertur huic etiamnum Cyprium. accessit his Puteolanum et Hispa-
niense, harena ibi confici coepta. tinguitur autem
omne et in sua coquitur herba bibitque sucum.
reliqua confectura eadem quae chrysocollae.

---

[1] *V.ll.* ommutuit, omittunt.
[2] *V.l.* quod maxime.
[3] *V.l.* probatum.
[4] mox *Mayhoff* (*coni. et* umore): hoc.

---

[a] Azurite, a basic copper carbonate.

which costs half the price of Attic. The third kind
is dark ochre, which other people call Scyric ochre,
as it comes from the island of Scyros, and nowadays
also from Achaia, which they use for the shadows of a
painting, price two sesterces a pound, while that
called clear ochre, coming from Gaul, costs two
asses less. This and the Attic kind they use for
painting different kinds of light, but only marbled
ochre for squared panel designs, because the marble
in it resists the acridity of the lime. This ochre is
also dug up in the mountains 20 miles from Rome.
It is afterwards burnt, and by some people it is
adulterated and passed off as dark ochre; but the
fact that it is not genuine and has been burnt is
shown by its acridity and by its crumbling into dust.

The custom of using yellow ochre for painting was
first introduced by Polygnotus and Micon, but they
only used the kind from Attica. The following
period employed this for representing lights but
ochre from Scyros and Lydia for shadows. Lydian
ochre used to be sold at Sardis, but now it has quite
gone out.

LVII. The blue pigment [a] is a sand. In old days *Azurite.*
there were three varieties: the Egyptian is thought
most highly of; next the Scythian mixes easily with
water, and changes into four colours when ground,
lighter or darker and coarser or finer; to this blue
the Cyprian is now preferred. To these were added
the Pozzuoli blue, and the Spanish blue, when blue
sand-deposits began to be worked in those places.
Every kind however undergoes a dyeing process,
being boiled with a special plant and absorbing its
juice; but the remainder of the process of manufac- *§§ 86 sqq.*
ture is the same as with gold-solder.

162   Ex caeruleo fit quod vocatur lomentum, perficitur id lavando terendoque. hoc est caeruleo candidius. pretia eius X x[1] in libras, caerulei X viii. usus in creta; calcis inpatiens. nuper accessit et Vestorianum, ab auctore appellatum. fit ex Aegyptii levissima parte; pretium eius in libras X xi. idem et Puteolani usus, praeterque ad fenestras; cyanon[2]
163 vocant. non pridem adportari et Indicum coeptum est, cuius pretium X vii. ratio in pictura ad incisuras, hoc est umbras dividendas ab lumine. est et vilissimum[3] genus lomenti, quod[4] tritum vocant, quinis assibus aestimatum.

    Caerulei sinceri experimentum in carbone ut flagret; fraus viola arida decocta in aqua sucoque per linteum expresso in cretam Eretriam. vis in medicina ut purget ulcera; itaque et emplastris adiciunt, item
164 causticis. teritur autem difficillime. sil in medendo leniter mordet adstringitque et explet ulcera. uritur in fictilibus, ut prosit.

    Pretia rerum, quae usquam posuimus, non ignoramus alia aliis locis esse et omnibus paene mutari annis, prout navigatione constiterint aut ut quisque mercatus sit aut aliquis praevalens manceps annonam

---

[1] §§ 162–163 *numeri varie traduntur.*
[2] cyanon *Brotier ex coni. Durandi*: cylon *B*: *V.ll.* cyllon, cylonon, cynolon.
[3] *V.ll.* utilissimum, subtilissimum.
[4] *V.l.* quondam: quod dant *edd. vett*: quidam *Hermolaus Barbarus.*

From blue is made the substance called blue wash, which is produced by washing and grinding it. Blue wash is of a paler colour than blue, and it costs 10 denarii per pound, while blue costs 8 denarii. Blue is used on a surface of clay, as it will not stand lime. A recent addition has been Vestorian blue, called after the man Vestorius who invented it; it is made from the finest part of Egyptian blue, and costs 11 denarii per pound. Pozzuoli blue is employed in the same way, and also near windows [a]; it is called cyanos. Not long ago Indian blue or *Indigo.* indigo began to be imported, its price being 7 denarii; painters use it for dividing-lines, that is, for separating shadows from light. There is also a blue wash of a very inferior kind, called ground blue, valued at 5 asses.

The test of genuine Indian blue is that when laid on burning coal it should blaze; it is adulterated by boiling dried violets in water and straining the liquor through linen on to Eretrian earth.[b] Its use as a medicament is to clean out ulcers; consequently it is employed as an ingredient in plasters, and also in cauteries, but it is extremely difficult to pound up. Yellow ochre used as a drug has a gently mordant and astringent effect, and fills up ulcers. To make it beneficial it is burnt in earthenware vessels.

We are not unaware that the prices of articles which we have stated at various points differ in different places and alter nearly every year, according to the shipping costs or the terms on which a particular merchant has bought them, or as some dealer dominating the market may whip up the

[a] *I.e.* it does not lose its colour in the light.
[b] See p. 283, note *k*.

flagellet, non obliti Demetrium a tota Seplasia
Neronis principatu accusatum apud consules; poni
tamen necessarium fuit quae plerumque erant Romae,
ut exprimeretur auctoritas rerum.

---

[a] A district in the city of Capua occupied by druggists and
perfumers and sellers of pigments.

selling price; we have not forgotten that, under the
emperor Nero, Demetrius was prosecuted before the A.D. 54–68.
Consuls by the entire Seplasia.^{a} Nevertheless I
have found it necessary to state the prices usual
at Rome, in order to give an idea of a standard value
of commodities.

BOOK XXXIV

I. Proxime dicantur aeris metalla, cui et in usu
proximum est pretium, immo vero ante argentum
ac paene etiam ante aurum Corinthio, stipis quoque
auctoritas, ut diximus. hinc aera militum, tribuni
aerarii et aerarium, obaerati, aere diruti. docuimus
quamdiu populus Romanus aere tantum signato usus
esset : et alia re [1] vetustas aequalem urbi auctoritatem eius declarat, a rege Numa collegio tertio aerarium
fabrum instituto.

2 II. Vena quo dictum est modo foditur ignique
perficitur. fit et e lapide aeroso, quem vocant
cadmean, celebri trans maria et quondam in Campania,
nunc et in Bergomatium agro extrema parte Italiae ;
ferunt nuper etiam in Germania provincia repertum.
fit et ex alio lapide, quem chalcitim appellant, in

[1] alia re *Mayhoff* (*qui et* alio *coni.*) : alta *coni. Ian* : alia.

---

[a] The word *aes* usually means a prepared alloy of copper
and tin, that is, bronze ; it also included brass (alloy of copper
and zinc). Much ancient *aes* contained lead. Pure copper
was properly called *aes Cyprium*. Note that native copper ore
or metallic copper is not common in the Old World, but the
ancients discovered that it could be produced artificially by
heating the much more plentiful oxides, sulphides, silicates,
and carbonates of copper.

[b] The others were the College of Priests and the College of
Augurs.

[c] Apparently mineral calamine and smithsonite = silicate
and carbonate of zinc ; cf. § 100.

BOOK XXXIV

I. Let our next subject be ores, etc., of copper *Base metals.*
and bronze [a] the metals which in point of utility
have the next value; in fact Corinthian bronze is *Bronze and*
valued before silver and almost even before gold; *copper.*
and bronze is also the standard of payments in
money as we have said: hence *aes* is embodied in the XXXIII.
terms denoting the pay of soldiers, the treasury 43, 138.
paymasters and the public treasury, persons held
in debt, and soldiers whose pay is stopped. We
have pointed out for what a long time the Roman XXXIII.
nation used no coinage except bronze; and by 42 *sqq.*
another fact antiquity shows that the importance
of bronze is as old as the city—the fact that the
third corporation [b] established by King Numa was *Trad. date*
the Guild of Coppersmiths. *715–672 B.C.*

II. The method followed in mining deposits of *Copper.*
copper and purifying the ore by firing is that which
has been stated. The metal is also got from a XXXII.
coppery stone called by a Greek name *cadmea*,[c] a 95 *sqq.*
kind in high repute coming from overseas and also
formerly found in Campania and at the present day
in the territory of Bergamo on the farthest confines
of Italy; and it is also reported to have been
recently found in the province [d] of Germany. In
Cyprus, where copper was first discovered, it is also

---

[d] Only the region of the left bank of the Rhine is meant.

127

Cypro, ubi prima aeris inventio, mox vilitas praecipua reperto in aliis terris praestantiore maximeque aurichalco, quod praecipuam bonitatem admirationemque diu optinuit nec reperitur longo iam
3 tempore effeta tellure. proximum bonitate fuit Sallustianum in Ceutronum Alpino tractu, non longi et ipsum aevi, successitque ei Livianum in Gallia. utrumque a metallorum dominis appellatum, illud ab
4 amico divi Augusti, hoc a coniuge. velocis defectus Livianum quoque; certe admodum exiguum invenitur. summa gloriae nunc in Marianum conversa, quod et Cordubense dicitur. hoc a Liviano cadmean maxime sorbet et aurichalci bonitatem imitatur in sestertiis dupondiariisque, Cyprio suo assibus contentis. et hactenus nobilitas in aere naturali[1] se habet.

5 III. Reliqua genera cura constant, quae suis locis reddentur, summa claritate ante omnia indicata. quondam aes confusum auro argentoque miscebatur, et tamen ars pretiosior erat; nunc incertum est, peior haec sit an materia, mirumque, cum ad infinitum operum pretia creverint, auctoritas artis extincta est.

---

[1] naturali *coni. Mayhoff*: naturalis.

---

[a] See § 117, note.
[b] *Aurichalcum*, the right word being probably *orichalcum* = ὀρείχαλκος, ' mountain-copper,' that is yellow copper ore and the brass made from it.
[c] Named after the great Marius (155–86 B.C.).

obtained from another stone also, called chalcitis,[a]
copper ore; this was however afterwards of exceptionally low value when a better copper was
found in other countries, and especially gold-copper,[b]
which long maintained an outstanding quality and
popularity, but which for a long time now has not
been found, the ground being exhausted. The next
in quality was the Sallustius copper, occurring in the
Alpine region of Haute Savoie, though this also only
lasted a short time; and after it came the Livia
copper in Gaul: each was named from the owners
of the mines, the former from the friend of Augustus
and the latter from his wife. Livia copper also
quickly gave out: at all events it is found in very
small quantity. The highest reputation has now
gone to the Marius[c] copper, also called Cordova
copper; next to the Livia variety this kind most
readily absorbs *cadmea* and reproduces the excellence
of gold-copper in making sesterces and double-*as*
pieces, the single *as* having to be content with its
proper Cyprus copper. That is the extent of the
high quality contained in natural bronze and copper.

III. The remaining kinds are made artificially,
and will be described in their proper places, the
most distinguished sorts being indicated first of all.
Formerly copper used to be blended with a mixture
of gold and silver, and nevertheless artistry was
valued more highly than the metal; but nowadays
it is a doubtful point whether the workmanship or
the material is worse, and it is a surprising thing
that, though the prices paid for these works of art
have grown beyond all limit, the importance attached
to this craftsmanship of working in metals has
quite disappeared. For this, which formerly used

129

quaestus enim causa, ut omnia, exerceri coepta est
quae gloriae solebat—ideo etiam deorum adscripta
operi, cum proceres gentium claritatem et hac via
quaererent—, adeoque exolevit fundendi aeris pre-
tiosi ratio, ut iam diu ne fortuna quidem in ea re ius
artis habeat.

6 Ex illa autem antiqua gloria Corinthium maxime
laudatur. hoc casus miscuit Corintho, cum caperetur,
incensa, mireque circa id multorum adfectatio
furuit,[1] quippe cum tradatur non alia de causa
Verrem, quem M. Cicero damnaverat, proscriptum
cum eo ab Antonio, quoniam [2] Corinthiis cessurum
se ei negavisset. ac mihi maior pars eorum simulare
eam scientiam videtur ad segregandos sese a ceteris
magis quam intellegere aliquid ibi suptilius; et
7 hoc paucis docebo. Corinthus capta est olympiadis
CLVIII anno tertio, nostrae urbis DCVIII, cum ante
haec saecula [3] fictores nobiles esse desissent, quorum
isti omnia signa hodie Corinthia appellant. quapropter ad coarguendos eos ponemus artificum
aetates; nam urbis nostrae annos ex supra dicta
comparatione olympiadum colligere facile erit.

---

[1] furuit *Warmington*: furit *Sillig*, *Ian*: fuerit *cdd.* (fuit
cd. Par. 6801).

[2] quoniam *Ian*: quam quoniam *coni. Sillig*: q̄ūm *B*:
quam *rell.* (quam quod *cd. Par.* 6801).

[3] saeculo *quid. ap. Dalecamp.*

---

[a] *I.e.* Hephaestus or Vulcan.

[b] Or ' has won praise normally due to art.'

[c] Corinth was destroyed by the Roman forces under
Mummius 146 B.C.

to be practised for the sake of glory—consequently it was even attributed to the workmanship of gods,[a] and the leading men of all the nations used to seek for reputation by this method also—has now, like everything else, begun to be practised for the sake of gain; and the method of casting costly works of art in bronze has so gone out that for a long time now not even luck in this matter has had the privilege of producing art.[b]

Of the bronze which was renowned in early days, the Corinthian is the most highly praised. This is a compound that was produced by accident, when Corinth was burned[c] at the time of its capture; and there has been a wonderful mania among many people for possessing this metal—in fact it is recorded that Verres, whose conviction Marcus Cicero had procured, was, together with Cicero, proscribed by Antony for no other reason than because he had refused to give up to Antony some pieces of Corinthian ware; and to me the majority of these collectors seem only to make a pretence of being connoisseurs, so as to separate themselves from the multitude, rather than to have any exceptionally refined insight in this matter; and this I will briefly show. Corinth was taken in the third year of the 158th Olympiad, which was the 608th year of our city, when for ages there had no longer been any famous artists in metalwork; yet these persons designate all the specimens of their work as Corinthian bronzes. In order therefore to refute them we will state the periods to which these artists belong; of course it will be easy to turn the Olympiads into the years since the foundation of our city by referring to the two corresponding dates given

Corinthian bronze.

70 B.C.

43 B.C.

146 B.C.

131

sunt ergo vasa tantum Corinthia, quae isti elegan-
tiores modo ad esculenta transferunt, modo in
lucernas aut trulleos nullo munditiarum dispectu.[1]
8 eius aeris [2] tria genera: candidum argento nitore
quam proxime accedens, in quo illa mixtura prae-
valuit; alterum, in quo auri fulva natura; tertium,
in quo aequalis omnium temperies fuit. praeter
haec est cuius ratio non potest reddi, quamquam
hominis manu est,[3] at fortuna temperatur [4] in
simulacris signisque illud suo colore pretiosum ad
iocineris imaginem vergens, quod ideo hepatizon
appellant, procul a Corinthio, longe tamen ante
Aegineticum atque Deliacum, quae diu optinuere
principatum.

9   IV. Antiquissima aeris gloria Deliaco fuit, mercatus in Delo celebrante toto orbe, et ideo cura
officinis. tricliniorum pedibus fulcrisque ibi prima
aeris nobilitas, pervenit deinde et ad deum simulacra
effigiemque hominum et aliorum animalium.

10   V. Proxima laus Aeginetico fuit, insula et ipsa
eo,[5] nec quod ibi gigneretur,[6] sed officinarûm temperatura, nobilitata. bos aereus inde captus in foro
boario est Romae. hoc erit exemplar Aeginetici

[1] V.l. despectu.
[2] eius aeris *Warmington*: aeris *coni. Mayhoff*: eius.
[3] manus est *Mayhoff*: manu facta *edd. vett.*: manus et *B*: manu sed *aut* manus sed *rell.*
[4] V.l. temperamentum: temperatum *C. F. W. Müller.*
[5] eo *Mayhoff*: est.
[6] gigneretur *Sillig*: signetur *aut* gignens *aut* gimens.

---

a Or ' for the neatness of the workmanship.'

above. The only genuine Corinthian vessels are then those which your connoisseurs sometimes convert into dishes for food and sometimes into lamps or even washing basins, without nice regard for decency.[a] There are three kinds of this sort of bronze: a white variety, coming very near to silver in brilliance, in which the alloy of silver predominates; a second kind, in which the yellow quality of gold predominates, and a third kind in which all the metals were blended in equal proportions. Besides these there is another mixture the formula for which cannot be given, although it is man's handiwork; but the bronze valued in portrait statues and others for its peculiar colour, approaching the appearance of liver and consequently called by a Greek name 'hepatizon' meaning 'liverish,' is a blend produced by luck; it is far behind the Corinthian blend, yet a long way in front of the bronze of Aegina and that of Delos which long held the first rank.

IV. The Delian bronze was the earliest to become *Delian* famous, the whole world thronging the markets *bronze.* in Delos; and hence the attention paid to the processes of making it. It was at Delos that bronze first came into prominence as a material used for the feet and framework of dining-couches, and later it came to be employed also for images of the gods and statues of men and other living things.

V. The next most famous bronze was the *Aeginetan* Aeginetan; and the island of Aegina itself became *bronze.* celebrated for it, though not because the metal copper was mined there but because of the compounding done in the workshops. A bronze ox looted from Aegina stands in the cattle-market at Rome, and will serve as a specimen of Aegina bronze,

aeris, Deliaci autem Iuppiter in Capitolio in Iovis Tonantis aede.[a] illo aere Myron usus est, hoc Polycletus, aequales atque condiscipuli; sic [1] aemulatio et in materia fuit.[b]

VI. Privatim Aegina candelabrorum superficiem
11 dumtaxat elaboravit, sicut Tarentum scapos. in iis ergo iuncta commendatio officinarum est. nec pudet tribunorum militarium salariis emere, cum ipsum nomen a candelarum lumine inpositum appareat.[c] accessio candelabri talis fuit Theonis iussu praeconis Clesippus fullo gibber et praeterea et alio foedus aspectu, emente id Gegania HS L̄. eadem ostentante in convivio empta ludibrii causa
12 nudatus atque inpudentia [2] libidinis receptus in torum, mox in testamentum, praedives numinum vice illud candelabrum coluit et hanc Corinthiis fabulam adiecit, vindicatis tamen moribus nobili sepulchro, per quod aeterna supra terras Geganiae dedecoris memoria duraret. sed cum esse nulla

1 sic *Mayhoff*: sit B^1: sed.
2 impotentia *Gelen.*

---

a Built in 22 B.C.
b Myron fl. *c.* 475 B.C., Polyclitus *c.* 435. See pp. 168–171.
c *I.e.* the sockets holding the candles, the other parts being the stems and the feet.

while that of Delos is seen in the Zeus or
Jupiter in the temple [a] of Jupiter the Thunderer
on the Capitol. Aegina bronze was used by
Myron and that from Delos by Polyclitus, who
were contemporaries [b] and fellow-pupils; thus there
was rivalry between them even in their choice of
materials.

VI. Aegina specialized in producing only the *Chandeliers.*
upper parts [c] of chandeliers, and similarly Taranto
made only the stems, and consequently credit
for manufacture is, in the matter of these articles,
shared between these two localities. Nor are people
ashamed to buy these at a price equal to the
pay of a military tribune, although they clearly
take even their name from the lighted candles they
carry. At the sale of a chandelier of this sort by the
instructions of the auctioneer (named Theon) selling it
there was thrown in as part of the bargain the fuller
Clesippus a humpback and also of a hideous appearance in other respects besides, the lot being bought by
a woman named Gegania for 50,000 sesterces. This
woman gave a party to show off her purchases, and
for the mockery of the guests the man appeared
with no clothes on; his mistress conceiving an
outrageous passion for him admitted him to her bed
and later gave him a place in her will. Thus becoming excessively rich he worshipped the lamp-
stand in question as a divinity and so caused this
story to be attached to Corinthian lampstands in
general, though the claims of morality were
vindicated by his erecting a noble tombstone to
perpetuate throughout the living world for all time
the memory of Gegania's shame. But although it is
admitted that there are no lampstands made of

Corinthia candelabra constet, nomen id praecipue in his celebratur, quoniam Mummi victoria Corinthum quidem diruit, sed e compluribus Achaiae oppidis simul aera dispersit.

13 VII. Prisci limina etiam ac valvas in templis ex aere factitavere. invenio et a Cn. Octavio, qui de Perseo rege navalem triumphum egit, factam porticum duplicem ad circum Flaminium, quae Corinthia sit appellata a capitulis aereis columnarum, Vestae quoque aedem ipsam Syracusana superficie tegi placuisse. Syracusana sunt in Pantheo capita columnarum a M. Agrippa posita. quin etiam privata opulentia eo modo usurpata est. Camillo inter crimina obiecit Spurius Carvilius quaestor, ostia quod aerata haberet in domo.

14 VIII. Nam triclinia aerata abacosque et mono-podia Cn. Manlium Asia devicta primum invexisse triumpho suo, quem duxit anno urbis DLXVII, L. Piso auctor est, Antias quidem heredes L. Crassi oratoris multa etiam triclinia aerata vendidisse. ex aere factitavere et cortinas tripodum nomine [et] [1] Delphi-cas, quoniam donis [2] maxime Apollini Delphico dicabantur. placuere et lychnuchi pensiles in delubris

---

[1] et *Mayhoff*: ac *B*: a *cd. Leid. Voss. m.* 1: *om. rell.*
[2] donis *cdd. pler.*: ludis *cd. Poll.*: erat *B*: aeratae *Urlichs*: eae *Ian*: dono *coni. Warmington.*

Corinthian metal, yet this name specially is commonly attached to them, because although Mummius's victory destroyed Corinth, it caused the dispersal of bronzes from a number of the towns of Achaia at the same time.

VII. In early times the lintels and folding doors *Various uses of bronze.* of temples as well were commonly made of bronze. I find that also Gnaeus Octavius, who was granted a 169 B.C. triumph after a sea-fight against King Perseus, constructed the double colonnade at the Flaminian 167 B.C. circus which owing to the bronze capitals of its columns has received the name of the Corinthian portico, and that a resolution was passed that even the temple of Vesta should have its roof covered with an outer coating of Syracusan metal. The capitals of the pillars in the Pantheon which were put up by Marcus 27 B.C. Agrippa are of Syracusan metal. Moreover even private opulence has been employed in similar uses: one of the charges brought against Camillus 391 B.C. by the quaestor Spurius Carvilius was that in his house he had doors covered with bronze.

VIII. Again, according to Lucius Piso dinner-couches and panelled sideboards and one-leg tables decorated with bronze were first introduced by Gnaeus Manlius at the triumph which he celebrated 187 B.C. in the 567th year of the city after the conquest of Asia; and as a matter of fact Antias states that the heirs of Lucius Crassus the orator also sold a number of dinner couches decorated with bronze. It was even customary for bronze to be used for making the cauldrons on tripods called Delphic cauldrons because they used to be chiefly dedicated as gifts to Apollo of Delphi; also lamp-holders were popular suspended from the ceiling in temples or with their

aut arborum mala ferentium modo lucentes, quale
est in templo Apollinis Palatini quod Alexander
Magnus Thebarum expugnatione captum in Cyme
dicaverat eidem deo.

15 IX. Transiit deinde ars vulgo ubique ad effigies
deorum. Romae simulacrum ex aere factum Cereri
primum reperio ex peculio Sp. Cassi, quem regnum
adfectantem pater ipsius interemerit. transiit et a
diis ad hominum statuas atque imagines multis
modis. bitumine antiqui tinguebant eas, quo magis
mirum est placuisse auro integere. hoc nescio an
Romanum fuerit inventum; certe etiam nomen non [1]
16 habet vetustum.[2] effigies hominum non solebant
exprimi nisi aliqua inlustri causa perpetuitatem
merentium, primo sacrorum certaminum victoria
maximeque Olympiae, ubi omnium, qui vicissent,
statuas dicari mos erat, eorum vero, qui ter ibi
superavissent, ex membris ipsorum similitudine
17 expressa, quas iconicas vocant. Athenienses nescio
an primis omnium Harmodio et Aristogitoni tyran-
nicidis publice posuerint statuas. hoc actum est
eodem anno, quo et Romae reges pulsi. excepta
deinde res est a toto orbe terrarum humanissima

[1] nomen non *B* : romae non *rell*. : Romae nomen *Sillig*.
[2] vetustum *B* : vetustatem *rell*.

[a] Dedicated by Augustus in 27 B.C.
[b] Probably to provide a protective polish.
[c] From the Greek εἰκών, εἰκωνικός.
[d] By Antenor. The conspiracy of Harmodius and Aristogeiton took place in 514–513 B.C. A marble copy of Critius' and
Nesiotes' later substitute still exists. See note on pp. 256–257.

lights arranged to look like apples hanging on trees, like the specimen in the temple [a] of Apollo of the Palatine which had been part of the booty taken by Alexander the Great at the storming of Thebes and dedicated by him to the same deity at Cyme. *335–4 B.C.*

IX. But after a time this art in all places came to be usually devoted to statues of gods. I find that the first image of a god made of bronze at Rome was that dedicated to Ceres and paid for out of the property of Spurius Cassius who was put to death by his own father when trying to make himself king. *485 B.C.* The practice passed over from the gods to statues and representations of human beings also, in various forms. In early days people used to stain statues with bitumen,[b] which makes it the more remarkable that they afterwards became fond of covering them with gold. This was perhaps a Roman invention, but it certainly has a name of no long standing at Rome. It was not customary to make effigies of human beings unless they deserved lasting commemoration for some distinguished reason, in the first case victory in the sacred contests and particularly those at Olympia, where it was the custom to dedicate statues of all who had won a competition; these statues, in the case of those who had been victorious there three times, were modelled as exact personal likenesses of the winners—what are called iconicae,[c] portrait statues. I rather believe that the first portrait statues [d] officially erected at Athens were those of the tyrannicides Harmodius and Aristogeiton. This happened in the same year *510 B.C.* as that in which the Kings were also driven out at Rome. The practice of erecting statues from a most civilized sense of rivalry was afterwards

Statues o bronze at Rome.

Greek portrait-statues.

ambitione, et in omnium municipiorum foris statuae
ornamentum esse coepere propagarique[1] memoria
hominum et honores legendi aevo basibus inscribi,
ne in sepulcris tantum legerentur. mox forum et in
domibus privatis factum atque in atriis: honos
clientium instituit sic colere patronos.

18 X. Togatae effigies antiquitus ita dicabantur.
placuere et nudae tenentes hastam ab epheborum e
gymnasiis exemplaribus; quas Achilleas vocant.
Graeca res nihil velare, at contra Romana ac militaris
thoraces addere. Caesar quidem dictator loricatam
sibi dicari in foro suo passus est. nam Lupercorum
habitu tam noviciae sunt quam quae nuper prodiere
paenulis indutae. Mancinus eo habitu sibi statuit,
19 quo deditus fuerat. notatum ab auctoribus et
L. Accium poetam in Camenarum aede maxima
forma statuam sibi posuisse, cum brevis admodum
fuisset. equestres utique statuae Romanam cele-
brationem habent, orto sine dubio a Graecis exemplo.
sed illi celetas tantum dicabant in sacris victores,

---

[1] propagarique *B* : prorogarique *rell.*

---

[a] *Forum Caesaris* or *Forum Iulium* was built by Julius
Caesar in the eighth region of the city, near the temple of
Janus and the old forum, which had become too small.

[b] Mancinus, consul 137 B.C., in a war with the Numantines
made a treaty which the senate refused to ratify, and he was
handed over to the enemy though they refused to receive
him. He seems to have regarded this as discreditable to the
senate but not to himself.

taken up by the whole of the world, and the custom proceeded to arise of having statues adorning the public places of all municipal towns and of perpetuating the memory of human beings and of inscribing lists of honours on the bases to be read for all time, so that such records should not be read on their tombs only. Soon after a publicity centre was established even in private houses and in our own halls: the respect felt by clients inaugurated this method of doing honour to their patrons.

X. In old days the statues dedicated were simply clad in the toga. Also naked figures holding spears, made from models of Greek young men from the gymnasiums—what are called figures of Achilles— became popular. The Greek practice is to leave the figure entirely nude, whereas Roman and military statuary adds a breastplate: indeed the dictator Caesar gave permission for a statue wearing a cuirass to be erected in his honour in his Forum.[a] As for the statues in the garb of the Luperci, they are modern innovations, just as much as the portrait-statues dressed in cloaks that have recently appeared. Mancinus[b] set up a statue of himself in the dress that he had worn when surrendered to the enemy. It has been remarked by writers that the poet Lucius Accius also set up a very tall statue of himself in the shrine of the Latin Muses, although he was a very short man. Assuredly equestrian statues are popular at Rome, the fashion for them having no doubt been derived from Greece; but the Greeks used only to erect statues of winners of races on horse-back at their sacred contests, although subsequently they also erected statues of

Greek and Roman styles.

49–44 B.C.

170–c. 85 B.C.

141

postea vero et qui bigis vel quadrigis vicissent; unde et nostri currus nati in iis, qui triumphavissent. serum hoc, et in iis non nisi a divo Augusto seiuges, sicut [1] elephanti.

20 XI. Non vetus et bigarum celebratio in iis, qui praetura functi curru vecti essent per circum; antiquior columnarum, sicuti C. Maenio, qui devicerat priscos Latinos, quibus ex foedere tertias praedae populus Romanus praestabat, eodemque in consulatu in suggestu rostra devictis Antiatibus fixerat anno urbis ccccxvi, item C. Duillio, qui primus navalem triumphum egit de Poenis, quae est etiam nunc in 21 foro, item L. Minucio praefecto annonae extra portam Trigeminam unciaria stipe conlata—nescio an primo honore tali a populo, antea enim a senatu erat,—praeclara res, ni frivolis coepisset initiis. namque et Atti Navi statua fuit ante curiam—basis eius conflagravit curia incensa P. Clodii funere—;

---

[1] sicut et *edd. vett.*: aut *Eugénie Sellers*.

---

a For purposes which were not religious. For a dedication of a six-horsed chariot in 189 B.C., cf. Livy, XXXVIII. 35, 4.

b For the surviving inscription of the *Columna Rostrata* see *Remains of Old Latin*, IV, pp. 128–131.

c Famous augur under King Tarquinius Priscus (traditional dates 616–579 B.C.).

winners with two-horse or four-horse chariots; and this is the origin of our chariot-groups in honour of those who have celebrated a triumphal procession. But this belongs to a late date, and among those monuments it was not till the time of his late lamented Majesty Augustus that chariots with six horses occurred,[a] and likewise elephants.

XI. The custom of erecting memorial chariots *Roman* with two horses in the case of those who held the *statues.* office of praetor and had ridden round the Circus in a chariot is not an old one; that of statues on pillars is of earlier date, for instance the statue of honour of Gaius Maenius who had vanquished the Old 338 B.C. Latins to whom the Roman nation gave by treaty a third part of the booty won from them. It was in the same consulship that the nation, after defeating the people of Antium, had fixed on the platform the beaked prows of ships taken in the victory over the people of Antium, in the 416th year of the city of Rome; and similarly the statue to Gaius Duillius, who was the first to obtain a naval triumph over the 260 B.C. Carthaginians—this statue still stands in the forum [b]— and likewise that in honour of the praefect of markets Lucius Minucius outside the Triplets Gate, defrayed 439 B.C. by a tax of one-twelfth of an *as* per head. I rather think this was the first time that an honour of this nature came from the whole people; previously it had been bestowed by the senate: it would be a very distinguished honour had it not originated on such unimportant occasions. In fact also the statue of Attus Navius [c] stood in front of the senate-house— when the senate-house was set on fire at the funeral of Publius Clodius the base of the statue was burnt 52 B.C.

fuit et Hermodori Ephesii in comitio, legum, quas
decemviri scribebant, interpretis, publice dicata.
22 alia causa, alia auctoritas M. Horati Coclitis statuae—
quae durat hodieque—, cum hostes a ponte sublicio
solus arcuisset. equidem et Sibyllae iuxta rostra
esse non miror, tres sint licet: una quam Sextus
Pacuius Taurus aed. pl. restituit[1]; duae quas
M. Messalla. primas putarem has et Atti Navi,
positas aetate Tarquinii Prisci, ni regum anteceden-
23 tium essent in Capitolio, ex iis Romuli et Tatii sine
tunica, sicut et Camilli in rostris. et ante aedem
Castorum fuit Q. Marci Tremuli equestris togata,
qui Samnites bis devicerat captaque Anagnia populum stipendio liberaverat. inter antiquissimas sunt
et Tulli Cloeli, L. Rosci, Sp. Nauti,[2] C. Fulcini in
rostris, a Fidenatibus in legatione interfectorum.
24 hoc a re p. tribui solebat iniuria caesis, sicut aliis et
P. Iunio, Ti. Coruncanio, qui ab Teuta Illyriorum
regina interfecti erant. non omittendum videtur,
quod annales adnotavere, tripedaneas iis statuas in

---

[1] *V.l.* instituit.
[2] nautii *B*: Antii *Caesarius coll. Liv.* IV.17.

[a] *I.e.* Castor and Pollux : Suetonius records (*Div. Iul.*, 10)
that it was dedicated to both, but usually spoken of as the
Temple of Castor.

with it; and the statue of Hermodorus of Ephesus
the interpreter of the laws drafted by the decemvirs, 451-450 B.C.
dedicated at the public cost, stood in the Assembly-
place of Rome. There was a different motive and
another reason—an important one—for the statue of
Marcus Horatius Cocles, which has survived even to 508 B.C.
the present day; it was erected because he had
single-handed barred the enemy's passage of the
Bridge on Piles. Also, it does not at all surprise me
that statues of the Sibyl stand near the Beaked
Platform though there are three of them—one
restored by Sextus Pacuvius Taurus, aedile of the
plebs, and two by Marcus Messalla. I should think
these statues and that of Attus Navius, all erected
in the period of Tarquinius Priscus, were the first, (616-579
if it were not for the statues on the Capitol of the B.C.)
kings who reigned before him, among them the
figures of Romulus and Tatius without the tunic,
as also that of Camillus on the Beaked Platform.
Also there was in front of the temple of the Castors [a]
an equestrian statue of Quintus Marcius Tremulus,
wearing a toga; he had twice vanquished the
Samnites, and by taking Anagni delivered the nation c. 305 B.C.
from payment of war-tax. Among the very old
statues are also those at the Platform of Tullus
Cloelius, Lucius Roscius, Spurius Nautius, and Gaius
Fulcinius, all assassinated by the people of Fidenae 438 B.C.
when on an embassy to them. It was the custom
for the state to confer this honour on those who had
been wrongfully put to death, as among others
Publius Junius and Titus Coruncanius, who had
been killed by Teuta the Queen of the Illyrians. 230 B.C.
It would seem not to be proper to omit the fact
noted by the annals that the statues of these persons,

foro statutas; haec videlicet mensura honorata tunc erat. non praeteribo et Cn. Octavium ob unum SC.[1] verbum. hic regem Antiochum daturum se responsum dicentem virga, quam tenebat forte, circumscripsit priusque, quam egrederetur circulo illo, responsum dare coegit. in qua legatione interfecto senatus statuam poni iussit quam oculatissimo loco, eaque est in rostris. invenitur statua decreta et Taraciae Gaiae sive Fufetiae virgini Vestali, ut poneretur ubi vellet, quod adiectum non minus honoris habet quam feminae esse decretam. meritum eius ipsis ponam annalium verbis: quod campum Tiberinum gratificata esset ea populo.

26   XII. Invenio et Pythagorae et Alcibiadi in cornibus comitii positas, cum bello Samniti Apollo Pythius iussisset fortissimo Graiae gentis et alteri sapientissimo simulacra celebri loco dicari. eae stetere, donec Sulla dictator ibi curiam faceret. mirumque est, illos patres Socrati cunctis ab eodem deo sapientia praelato Pythagoran praetulisse aut tot aliis virtute Alcibiaden et quemquam utroque [2] Themistocli.

[1] sc *B, cd. Leid. Voss.*: scilicet *rell.*
[2] utraque *coni. Mayhoff.*

---

[a] This is a mistake. This act was performed by C. Popillius Laenas when Antiochus IV was on his 4th campaign against Egypt.

[b] In fact on an embassy to Syria connected with troubles on the accession of Antiochus V (not IV).

[c] *Oculatissimus* is the 'single word' to which Pliny alludes above, meaning 'most visible to the eye.'

[d] It was in fact an enlargement of the original Senate-house.

[e] The Athenian chiefly responsible for Athens' sea-power and the defeat of Xerxes by sea at Salamis in 480 B.C.

erected in the forum, were three feet in height,
showing that this was the scale of these marks of
honour in those days. I will not pass over the case
of Gnaeus Octavius also, because of a single word
that occurs in a Decree of the Senate. When King
Antiochus IV said he intended to answer him,
Octavius [a] with the stick he happened to be holding
in his hand drew a line all round him and compelled
him to give his answer before he stepped out of the 168 B.C.
circle. And as Octavius was killed while on this 162 B.C.
embassy,[b] the senate ordered a statue to be erected
to him ' in the spot most eyed '[c] and that statue
stands on the Platform. We also find that a decree
was passed to erect a statue to a Vestal Virgin named
Taracia Gaia or Fufetia ' to be placed where she
wished,' an addition that is as great a compliment as
the fact that a statue was decreed in honour of a woman.
For the Vestal's merits I will quote the actual words
of the Annals : ' because she had made a gratuitous
present to the nation of the field by the Tiber.'

XII. I also find that statues were erected to *Greek*
Pythagoras and to Alcibiades, in the corners of the *statues.*
Place of Assembly, when during one of our Samnite
Wars Pythian Apollo had commanded the erection 343 B.C.
in some conspicuous position of an effigy of the
bravest man of the Greek race, and likewise, one
of the wisest man; these remained until Sulla the
dictator made [d] the Senate-house on the site. It 80 B.C.
is surprising that those illustrious senators of ours
rated Pythagoras above Socrates, whom the same
deity had put above all the rest of mankind in respect
of wisdom, or rated Alcibiades above so many other
men in manly virtue, or anybody above Themistocles [e]
for wisdom and manly virtue combined.

27   Columnarum ratio erat attolli super ceteros mor-
tales, quod et arcus significant novicio invento.
primus tamen honos coepit a Graecis, nullique
arbitror plures statuas dicatas quam Phalereo
Demetrio Athenis, siquidem cccLx statuere, nondum
anno hunc numerum dierum excedente, quas mox
laceravere. statuerunt et Romae in omnibus vicis
tribus Mario Gratidiano, ut diximus, easdemque
subvertere Sullae introitu.

28   XIII. Pedestres sine dubio Romae fuere in
auctoritate longo tempore; et equestrium tamen
origo perquam vetus est, cum feminis etiam honore
communicato Cloeliae statua equestri, ceu parum
esset toga eam cingi, cum Lucretiae ac Bruto, qui
expulerant reges, propter quos Cloelia inter obsides

29   fuerat, non decernerentur. hanc primam cum
Coclitis publice dicatam crediderim—Atto enim ac
Sibyllae Tarquinium, ac reges sibi ipsos posuisse
verisimile est—, nisi Cloeliae quoque Piso traderet
ab iis positam, qui una opsides fuissent, redditis a
Porsina[1] in honorem eius.   e diverso Annius Fetialis

---

[1] porsina B^1: porsena cd. Par. 6801: porsenna rell.

---

[a] An Attic orator and statesman who lived c. 345–282 b.c.,
and was exiled in 307 b.c. after a ten years' tyranny.
[b] The last king of Rome, Tarquinius Superbus, was supported
against his republican enemies at Rome by Lars Porsena, the
Tuscan king of Clusium, who invaded Rome and seized the
Capitol, but withdrew after receiving twenty hostages.
Among them was a maiden Cloelia, who escaped, swam across
the Tiber and reached Rome. She was sent back to Porsena,
but he was so struck with her gallantry that he set her free
and allowed her to take back with her some of the other
hostages. The rape of Lucretia by Sextus Tarquinius led to

The purport of placing statues of men on columns was to elevate them above all other mortals; which is also the meaning conveyed by the new invention of arches. Nevertheless the honour originally began with the Greeks, and I do not think that any person ever had more statues erected to him than Demetrius [a] of Phalerum had at Athens, inasmuch as they set up 360, at a period when the year did not yet exceed that number of days, statues however the Athenians soon shattered in pieces. At Rome also the tribes in all the districts set up statues to Marius Gratidianus, as we have stated, and likewise threw them down again at the entrance of Sulla.

Roman statues.

XXXIII. 132.

XIII. Statues of persons on foot undoubtedly held the field at Rome for a long time; equestrian statues also however are of considerable antiquity, and this distinction was actually extended to women with the equestrian statue of Cloelia, as if it were not enough for her to be clad in a toga, although statues were not voted to Lucretia and Brutus, who had driven out the kings owing to whom Cloelia had been handed over with others as a hostage.[b] I should have held the view that her statue and that of Cocles were the first erected at the public expense— for it is probable that the monuments to Attus and the Sibyl were erected by Tarquin and those of the kings by themselves—were it not for the statement of Piso that the statue of Cloelia also was erected by the persons who had been hostages with her, when they were given back by Porsena, as a mark of honour to her; whereas on the other hand Annius Fetialis states that an equestrian figure which once

510–509 B.C.

508 B.C.

§ 22.

the expulsion of the Tarquins by Brutus and his companions and the establishment of the republican government.

equestrem, quae fuerit contra Iovis Statoris aedem
in vestibulo Superbi domus, Valeriae fuisse, Publi-
colae consulis filiae, eamque solam refugisse Tiberim-
que transnatavisse ceteris opsidibus, qui [1] Porsinae [2]
mittebantur, interemptis Tarquinii insidiis.

30   XIV. L. Piso prodidit M. Aemilio C. Popilio
iterum cos. a censoribus P. Cornelio Scipione M.
Popilio statuas circa forum eorum, qui magistratum
gesserant, sublatas omnes praeter eas, quae populi
aut senatus sententia statutae essent, eam vero,
quam apud aedem Telluris statuisset sibi Sp. Cassius,
qui regnum adfectaverat, etiam conflatam a censoribus.   nimirum in ea quoque re ambitionem provide-
31 bant illi viri.   exstant Catonis in censura vocifera-
tiones mulieribus statuas Romanis [3] in provinciis
poni; nec tamen potuit inhibere, quo minus Romae
quoque ponerentur, sicuti Corneliae Gracchorum
matri, quae fuit Africani prioris filia.   sedens huic
posita soleisque sine ammento insignis in Metelli
publica porticu, quae statua nunc est in Octaviae
operibus.[a]

32   XV. Publice autem ab exteris posita est Romae
C. Aelio tr. pl. lege perlata in Sthennium Stallium

---

[1] *V.l.* quae.
[2] porsinae *B*[1] : porsennae.
[3] statuas romanis *B*: romanis statuas *rell.* (r. in p. statuas
cl. Par. 6801).

---

[a] Public buildings in Rome erected by Augustus on the
site of Metellus' colonnade built in 146 B.C. and named after
his sister Octavia.   The *basis* of Cornelia's statue survives.

stood opposite the temple of Jupiter Stator in the forecourt of Tarquinius Superbus's palace was the statue of Valeria, daughter of Publicola, the consul, and that she alone had escaped and had swum across the Tiber, the other hostages who were being sent to Porsena having been made away with by a stratagem of Tarquin.

XIV. Lucius Piso has recorded that, in the second consulship of Marcus Aemilius and Gaius 158 B.C. Popilius, the censors Publius Cornelius Scipio and Marcus Popilius caused all the statues round the forum of men who had held office as magistrates to be removed excepting those that had been set up by a resolution of the people or the Senate, while the statue which Spurius Cassius, who had aspired to 485 B.C. monarchy, had erected in his own honour before the temple of the Earth was actually melted down by censors: obviously the men of those days took precautions against ambition in the matter of statues also. Some declamatory utterances made by Cato during his censorship are extant protesting against 184 B.C. the erection in the Roman provinces of statues to women; yet all the same he was powerless to prevent this being done at Rome also: for instance there is the statue of Cornelia the mother of the Gracchi and daughter of the elder Scipio Africanus. This represents her in a sitting position and is remarkable because there are no straps to the shoes; it stood in the public colonnade of Metellus, but is now in Octavia's Buildings.[a]

XV. The first statue publicly erected at Rome by foreigners was that in honour of the tribune of the people Gaius Aelius, for having introduced a law against Sthennius Stallius the Lucanian who had

Lucanum, qui Thurinos bis infestaverat. ob id
Aelium Thurini statua et corona aurea donarunt.
iidem postea Fabricium donavere statua liberati
obsidione, passimque gentes[1] in clientelas ita[2]
receptae, et adeo discrimen omne sublatum, ut
Hannibalis etiam statuae tribus locis visantur in ea
urbe, cuius intra muros solus hostium emisit hastam.

33 XVI. Fuisse autem statuariam artem familiarem
Italiae quoque et vetustam, indicant Hercules ab
Euandro sacratus, ut produnt, in foro boario, qui
triumphalis vocatur atque per triumphos vestitur
habitu triumphali, praeterea Ianus geminus a Numa
rege dicatus, qui pacis bellique argumento colitur
digitis ita figuratis, ut CCCLV[3] dierum nota[4] et aevi
34 esse deum indicent.[5] signa quoque Tuscanica per
terras dispersa quin[6] in Etruria factitata sint,[7] non
est dubium. deorum tantum putarem ea fuisse, ni
Metrodorus Scepsius, cui cognomen a Romani
nominis odio inditum est, propter MM statuarum
Volsinios expugnatos obiceret. mirumque mihi

---

[1] gentes *cdd.* : clientes *Gelen* : statuae *edd. vett.*
[2] *V.l.* clientela sua : sunt *edd. vett.*
[3] CCCLV *edd. vett.* : CCCLXV.
[4] nota aut per significationem anni temporis *cdd.* : *seclud.*
aut . . . temporis *Mayhoff.*
[5] indicent *B* : indicet *rell.* : indicaret *edd. vett.*
[6] quin *Detlefsen* : quae quin *Urlichs* : quae.
[7] sint *B* : *om. rell.*

---

[a] When he came up to the walls of Rome in 211 B.C. without
attacking the city.
[b] Presumably three fingers of one hand made III, the first
finger and thumb of that hand V, and the first finger and
second finger of the other hand V, the thumb and the third
and little finger of that hand being bent and not counting.
The MSS. have 365 (which number was not valid until Caesar's

twice made an attack upon Thurii; for this the 289, 285 B.C. inhabitants of that place presented Aelius with a statue and a crown of gold. The same people afterwards presented Fabricius with a statue for having rescued them from a state of siege; and various 283 B.C. races successively in some such way placed themselves under Roman patronage, and all discrimination was so completely abrogated that even a statue of Hannibal may be seen in three places in the city within the walls of which he alone of its national foes had hurled a spear.[a]

XVI. That the art of statuary was familiar to *Italian* Italy also and of long standing there is indicated by *statues.* the statue of Hercules in the Cattle Market said to have been dedicated by Evander, which is called ' Hercules Triumphant,' and on the occasion of triumphal processions is arrayed in triumphal vestments; and also by the two-faced Janus, dedicated by King Numa, which is worshipped as indicating war and peace, the fingers of the statue being so arranged as to indicate the 355 [b] days of the year, and to betoken that Janus is the god of the duration of time. Also there is no doubt that the so-called Tuscanic images scattered all over the world were regularly made in Etruria. I should have supposed these to have been statues of deities only, were it not that Metrodorus of Scepsis, who received his surname [c] from his hatred of the very name of Rome, reproached us with having taken by storm the city of Volsinii for the sake of the 2000 statues which it 264 B.C. contained. And it seems to me surprising that

time). In such a case the VI could be represented by the thumb and the first and second fingers of one hand.

[c] Misoromaeus, μισορωμαῖος, ' Roman-hater.'

videtur, cum statuarum origo tam vetus Italiae
sit, lignea potius aut fictilia deorum simulacra in
delubris dicata usque ad devictam Asiam, unde
luxuria.

35 Similitudines exprimendi quae prima fuerit origo,
in ea, quam plasticen Graeci vocant, dici convenientius
erit; etenim prior quam statuaria fuit. sed haec ad
infinitum effloruit, multorum voluminum opere, si
quis plura persequi velit; omnia enim quis possit?

36 (XVII.) M. Scauri aedilitate signorum MMM in scaena
tantum fuere temporario theatro. Mummius Achaia
devicta replevit urbem, non relicturus filiae dotem;
cur enim non cum excusatione ponatur? multa et
Luculli invexere. Rhodi etiamnum $\overline{\text{III}}$ [1] signorum
esse Mucianus ter cos. prodidit, nec pauciora Athenis,
Olympiae, Delphis superesse creduntur. quis ista

37 mortalium persequi possit aut quis usus noscendi
intellegatur? insignia maxime et aliqua de causa
notata voluptarium sit attigisse artificesque cele-
bratos nominavisse, singulorum quoque inexplicabili
multitudine, cum Lysippus MD opera fecisse prodatur,
tantae omnia artis, ut claritatem possent dare vel
singula: numerum apparuisse defuncto eo, cum

---

[1] $\overline{\text{III}}$ *Mayhoff*: tria milia *cd. Par.* 6801 : $\overline{\text{LXXIII}}$ *aut* LXXIII
rell.

---

[a] See p. 110, notes.
[b] L. Licinius Lucullus, consul 74 B.C., and his brother M.,
consul 73 B.C.

although the initiation of statuary in Italy dates
so far back, the images of the gods dedicated in the
shrines should have been more usually of wood or
terracotta right down to the conquest of Asia,[a] which
introduced luxury here.

What was the first origin of representing likenesses *Popularity*
in the round will be more suitably discussed when *of the art of bronze*
we are dealing with the art for which the Greek *statuary.*
term is 'plasticē' *plastic*, as that was earlier than *XXXV. 151 sqq.*
the art of bronze statuary. But the latter has
flourished to an extent passing all limit and offers a
subject that would occupy many volumes if one
wanted to give a rather extensive account of it—
for as for a completely exhaustive account, who
could achieve that? (XVII.) In the aedileship of *58 B.C.*
Marcus Scaurus there were 3000 statues on the stage
in what was only a temporary theatre. Mummius
after conquering Achaia filled the city with statues, *146 B.C.*
though destined not to leave enough at his death to
provide a dowry for his daughter—for why not
mention this as well as the fact that excuses it? A
great many were also imported by the Luculli.[b]
Yet it is stated by Mucianus who was three times *A.D. 52, 70,*
consul that there are still 3000 statues at Rhodes, *75.*
and no smaller number are believed still to exist at
Athens, Olympia and Delphi. What mortal man
could recapitulate them all, or what value can be
felt in such information? Still it may give pleasure
just to allude to the most remarkable and to name
the artists of celebrity, though it would be impossible
to enumerate the total number of the works of each, *Lysippus.*
inasmuch as Lysippus is said to have executed 1500 *Latter part of*
works of art, all of them so skilful that each of them *4th century*
by itself might have made him famous; the number *B.C.*

thesaurum effregisset heres; solitum enim ex mani-
pretio cuiusque signi denarios seponere aureos
singulos.

38 Evecta supra humanam fidem ars est successu,
mox et audacia. in argumentum successus unum
exemplum adferam, nec deorum hominumve similitudinis expressae. aetas nostra vidit in Capitolio,
priusquam id novissime conflagraret a Vitellianis
incensum, in cella Iunonis canem ex aere volnus
suum lambentem, cuius eximium miraculum et
indiscreta veri similitudo non eo solum intellegitur,
quod ibi dicata fuerat, verum et satisdatione; nam
quoniam summa nulla par videbatur, capite tutelarios

39 cavere pro ea institutum publice fuit. (XVIII.)
audaciae innumera sunt exempla. moles quippe
excogitatas videmus statuarum, quas colossaeas
vocant, turribus pares. talis est in Capitolio Apollo,
tralatus a M. Lucullo ex Apollonia Ponti urbe, xxx

40 cubitorum, D [1] talentis factus; talis in campo Martio
Iuppiter, a [2] Claudio Caesare dicatus, qui devoratur
Pompeiani theatri vicinitate; talis et Tarenti
factus a Lysippo, xl cubitorum. mirum in eo quod
manu, ut ferunt, mobilis ea ratio libramenti est, ut
nullis convellatur procellis. id quidem providisse
et artifex dicitur modico intervallo, unde maxime

[1] D *cdd.* : L *Overbeck* : CL *edd. vett.* (*recte ?*).
[2] a *B, cd. Par.* 6801 : aulo *rell.* : a divo *Gronov.*

---

[a] No doubt a στατήρ.
[b] This figure seems too large.
[c] Lit. ' is swallowed up by.'

is said to have been discovered after his decease, when his heir broke open his coffers, it having been his practice to put aside a coin [a] of the value of one gold denarius out of what he got as reward for his handicraft for each statue.

The art rose to incredible heights in success and afterwards in boldness of design. To prove its success I will adduce one instance, and that not of a representation of either a god or a man: our own generation saw on the Capitol, before it last went up A.D. 69. in flames burnt at the hands of the adherents of Vitellius, in the shrine of Juno, a bronze figure of a hound licking its wound, the miraculous excellence and absolute truth to life of which is shown not only by the fact of its dedication in that place but also by the method taken for insuring it; for as no sum of money seemed to equal its value, the government enacted that its custodians should be answerable for its safety with their lives. (XVIII.) Of boldness of design the examples are innumerable. We see enormously huge statues devised, what are called Colossal Colossi, as large as towers. Such is the Apollo on statues. the Capitol, brought over by Marcus Lucullus from 73 B.C. Apollonia, a city of Pontus, 45 ft. high, which cost 500 [b] talents to make; or the Jupiter which the Emperor Claudius dedicated in the Campus Martius, which is dwarfed [c] by the proximity of the theatre of Pompey; or the 60 ft. high statue at Taranto made by Lysippus. The remarkable thing in the case of the last is that though it can be moved by the hand, it is so nicely balanced, so it is said, that it is not dislodged from its place by any storms. This indeed, it is said, the artist himself provided against by erecting a column a short distance from

flatum opus erat frangi, opposita columna. itaque
magnitudinem propter difficultatemque moliendi [1]
non attigit eum Fabius Verrucosus, cum Herculem,
41 qui est in Capitolio, inde transferret. ante omnes
autem in admiratione fuit Solis colossus Rhodi,
quem fecerat Chares Lindius, Lysippi supra dicti
discipulus. LXX cubitorum altitudinis fuit hoc simulacrum, post LXVI [2] annum terrae motu prostratum,
sed iacens quoque miraculo est. pauci pollicem eius
amplectuntur, maiores sunt digiti quam pleraeque
statuae. vasti specus hiant defractis membris;
spectantur intus magnae molis saxa, quorum pondere
stabiliverat eum constituens. duodecim annis tradunt effectum ccc talentis, quae contigerant ex
apparatu regis Demetrii relicto morae taedio
42 obsessa [3] Rhodo. sunt alii centum numero in eadem
urbe colossi minores hoc, sed ubicumque singuli
fuissent, nobilitaturi locum, praeterque hos deorum
quinque, quos fecit Bryaxis.

43 Factitavit colossos et Italia. videmus certe Tus-
canicum Apollinem in bibliotheca templi Augusti
quinquaginta pedum a pollice, dubium aere mira-
biliorem an pulchritudine. fecit et Sp. Carvilius
Iovem, qui est in Capitolio, victis Samnitibus sacrata

---

[1] movendi *B*[2].
[2] LXVI *B*[1] : LVI.
[3] obsessa *B* : obesse à *rell.* (obsesso *cd. Vind.*) : obsessae
Sillig : *del. edd. vett.* (*item* rhodo).

---

[a] Demetrius Poliorcetes.   [b] Cf. § 34.

it to shelter it on the side where it was most necessary to break the force of the wind. Accordingly, because of its size, and the difficulty of moving it with great labour, Fabius Verrucosus left it alone when he transferred the Heracles from that place 209 B.C. to the Capitol where it now stands. But calling for admiration before all others was the colossal Statue *Chares.* of the Sun at Rhodes made by Chares of Lindus, *Colossal* the pupil of Lysippus mentioned above. This *Rhodes.* statue was 105 ft. high; and, 66 years after its erection, was overthrown by an earthquake, but *c.* 226 B.C. even lying on the ground it is a marvel. Few people can make their arms meet round the thumb of the figure, and the fingers are larger than most statues; and where the limbs have been broken off enormous cavities yawn, while inside are seen great masses of rock with the weight of which the artist steadied it when he erected it. It is recorded that it took twelve years to complete and cost 300 talents, money realized from the engines of war belonging to King Demetrius[a] which he had abandoned when he got tired of the protracted siege of Rhodes. There 305-4 B.C. are a hundred other colossal statues in the same city, which though smaller than this one would have each of them brought fame to any place where it might have stood alone; and besides these there were five colossal statues of gods, made by Bryaxis.

Italy also was fond of making colossal statues. *Other* At all events we see the Tuscanic[b] Apollo in the *colossal* library of the Temple of Augustus, 50 ft. in height *statues.* measuring from the toe; and it is a question whether it is more remarkable for the quality of the bronze or for the beauty of the work. Spurius Carvilius also made the Jupiter that stands in the Capitol,

lege pugnantibus e pectoralibus eorum ocreisque et
galeis. amplitudo tanta est, ut conspiciatur a
Latiari Iove. e reliquiis limae suam statuam
44 fecit, quae est ante pedes simulacri eius. habent in
eodem Capitolio admirationem et capita duo, quae
P. Lentulus cos. dicavit, alterum a Charete supra
dicto factum, alterum fecit . . . dicus [1] comparatione in tantum victus, ut artificum [2] minime pro-
45 babilis videatur. verum omnem amplitudinem statuarum eius generis vicit aetate nostra Zenodorus
Mercurio facto in civitate Galliae Arvernis per
annos decem, HS ⌈CCCC⌉ [3] manipretii, postquam satis
artem ibi adprobaverat, Romam accitus a Nerone,
ubi destinatum illius principis simulacro [4] colossum
fecit CVIS [5] pedum in [6] longitudinem, qui dicatus
Soli [7] venerationi est damnatis sceleribus illius
46 principis. mirabamur in officina non modo ex
argilla similitudinem insignem, verum et de parvis
admodum surculis [8] quod primum operis instaurati
fuit. ea statua indicavit interisse fundendi aeris
scientiam, cum et Nero largiri aurum argentumque
paratus esset et Zenodorus scientia fingendi cae-
47 landique nulli veterum postponeretur. statuam Ar-

[1] Prodicus *coni. Sillig* : Pythodicus *Urlichs coll.* § 85.
[2] *V.l.* artificium.
[3] |CCCC| *Ian* : CCCC.
[4] simulacro *B* : simulacrum *rell.*
[5] CVIS *Detlefsen* : CXIX *Ian* : CXIXS *Urlichs* : cui
nonaginta (= CVIXC) *B* : cui x *aut alia rell.*
[6] in *add. Mayhoff.*
[7] soli *B* : solis *rell.*
[8] sublicis *coni. Warmington.*

---

[a] On the Alban Mount, ten miles from Rome.
[b] Only the last five letters of the name survive in MSS.
Another conjectured restoration is Pythodicus, cf. § 85.

after defeating the Samnites in the war which they 293 B.C. fought under a most solemn oath; the metal was obtained from their breastplates, greaves and helmets, and the size of the figure is so great that it can be seen from the temple of Jupiter Latiaris.[a] Out of the bronze filings left over Carvilius made the statue of himself that stands at the feet of the statue of Jupiter. The Capitol also contains two much admired heads dedicated by the consul Publius 57 B.C. Lentulus, one made by Chares above-mentioned and the other by Prodicus,[b] who is so outdone by comparison as to seem the poorest of artists. But all the gigantic statues of this class have been beaten in our period by Zenodorus with the Hermes or Zenodorus. Mercury which he made in the community of the Arverni in Gaul; it took him ten years and the sum paid for its making was 40,000,000 sesterces. Having given sufficient proof of his artistic skill in Gaul he was summoned to Rome by Nero, and there made the A.D. 54-68. colossal statue, $106\frac{1}{2}$ ft. high, intended to represent that emperor but now, dedicated to the sun after the condemnation of that emperor's crimes, it is an object of awe. In his studio we used not only to admire the remarkable likeness of the clay model but also to marvel at the frame of quite small timbers [c] which constituted the first stage of the work put in hand. This statue has shown that skill in bronze-founding has perished, since Nero was quite ready to provide gold and silver, and also Zenodorus was counted inferior to none of the artists of old in his knowledge of modelling and chasing. When he

[c] A skeleton for the model; or, according to Eugénie Sellers, slender wax tubes covering a wax model, which was then cased in loam before bronze was poured in.

161

vernorum cum faceret provinciae Dubio [1] Avito
praesidente, duo pocula Calamidis manu caelata,
quae Cassio Salano avunculo eius, praeceptori suo,
Germanicus Caesar adamata donaverat, aemulatus
est, ut vix ulla differentia esset artis. quanto maior
Zenodoro praestantia fuit, tanto magis deprehenditur
aeris obliteratio.

48   Signis, quae vocant Corinthia, plerique in tantum
capiuntur, ut secum circumferant, sicut Hortensius
orator sphingem Verri reo ablatam, propter quam
Cicero illo iudicio in altercatione neganti ei, aenig-
mata se intellegere, respondit debere, quoniam
sphingem domi haberet. circumtulit et Nero princeps Amazonem, de qua dicemus, et paulo ante C.
Cestius consularis [2] signum,[3] quod secum etiam in
proelio habuit. Alexandri quoque Magni taber-
naculum sustinere traduntur solitae statuae, ex
quibus duae ante Martis Ultoris aedem dicatae sunt,
totidem ante regiam.

49   XIX. Minoribus simulacris signisque innumera
prope artificum multitudo nobilitata est, ante omnes
tamen Phidias Atheniensis Iove Olympio [4] facto ex
ebore quidem et auro, sed et ex aere signa fecit.

---

[1] Vibio *cd. Par.* 6801 : Duuio *J. Klein.*
[2] consularis Laris *Fröhner.*
[3] sphingem *coni. Mayhoff*: *seclud. Urlichs.*
[4] Olympio *B* : Olympiae *rell.*

---

[a] The reference is probably to statuettes, not medallions or
signet rings or brooches.
[b] Apparently Pliny has made a mistake, because Alexander's
σκηνή was the canopy (supported by four golden statues of
Victory) of the chariot which carried Alexander's dead body
to Alexandria.
[c] In the forum of Augustus at Rome.
[d] Near the temple of Vesta.

was making the statue for the Arverni, when the
governor of the province was Dubius Avitus, he
produced facsimiles of two chased cups, the handiwork
of Calamis, which Germanicus Caesar had prized
highly and had presented to his tutor Cassius Salanus,
Avitus's uncle; the copies were so skilfully made that
there was scarcely any difference in artistry between
them and the originals. The greater was the eminence of Zenodorus, the more we realize how the
art of working bronze has deteriorated.

Owners of the figurines [a] called Corinthian are *Figurines.*
usually so enamoured of them that they carry them
about with them; for instance the orator Hortensius
was never parted from the sphinx which he had got
out of Verres when on trial; this explains Cicero's 70 B.C.
retort when Hortensius in the course of an altercation
at the trial in question said he was not good at
riddles. ' You ought to be,' said Cicero, ' as you keep
a figurine in your pocket.' The emperor Nero also A.D. 54–68.
used to carry about with him an Amazon which we
shall describe later, and a little before Nero, the § 82.
ex-consul Gaius Cestius used to go about with a
sphinx, which he had with him even on the battle-
field. It is also said that the tent [b] of Alexander
the Great was regularly erected with four statues
as tent-poles, two of which have now been dedicated
to stand in front of the temple [c] of Mars the Avenger
and two in front of the Royal Palace.[d]

XIX. An almost innumerable multitude of artists *Small*
have been rendered famous by statues and figures *statues.*
of smaller size; but before them all stands the *Greek*
Athenian Pheidias, celebrated for the statue of *statuaries.*
Olympian Zeus, which in fact was made of ivory and *Born c. 500*
gold, although he also made figures of bronze. He *c. 450 B.C.*

floruit autem olympiade LXXXIII, circiter CCC urbis
nostrae annum,[1] quo eodem tempore aemuli eius
fuere Alcamenes, Critias, Nesiotes, Hegias, et deinde
olympiade LXXXVII Hagelades, Callon, Gorgias Lacon ;
rursus LXXXX Polyclitus, Phradmon, Myron, Pytha-
50 goras, Scopas, Perellus.[2] ex iis Polyclitus discipulos
habuit Argium, Asopodorum, Alexim, Aristidem,
Phrynonem, Dinonem,[3] Athenodorum, Demean Cli-
torium, Myron Lycium. LXXXXV olympiade floruere
Naucydes, Dinomenes, Canachus, Patroclus ; CII
Polycles, Cephisodotus, Leochares,[4] Hypatodorus ;[5]
CIIII Praxiteles, Euphranor ; CVII Aetion, Theri-
51 machus. CXIII Lysippus fuit, cum et Alexander
Magnus, item Lysistratus frater eius, Sthennis,[6]
Euphron, Sofocles,[7] Sostratus, Ion, Silanion—in hoc
mirabile quod nullo doctore nobilis fuit ; ipse disci-
pulum habuit Zeuxiaden— ; CXXI Eutychides, Euthycrates, Laippus,[8] Cephisodotus,[9] Timarchus, Pyro-
52 machus.[10] cessavit deinde ars ac rursus olympiade
CLVI revixit, cum fuere longe quidem infra praedictos,

[1] V.l. anno.
[2] Perellus B : Perelius rell. : Perileus Thiersch.
[3] om. B.
[4] Leochares Hermolaus Barbarus (cf. § 79) : leuchares B :
leuihares aut sim. rell.
[5] Epatodorus Hermolaus Barbarus.
[6] Sthennis Hermolaus Barbarus : thenis.
[7] Euphron, Sofocles Loewy : E., Eucles Ian : euphron
fucles B : euphronicles aut -ides rell.
[8] Dahippus Hardouin.
[9] Cephisodorus Gelen.
[10] Phyromachus Keil.

[a] More exactly the 306th to the 309th year of the city of
Rome = 448–445 B.C.
[b] In merit. For Critias the Marmor Parium has Critios.

flourished in the 83rd Olympiad, about [a] the 300th   448-445 B.C.
year of our city, at which same period his rivals [b]   454 B.C.
were Alcamenes, Critias, Nesiotes and Hegias; and
later, in the 87th Olympiad there were Hagelades,[c]   432-429 B.C.
Callon and the Spartan Gorgias, and again in the
90th Olympiad Polycleitus, Phradmon, Myron,   420-417 B.C.
Pythagoras,[d] Scopas[e] and Perellus. Of these
Polycleitus had as pupils Argius, Asopodorus, Alexis,
Aristides, Phryno, Dino, Athenodorus, and Demeas
of Clitor; and Myron had Lycius. In the 95th   400-397 B.C.
Olympiad flourished Naucydes, Dinomenes, Cana-
chus and Patroclus; and in the 102nd Polycles, Cephi-   372-369 B.C.
sodotus, Leochares and Hypatodorus; in the 104th   364-361 B.C.
Praxiteles and Euphranor; in the 107th Aetion   352-349 B.C.
and Therimachus. Lysippus [f] was in the 113th, the   327-324 B.C.
period of Alexander the Great, and likewise his
brother Lysistratus, Sthennis, Euphron, Sophocles,
Sostratus, Ion and Silanion—a remarkable fact in
the case of the last named being that he became
famous without having had any teacher; he him-
self had Zeuxiades as his pupil—and in the 121st   295-292 B.C.
Eutychides, Euthycrates, Laippus,[g] Cephisodotus,
Timarchus and Pyromachus. After that the art
languished, and it revived again in the 156th   156-153 B.C.
Olympiad, when there were the following, far
inferior it is true to those mentioned above, but

---

[c] The Greek form is Hagelaidas. He really flourished
c. 515-485 B.C.

[d] In fact Myron's best work was done before 450 B.C.,
Pythagoras' before 475 B.C.

[e] In fact Scopas was still working in 350 B.C. unless we have
here an elder Scopas

[f] He was apparently working soon after 369 B.C.

[g] Probably this should be Daippus as in § 87, cf. Paus. VI.
12, 6; 16, 35, Δάιππος.

probati tamen, Antaeus,[1] Callistratus, Polycles
Athenaeus, Callixenus, Pythocles, Pythias, Timocles.

53   Ita distinctis celeberrimorum aetatibus insignes
raptim transcurram, reliqua multitudine passim
dispersa. venere autem et in certamen laudatissimi,
quamquam diversis aetatibus geniti, quoniam fecerant
Amazonas, quae cum in templo Dianae Ephesiae
dicarentur, placuit eligi probatissimam ipsorum
artificum, qui praesentes erant, iudicio, cum [2]
apparuit eam esse, quam omnes secundam a sua
quisque iudicassent. haec est Polycliti, proxima
ab ea Phidiae, tertia Cresilae,[3] quarta Cydonis,
quinta Phradmonis.

54   Phidias praeter Iovem Olympium, quem nemo
aemulatur, fecit ex ebore auroque [4] Minervam
Athenis, quae est in Parthenone stans, ex aere vero
praeter Amazonem supra dictam Minervam tam
eximiae pulchritudinis, ut formae cognomen acceperit.
fecit et cliduchum et aliam Minervam, quam Romae
Paulus Aemilius ad aedem Fortunae Huiusce Diei
dicavit, item duo signa, quae Catulus in eadem aede,
palliata et alterum colossicon nudum, primusque
artem toreuticen aperuisse atque demonstrasse
merito iudicatur.

---

[1] Antheus *edd. vett.*
[2] tum *O. Jahn.*
[3] clesilae *B* : Ctesilae *Gelen* : Ctesilai *Hardouin.*
[4] auroque *coni. Mayhoff* : aeque.

---

[a] Some blunder has produced a new artist out of the name
Cydonia, Cresilas's birthplace.
[b] Perhaps Callimorphos, ' fair of form.'
[c] A priestess probably, but possibly Persephone.

nevertheless artists of repute: Antaeus, Callistratus, Polycles of Athens, Callixenus, Pythocles, Pythias and Timocles.

After thus defining the periods of the most famous artists, I will hastily run through those of out-standing distinction, throwing in the rest of the throng here and there under various heads. The most celebrated have also come into competition with each other, although born at different periods, because they had made statues of Amazons; when these were dedicated in the Temple of Artemis of Ephesus, it was agreed that the best one should be selected by the vote of the artists themselves who were present; and it then became evident that the best was the one which all the artists judged to be the next best after their own: this is the Amazon by Polycleitus, while next to it came that of Pheidias, third Cresilas's, fourth Cydon's [a] and fifth Phradmon's.

Pheidias, besides the Olympian Zeus, which *Pheidias.* nobody has ever rivalled, executed in ivory and gold § 49. the statue of Athene that stands erect in the Parthenon at Athens, and in bronze, besides the Amazon mentioned above, an Athene of such ex-quisite beauty that it has been surnamed the Fair.[b] He also made the Lady [c] with the Keys, and another Athene which Aemilius Paulus dedicated in Rome 167 B.C.? at the temple of Today's Fortune, and likewise a work consisting of two statues wearing cloaks which Catulus erected in the same temple, and another 101 B.C. work, a colossal statue undraped; and Pheidias is deservedly deemed to have first revealed the capa-bilities and indicated the methods of statuary.[d]

[d] Here perhaps all statuary as contrasted with painting; or else all metal-work only.

55 Polyclitus Sicyonius, Hageladae discipulus, dia-
dumenum fecit molliter iuvenem, centum talentis
nobilitatum, idem et doryphorum viriliter puerum.
fecit et [1] quem canona artifices vocant liniamenta
artis ex eo petentes veluti a lege quadam, solusque
hominum artem ipsam fecisse artis opere iudicatur.
fecit et destringentem se et nudum telo [2] incessentem
duosque pueros item nudos, talis ludentes, qui
vocantur astragalizontes et sunt in Titi imperatoris
atrio—quo [3] opere nullum absolutius plerique iudi-
cant;—item Mercurium qui fuit Lysimacheae,
56 Herculem, qui Romae, hagetera [4] arma sumentem,
Artemona, qui periphoretos appellatus est. hic
consummasse hanc scientiam iudicatur et toreuticen
sic erudisse, ut Phidias aperuisse. proprium eius
est, uno crure ut insisterent signa, excogitasse,
quadrata tamen esse ea ait Varro et paene ad
exemplum.
57 Myronem Eleutheris natum, Hageladae et ipsum
discipulum, bucula maxime nobilitavit celebratis

---

[1] *del.* et *Urlichs puncto post* puerum *sublato.*
[2] telo *Benndorf* : talo.
[3] quo *cd. Par.* 6801 : hoc *Ian* : duo. hoc *B* : *cm. rell.*
[4] hagetera *B* : agetera *rell.* : alexetera *Hardouin* : Anteum
e terra (*om.* arma) *ed. vett.*

---

[a] Of Argos, says Plato (*Protag.* 311c). But his family
moved to Sicyon. He cannot however have been a pupil of
Hagelades (§ 49). Copies of Polycleitus' *Diadumenos* and of
his *Doryphoros* are extant. Pliny confuses the great Polycleitus with P. the younger, likewise of Argos, who lived in
the 4th century B.C.
[b] We know however that this ' model statue ' or ' standard '
was the *Doryphoros* just mentioned.
[c] Or, ' in a single work embodied the principles of his art.'
P. wrote a treatise on art, called it Κανών, then made his

Polycleitus of Sicyon,[a] pupil of Hagelades, made a *Polycleitus.*
statue of the ' Diadumenos' or Binding his Hair—
a youth, but soft-looking—famous for having cost
100 talents, and also the ' Doryphoros ' or Carrying a Spear—a boy, but manly-looking. He also
made what artists call a ' Canon ' or Model Statue,[b]
as they draw their artistic outlines from it as
from a sort of standard; and he alone of mankind is deemed by means of one work of art to have
created the art itself.[c] He also made the statue of
the Man using a Body-scraper (' Apoxyomenos ')
and, in the nude, the Man Attacking with Spear,
and the Two Boys Playing Dice, likewise in the nude,
known by the Greek name of *Astragalizontes* and now
standing in the fore-court of the Emperor Titus— A.D. 79-81.
this is generally considered to be the most perfect
work of art in existence—and likewise the Hermes
that was once at Lysimachea ; Heracles ; the Leader
Donning his Armour, which is at Rome; and Artemon,[d]
called the Man in the Litter. Polycleitus is deemed
to have perfected this science of statuary and to
have refined the art of carving sculpture, just as
Pheidias is considered to have revealed it. A discovery that was entirely his own is the art of making
statues throwing their weight on one leg, although
Varro says these figures are of a square build and
almost all made on one model.

Myron, who was born at Eleutherae, was himself *Myron.*
also a pupil of Hagelades; he was specially famous *c.* 477 B.C.
for his statue of a heifer, celebrated in some well-

Doryphoros on his own principles, and called the sculptured
work also Κανών.

[d] A famous voluptuary (not the engineer of Pericles'
time).

versibus laudata, quando alieno plerique ingenio
magis quam suo commendantur. fecit et Ladam [1]
et discobolon et Perseum et pristas [2] et Satyrum
admirantem tibias et Minervam, Delphicos penta-
thlos, pancratiastas, Herculem, qui est apud circum
maximum in aede Pompei Magni. fecisse et
cicadae monumentum ac locustae carminibus suis
58 Erinna significat. fecit et Apollinem, quem ab
triumviro Antonio sublatum restituit Ephesiis divus
Augustus admonitus in quiete. primus hic multi-
plicasse veritatem videtur, numerosior in arte quam
Polyclitus et in [3] symmetria diligentior, et ipse tamen
corporum tenus curiosus animi sensus non expressisse,
capillum quoque et pubem non emendatius fecisse,
quam rudis antiquitas instituisset.

59 Vicit eum Pythagoras Reginus ex Italia pancratiaste
Delphis posito ; eodem vicit et Leontiscum. fecit
et stadiodromon Astylon, qui Olympiae ostenditur,
et Libyn [4] puerum tenentem tabellam [5] eodem loco,
et mala ferentem nudum, Syracusis autem claudi-
cantem, cuius ulceris dolorem sentire etiam spectantes

---

[1] Ladam *Benndorf* : canem.   [2] pyctas *Löschke.*
[3] et in *cdd.* : *del. et Sillig.*
[4] Libyn *Hermolaus Barbarus* : lybin *B* : lipin *aut* lympin
aut iolpum *rell.*   [5] tabellam *B* : tabellas *rell.*

---

[a] Ladas was a famous runner. But the MSS. give *canem*,
'dog.' Copies of Myron's *Discobolos* are extant.

[b] It is possible that Pliny wrote ' the Boxers.'

[c] Probably a group of Marsyas and Athene, of which copies
exist.

[d] Experts in both boxing and wrestling.

[e] This absurd statement is caused by a confusion of Μυρών
and a girl Μυρώ (*Anth. Pal.* VII. 190—Myro makes a tomb
for her pet insect).

[f] For another interpretation see E. Gardner, *Classical Re-
view*, II. 69.   [g] Leontiscos was an athlete, not an artist.

known sets of verses—inasmuch as most men owe their reputation more to someone else's talent than to their own. His other works include Ladas[a] and a 'Discobolos' or Man Throwing a Discus, and Perseus, and The Sawyers,[b] and The Satyr Marvelling at the Flute and Athene,[c] Competitors in the Five Bouts at Delphi, the All-round Fighters,[d] the Heracles now in the house of Pompey the Great at the Circus Maximus. Erinna[e] in her poems indicates that he even made a memorial statue of a tree-cricket and a locust. He also made an Apollo which was taken from the people of Ephesus by Antonius the Triumvir but restored to them by his late lamented Majesty Augustus in obedience to a warning given him in a dream. Myron is the first sculptor who appears to have enlarged the scope of realism, having more rhythms in his art than Polycleitus and being more careful in his proportions.[f] Yet he himself so far as surface configuration goes attained great finish, but he does not seem to have given expression to the feelings of the mind, and moreover he has not treated the hair and the pubes with any more accuracy than had been achieved by the rude work of olden days.

Myron was defeated by the Italian Pythagoras *Pythagoras* of Reggio with his All-round Fighter which stands at *of Rhegium.* Delphi, with which he also defeated Leontiscus[g]; Pythagoras also did the runner Astylos which is on show at Olympia; and, in the same place, the Libyan[h] as a boy holding a tablet; and the nude Man Holding Apples, while at Syracuse there is his Lame Man, which actually makes people looking at it feel a pain from his ulcer in their own leg, and

---

[h] Mnaseas of Cyrene. Paus. VI. 13, 7; 18, 1.

videntur, item Apollinem serpentemque eius sagittis
configi,[1] citharoedum, qui Dicaeus appellatus est,
quod, cum Thebae ab Alexandro caperentur, aurum
a fugiente conditum sinu eius celatum esset. hic
primus nervos et venas expressit capillumque
diligentius.

60 Fuit et alius Pythagoras Samius, initio pictor, cuius
signa ad aedem Fortunae Huiusce Diei septem nuda
et senis unum laudata sunt. hic supra dicto facie
quoque indiscreta similis fuisse traditur, Regini
autem discipulus et filius sororis fuisse Sostratus.

61 Lysippum Sicyonium Duris negat ullius fuisse
discipulum, sed primo aerarium fabrum audendi
rationem cepisse pictoris Eupompi responso. eum
enim interrogatum, quem sequeretur antecedentium,
dixisse monstrata hominum multitudine, naturam
62 ipsam imitandam esse, non artificem. plurima ex
omnibus signa fecit, ut diximus, fecundissimae artis,
inter quae destringentem se, quem M. Agrippa ante
Thermas suas dicavit, mire gratum Tiberio principi.
non quivit temperare sibi in eo, quamquam imperiosus
sui inter initia principatus, transtulitque in cubiculum
alio signo substituto, cum quidem tanta pop. R.

---

[1] configi *Ian* : configit *B* : confici *rell.*

---

[a] Named Cleon, a Theban poet. Athenae. I. 19b.
[b] *I.e.* the statue afterwards restored the deposit entrusted
to it.
[c] It is now known that he was the same as P. of Reggio.
Paus. VI. 4, 3–4 shows that P. of Reggio was the sculptor of
the statue of Euthynos; but we have the basis of that statue,
whereon P. signs himself as ' Samian.' He must therefore
have migrated to Reggio.
[d] ' Apoxyomenos.' The example in the Vatican is probably
by a 3rd century artist.

also Apollo shooting the Python with his Arrows,
a Man [a] playing the Harp, that has the Greek name
of The Honest Man [b] given it because when
Alexander took Thebes a fugitive successfully hid   335 B.C.
in its bosom a sum of gold. Pythagoras of Reggio
was the first sculptor to show the sinews and veins,
and to represent the hair more carefully.

There was also another [c] Pythagoras, a Samian,   *Pythagoras*
who began as a painter; his seven nude statues now   *of Samos.*
at the temple of To-day's Fortune and one of an old
man are highly spoken of. He is recorded to have
resembled the above mentioned Pythagoras so
closely that even their features were indistinguishable; but we are told that Sostratus was a pupil
of Pythagoras of Reggio and a son of this Pythagoras'
sister.

Lysippus of Sicyon is said by Duris not to have   *Lysippus.*
been the pupil of anybody, but to have been originally a copper-smith and to have first got the idea of
venturing on sculpture from the reply given by the
painter Eupompus when asked which of his predecessors he took for his model; he pointed to a crowd
of people and said that it was Nature herself, not an
artist, whom one ought to imitate. Lysippus as we
have said was a most prolific artist and made more   § 37.
statues than any other sculptor, among them the Man
using a Body-scraper [d] which Marcus Agrippa gave
to be set up in front of his Warm Baths and of
which the emperor Tiberius was remarkably fond.   A.D. 14–37.
Tiberius, although at the beginning of his principate
he kept some control of himself, in this case could
not resist the temptation, and had the statue removed to his bedchamber, putting another one in
its place at the baths; but the public were so

contumacia fuit, ut theatri clamoribus reponi apoxyo-
menon flagitaverit princepsque, quamquam adama-
63 tum, reposuerit. nobilitatur Lysippus et temulenta
tibicina et canibus ac venatione, in primis vero
quadriga cum Sole Rhodiorum. fecit et Alexandrum
Magnum multis operibus, a pueritia eius orsus,
quam statuam inaurari iussit Nero princeps delectatus
admodum illa; dein, cum pretio perisset gratia artis,
detractum est aurum, pretiosiorque talis existimabatur [1] etiam cicatricibus operis atque concisuris,
64 in quibus aurum haeserat,[2] remanentibus. idem
fecit Hephaestionem, Alexandri Magni amicum,
quem quidam Polyclito adscribant, cum is centum
prope annis ante fuerit; item Alexandri venationem,
quae Delphis sacrata est, Athenis Satyrum, turmam
Alexandri, in qua amicorum eius imagines summa
omnium similitudine expressit; hanc Metellus
Macedonia subacta transtulit Romam. fecit et
65 quadrigas multorum generum. statuariae arti plurimum traditur contulisse, capillum exprimendo,
capita minora faciendo quam antiqui, corpora
graciliora siccioraque, per quae proceritas signorum
maior videretur. non habet Latinum nomen sym-

[1] *V.l.* existimatur.
[2] fuerat *B* : haeserat *rell.*

---

[a] With the head encircled with rays.
[b] This would be right, perhaps, if they meant the younger
P., unknown to Pliny.
[c] Twenty-five officers who fell in the Battle of the Granicus.
Vellei. Paterc., I. 11, 3.

obstinately opposed to this that they raised an out-
cry at the theatre, shouting " Give us back the
' *Apoxyomenos* ' "—Man using a Body-scraper—and
the Emperor, although he had fallen quite in love
with the statue, had to restore it.   Lysippus is also
famous for his Tipsy Girl playing the Flute, and his
Hounds and Huntsmen in Pursuit of Game, but
most of all for his Chariot with the Sun belonging to
Rhodes.[a]   He also executed a series of statues of
Alexander the Great, beginning with one in 356-323 B.C.
Alexander's boyhood.   The emperor Nero was so A.D. 54-68.
delighted by this statue of the young Alexander
that he ordered it to be gilt; but this addition to its
money value so diminished its artistic attraction
that afterwards the gold was removed, and in that
condition the statue was considered yet more valuable, even though still retaining scars from the
work done on it and incisions in which the gold had
been fastened.   The same sculptor did Alexander
the Great's friend Hephaestio, a statue which some
people ascribe to Polycleitus,[b] although his date is
about a hundred years earlier; and also Alexander's
Hunt, dedicated at Delphi, a Satyr now at Athens,
and Alexander's Squadron of Horse, in which the
sculptor introduced portraits of Alexander's friends[c]
consummately lifelike in every case.   After the conquest of Macedonia this was removed to Rome by 148 B.C.
Metellus; he also executed Four-horse Chariots of
various kinds.   Lysippus is said to have contributed
greatly to the art of bronze statuary by representing
the details of the hair and by making his heads
smaller than the old sculptors used to do, and his
bodies more slender and firm, to give his statues
the appearance of greater height.   He scrupulously

metria, quam diligentissime custodiit [1] nova intacta-
que ratione quadratas veterum staturas permutando,
vulgoque dicebat ab illis factos quales essent homines,
a se quales viderentur esse. propriae huius videntur
esse argutiae operum custoditae in minimis quoque
rebus.

66 Filios et discipulos reliquit laudatos artifices
Laippum,[2] Boëdan, sed ante omnes Euthycraten,
quamquam is constantiam potius imitatus patris
quam elegantiam austero maluit genere quam
iucundo placere. itaque optume expressit Herculem
Delphis et Alexandrum Thespiis venatorem et
Thespiadas,[3] proelium equestre, simulacrum ipsum
Trophoni ad oraculum, quadrigas complures, equum [4]

67 cum fiscinis,[5] canes venantium. huius porro discipulus fuit Tisicrates, et ipse Sicyonius, sed Lysippi
sectae propior, ut vix discernantur complura signa,
ceu senex Thebanus et Demetrius rex, Peucestes,
Alexandri Magni servator, dignus tanta gloria.

68 Artifices, qui compositis voluminibus condidere
haec, miris laudibus celebrant Telephanen Pho-

---

[1] custodiit *Mayhoff* : custodit *aut* custodivit *cdd.*
[2] Dahippum *Hardouin.*
[3] et Thespiadas *cd. Par. Lat.* 6797 : *om.* thespiadas *B* :
om. et *cd. Leid. Voss., cd. Flor. Ricc.* : et thespiadum *cd. Par.*
6801.
[4] equitem *coni. T. B. L. Webster.* [5] fiscinis *B* : fuscinis *rell.*

---

[a] See note on Laippus in § 51.
[b] Or ' his Heracles made for Delphi, and his Alexander
Hunting, and his Thespiades (these two made for Thespiae).'
The Thespiades were the Muses.
[c] All MSS. except one give *fuscinis*, ' with Two-pronged
Spears.'

preserved the quality of 'symmetry' (for which there is no word in Latin) by the new and hitherto untried method of modifying the squareness of the figure of the old sculptors, and he used commonly to say that whereas his predecessors had made men as they really were, he made them as they appeared to be. A peculiarity of this sculptor's work seems to be the minute finish maintained in even the smallest details.

Lysippus left three sons who were his pupils, the celebrated artists Laippus,[a] Boëdas and Euthycrates, the last pre-eminent, although he copied the harmony rather than the elegance of his father, preferring to win favour in the severely correct more than in the agreeable style. Accordingly his Heracles, at Delphi, and his Alexander Hunting, at Thespiae, his group of Thespiades,[b] and his Cavalry in Action are works of extreme finish, and so are his statue of Trophonius at the oracular shrine of that deity, a number of Four-horse Chariots, a Horse with Baskets [c] and a Pack of Hounds. Moreover Tisicrates, another native of Sicyon, was a pupil of Euthycrates, but closer to the school of Lysippus—indeed many of his statues cannot be distinguished from Lysippus's work, for instance his Old Man of Thebes, his King Demetrius (Poliorcetes), and his Peucestes, the man who saved the life of Alexander the Great and so deserved the honour of this commemoration. *Lysippus' sons.* *Tisicrates.*

Artists [d] who have composed treatises recording these matters speak with marvellously high praise of Telephanes of Phocis, who is otherwise unknown, *Telephanes.*

[a] Pliny means the writers Xenocrates of Sicyon and Antigonus of Carystus, from whom, through Varro, much of Pliny's material about art comes.

caeum, ignotum alias, quoniam[1] . . . Thessaliae[2] habitaverit et ibi opera eius latuerint; alioqui suffragiis ipsorum aequatur Polyclito, Myroni, Pythagorae. laudant eius Larisam et Spintharum pentathlum et Apollinem. alii non hanc ignobilitatis fuisse causam, sed quod se regum Xerxis atque Darei officinis dediderit, existimant.

69 Praxiteles quoque, qui[3] marmore felicior, ideo et clarior fuit, fecit tamen et ex aere pulcherrima opera : Proserpinae raptum, item catagusam et Liberum patrem, Ebrietatem nobilemque una Satyrum, quem Graeci periboëton cognominant, et signa, quae ante Felicitatis aedem fuere, Veneremque, quae ipsa aedis incendio cremata est Claudii principatu, marmoreae illi suae per terras inclutae parem,

70 item stephanusam,[4] pseliumenen,[5] Oporan,[6] Harmodium et Aristogitonem tyrannicidas, quos a Xerxe Persarum rege captos victa Perside Atheniensibus remisit Magnus Alexander. fecit et puberem Apollinem subrepenti lacertae comminus sagitta

---

[1] *lac. C. F. W. Müller.*
[2] thessaliae *B* : in thessalia *rell.*
[3] qui *add. Mayhoff.*
[4] *Fortasse* ⟨se⟩ stephanusam *vel* stephanusam ⟨se⟩.
[5] pseliumenen *Urlichs, O. Jahn* : pseliumenen *cd. Leid. Voss.*: varia *rell.*
[6] Oporan *cd. Flor. Ricc.* : varia *rell.* (oporum *cd. Par.* 6801) : oenophorum *edd. vett.*: canephoram *Urlichs* (*immo* canephorum).

---

[a] Κατάγουσα, from κατάγω ' draw down,' ' spin.'
[b] The wreath would be one bestowed on an athlete by the city (personified) when he won his victory; *pseliumene* is from ψελιῶ, and ψέλιον means an armlet.
[c] Not the actual tyrant Hippias but his brother and assistant Hipparchus, at Athens, 514–13 B.C.

since he lived at . . . in Thessaly where his works
have remained in concealment, although these
writers' own testimony puts him on a level with
Polycleitus, Myron and Pythagoras. They praise
his Larisa, his Spintharus the Five-bout Champion,
and his Apollo. Others however are of opinion
that the cause of his lack of celebrity is not the
reason mentioned but his having devoted himself
entirely to the studios established by King Xerxes
and King Darius.

Praxiteles although more successful and therefore *Praxiteles.*
more celebrated in marble, nevertheless also made *fl. c. 370 B.C.*
some very beautiful works in bronze: the Rape of
Persephone, also The Girl Spinning,[a] and a Father
Liber or Dionysus, with a figure of Drunkenness and
also the famous Satyr, known by the Greek title
Periboëtos meaning ' Celebrated,' and the statues
that used to be in front of the Temple of Happiness,
and the Aphrodite, which was destroyed by fire when
the temple of that goddess was burnt down in the
reign of Claudius, and which rivalled the famous A.D. 41–54.
Aphrodite, in marble, that is known all over the
world; also A Woman Bestowing a Wreath, A
Woman Putting a Bracelet on her Arm,[b] Autumn,
Harmodius and Aristogeiton who slew the tyrant[c]—
the last piece[d] carried off by Xerxes King of the 480 B.C.
Persians but restored to the Athenians by Alexander 331 B.C.
the Great after his conquest of Persia. Praxiteles
also made a youthful Apollo called in Greek the
Lizard-Slayer[e] because he is waiting with an arrow

---

[d] But the group carried off was by Antenor, and its
restoration is attributed also to Seleucus I, and to Antiochus I.
See note on pp. 256–257.

[e] Degenerate copies of this still exist.

insidiantem, quem sauroctonon vocant. spectantur
et duo signa eius diversos adfectus exprimentia,
flentis matronae et meretricis gaudentis. hanc
putant Phrynen fuisse deprehenduntque in ea
amorem artificis et mercedem in vultu meretricis.
71 habet simulacrum et benignitas eius; Calamidis
enim quadrigae aurigam suum inposuit, ne melior
in equorum effigie defecisse in homine crederetur.
ipse Calamis et alias quadrigas bigasque fecit equis
semper [1] sine aemulo expressis; sed, ne videatur
in hominum effigie inferior, Alcmena [2] nullius est
nobilior.
72 Alcamenes, Phidiae discipulus, et marmorea fecit,
sed aereum pentathlum, qui vocatur encrinomenos;
at Polycliti discipulus Aristides quadrigas bigasque.
Amphicrates [3] Leaena laudatur. scortum haec,
lyrae cantu familiaris Harmodio et Aristogitoni.
consilia eorum de tyrannicidio usque in mortem
excruciata a tyrannis non prodidit; quam ob rem
Athenienses, et honorem habere ei volentes nec
tamen scortum celebrasse, animal nominis eius fecere
atque, ut intellegeretur causa honoris, in opere
linguam addi ab artifice vetuerunt.

---

[1] equis semper *cdd.* (sēm pari equis *B*) : se impari, equis
Traube.

[2] alcumena *cd. deperd. ap. Dalecamp* : Achamene *edd. vett.* :
Alcman poeta *Eugénie Sellers* : alchimena *aut* alcm- *cdd.*
(alcamenet *B*[1] : -me et *B*[2]).

[3] iphicrates *cd. Par.* 6801 : Tisicratis *Hardouin.*

---

[a] Or, 'received by her.' The exact meaning is not clear.

[b] Or perhaps : ' Undergoing the test ' for recognition as an
athlete.

[c] Hippias and Thessalus of Athens after the killing of their
brother, 514–13 B.C.; *cf.* § 70 above, and note.

for a lizard creeping towards him. Also two of his statues expressing opposite emotions are admired, his Matron Weeping and his Merry Courtesan. The latter is believed to have been Phryne and connoisseurs detect in the figure the artist's love of her and the reward promised him [a] by the expression on the courtesan's face. The kindness also of Praxiteles is represented in sculpture, as in the Chariot and Four of Calamis he contributed the *Calamis.* charioteer, in order that the sculptor might not be thought to have failed in the human figure although more successful in representing horses. Calamis himself also made other chariots, some with four horses and some with two, and in executing the horses he is invariably unrivalled: but—that it may not be supposed that he was inferior in his human figures—his Alcmena is as famous as that of any other sculptor.

Alcamenes a pupil of Pheidias made marble figures, *Alcamenes.* and also in bronze a Winner of the Five Bouts, known by the Greek term meaning Highly Commended,[b] but Polycleitus's pupil Aristides made four-horse and pair-horse chariots. Amphicrates is *Amphicrates.* praised for his Leaena; she was a harlot, admitted to the friendship of Harmodius and Aristogeiton because of her skill as a harpist, who though put to the torture by the tyrants [c] till she died refused to betray their plot to assassinate them. Consequently the Athenians wishing to do her honour and yet unwilling to have made a harlot famous, had a statue made of a lioness, as that was her name, and to indicate the reason for the honour paid her instructed the artist to represent the animal as having no tongue.

73   Bryaxis Aesculapium et Seleucum [1] fecit, Boëdas
     adorantem, Baton Apollinem et Iunonem, qui sunt
74   Romae in Concordiae templo, Cresilas [2] volneratum
     deficientem, in quo possit intellegi quantum restet
     animae, et Olympium Periclen dignum cognomine,
     mirumque in hac arte est quod nobiles viros nobiliores
     fecit.   Cephisodorus [3] Minervam mirabilem in portu
     Atheniensium et aram ad templum Iovis Servatoris
     in eodem portu, cui pauca comparantur, Canachus
75   Apollinem nudum, qui Philesius cognominatur, in
     Didymaeo Aeginetica aeris temperatura, cervumque
     una ita vestigiis suspendit, ut linum [4] subter pedes
     trahatur alterno morsu calce digitisque retinentibus
     solum, ita vertebrato ungue [5] utrisque in partibus,
     ut a repulsu per vices resiliat.   idem et celetizontas
     pueros, Chaereas Alexandrum Magnum et Philip-
     pum patrem eius fecit, Ctesilaus doryphoron et
76   Amazonem volneratam, Demetrius Lysimachen,
     quae sacerdos Minervae fuit LXIIII annis, idem et
     Minervam, quae mycetica [6] appellatur—dracones [7]
     in Gorgone eius ad ictus citharae tinnitu resonant;—

[1] Salutem *Hardouin.*
[2] ctesilas *cd. Leid. Voss.* : Ctesilaus *Hardouin.*
[3] Cephisodotus *Hardouin.*
[4] ut inlitum *B.*
[5] ungue *aut* pede *aut* vertebrata ungula *Warmington*:
dente.
[6] mycetica *Traube* : myctica *Ian* : mystica *Fröhner* :
myetica *B* : musica *rell.*
[7] dracones *B* : quoniam dracones *rell.*

---

[a] Probably Seleucus I, King 312–280 B.C.
[b] ' Olympian,' ' High and Mighty,' as Pericles himself was
called during his lifetime.   Copies of this, and the *basis* of the
original, still exist.

Bryaxis made statues of Asclepius and Seleucus,[a] *Other famous sculptors.*
Boëdas a Man Praying, Baton an Apollo and a Hera,
both now in the Temple of Concord at Rome.
Cresilas did a Man Fainting from Wounds, the
expression of which indicates how little life remains,
and the Olympian Pericles, a figure worthy of its
title [b]; indeed it is a marvellous thing about the art
of sculpture that it has added celebrity to men
already celebrated.   Cephisodorus made the wonderful Athene at the harbour of Athens and the
almost unrivalled altar at the temple of Zeus the
Deliverer at the same harbour, Canachus the naked
Apollo, surnamed Philesius, at Didyma, made of
bronze compounded at Aegina [c]; and with it he
made a stag so lightly poised in its footprints as to
allow of a thread being passed underneath its feet,
the ' heel ' and the ' toes ' holding to the base with
alternate contacts, the whole hoof being so jointed in
either part that it springs back from the impact
alternately.[d]   He also made a Boys Riding on
Race-horses.   Chæreas did Alexander the Great
and his father Philip, Ctesilaus a Man with a Spear
and a Wounded Amazon, Demetrius Lysimache
who was a priestess of Athene for 64 years, and also
the Athene called the Murmuring Athene [e]—
the dragons on her Gorgon's head sound with a
tinkling note when a harp is struck; he likewise did

---

[c] Or, 'compounded on the Aeginetan formula.' Cf. § 10.

[d] Pliny is not clear; the MSS. reading *dente* (' tooth ' not
' ivory ' ?) is altered here to *ungue* by conjecture.   Perhaps he
simply means that when the figure was rocked to and fro, a
thread could be slipped under two feet.   From coins we know
that the small stag was not on the ground but on the god's
hand.

[e] The right reading is unknown.

idem equitem Simonem, qui primus de equitatu
scripsit. Daedalus, et ipse inter fictores laudatus,
pueros duos destringentes se fecit, Dinomenes
77 Protesilaum et Pythodemum luctatorem. Euphranoris Alexander Paris est, in quo laudatur quod
omnia simul intellegantur, iudex dearum, amator
Helenae et tamen Achillis interfector. huius est
Minerva, Romae quae dicitur Catuliana,[1] infra
Capitolium a Q. Lutatio dicata, et simulacrum Boni
Eventus, dextra pateram, sinistra spicam ac papa-
vera tenens, item Latona puerpera Apollinem et
Dianam infantes sustinens in aede Concordiae.
78 fecit et quadrigas bigasque et cliduchon[2] eximia
forma et Virtutem et Graeciam, utrasque colossaeas,
mulierem admirantem et adorantem, item Alexandrum et Philippum in quadrigis; Eutychides Euro-
tam, in quo artem ipso amne liquidiorem plurimi
dixere. Hegiae Minerva Pyrrhusque rex laudatur
et celetizontes pueri et Castor ac Pollux ante aedem
Iovis Tonantis, Hagesiae in Pario colonia Hercules,
79 Isidoti[3] buthytes.[4] Lycius Myronis discipulus fuit,
qui fecit dignum praeceptore puerum sufflantem
languidos ignes et Argonautas; Leochares aquilam
sentientem, quid rapiat in Ganymede et cui ferat,

---

[1] Catulina *Manutius.*
[2] cliduchon *Hermolaus Barbarus* : cliticon *B* : cliticum *rell.*
[3] *V.l.* Isidori.
[4] Buthytes *B* : Buthyres *rell.* : Eleuthereus *Hardouin.*

---

[a] In Greek 'Αγαθὴ Τύχη. But it appears that the statue
was one of Triptolemus, re-named as a Roman rustic divinity.
[b] Cf. § 54.
[c] The river on which Sparta stood.
[d] *c.* 318–272 B.C. But perhaps *rex*, king, should be deleted.
[e] Parium was made a Roman colony by Augustus.

the mounted statue of Simon who wrote the first treatise on horsemanship. Dædalus (also famous as a modeller in clay) made Two Boys using a Body-Scraper, and Dinomenes did a Protesilaus and the wrestler Pythodemus. The statue of Alexander Paris is by Euphranor; it is praised because it conveys all the characteristics of Paris in combination—the judge of the goddesses, the lover of Helen and yet the slayer of Achilles. The Athene, called at Rome the Catuliana, which stands below the Capitol and was dedicated by Quintus Lutatius 78 B.C. Catulus, is Euphranor's, and so is the figure of Success,[a] holding a dish in the right hand and in the left an ear of corn and some poppies, and also in the temple of Concord a Leto as Nursing Mother, with the infants Apollo and Artemis in her arms. He also made four-horse and two-horse chariots, and an exceptionally beautiful Lady with the Keys,[b] and two colossal statues, one of Virtue and one of Greece, a Woman Wondering and Worshipping, and also an Alexander and a Philip in four-horse chariots. Eutychides did a Eurotas,[c] in which it has frequently been said that the work of the artist seems clearer than the water of the real river. The Athene and the King Pyrrhus [d] of Hegias are praised, and his Boys Riding on Race-horses, and his Castor and Pollux that stand before the temple of Jupiter the Thunderer; and so are Hagesias's Heracles in our colony [e] of Parium, and Isidotus's Man Sacrificing an Ox. Lycius who was a pupil of Myron did a Boy Blowing a Dying Fire that is worthy of his instructor, also a group of the Argonauts; Leochares an Eagle carrying off Ganymede in which the bird is aware of what his burden is and for whom he is

parcentemque unguibus etiam per vestem puero,[1]
Autolycum pancratii [2] victorem, propter quem
Xenophon symposium scripsit, Iovemque illum
Tonantem in Capitolio ante cuncta laudabilem, item
Apollinem diadematum, Lysiscum,[3] mangonem,[4]
puerum subdolae ac fucatae vernilitatis, Lycius
80 et ipse puerum suffitorem. Menaechmi vitulus
genu premitur replicata cervice. ipse Menaechmus
scripsit de sua arte. Naucydes Mercurio et dis-
cobolo et immolante arietem censetur, Naucerus [5]
luctatore anhelante,[6] Niceratus Aesculapio et
Hygia,[7] qui sunt in Concordiae templo Romae.
Pyromachi quadriga ab Alcibiade regitur. Polycles
Hermaphroditum nobilem fecit, Pyrrhus Hygiam
81 et Minervam, Phanis, Lysippi discipulus, epithyu-
san. Styppax Cyprius uno celebratur signo,
splanchnopte; Periclis Olympii vernula hic fuit
exta torrens ignemque oris pleni spiritu accendens.
Silanion Apollodorum fudit, fictorem et ipsum, sed
inter cunctos diligentissimum artis et iniquum
sui iudicem, crebro perfecta signa frangentem,
dum satiari cupiditate artis non quit, ideoque
82 insanum cognominatum—hoc in eo expressit, nec

---

[1] puero B : puerum rell.
[2] pancrati B : pancratio rell.
[3] lysiscum B : lusiscus rell. : Lyciscus Gelen.
[4] mangonem B : langonem vel lagonem rell.
[5] Nauclerus coni. Hardouin.
[6] V.l. luctatorem anhelantem (fecit add. edd. vett.).
[7] Aesculapio et Hygia coni. Ian : aesculapium et hygiam aut a.h. cdd. : Hygiam fecit Detlefsen.

---

[a] The banquet described in Xenophon's *Symposium* was
given by Callias in honour of Autolycus's victory in the
pentathlum at the Great Panathenaea in 422 B.C.
[b] See § 74, note.

carrying it, and is careful not to let his claws hurt the boy even through his clothes, and Autolycus Winner of the All-round Bout, being also the athlete in whose honour Xenophon wrote his *Banquet*,[a] and the famous Zeus the Thunderer now on the Capitol, of quite unrivalled merit, also an Apollo crowned with a Diadem; also Lyciscus, the Slave-dealer, and a Boy, with the crafty cringing look of a household slave. Lycius also did a Boy Burning Perfumes. There is a Bull-calf by Menæchmus, on which a man is pressing his knee as he bends its neck back; Menæchmus has written a treatise about his own work. The reputation of Naucydes rests on his Hermes and Man throwing a Disc and Man Sacrificing a Ram, that of Naucerus on his Wrestler Winded, that of Niceratus on his Asclepius and his Goddess of Health, which are in the Temple of Concord at Rome. Pyromachus has an Alcibiades Driving a Chariot and Four; Polycles made a famous Hermaphrodite, Pyrrhus, a Goddess of Health and an Athene, Phanis, who was a pupil of Lysippus, a Woman Sacrificing. Styppax of Cyprus is known for a single statue, his Man Cooking Tripe, which represented a domestic slave of the Olympian [b] Pericles roasting inwards and puffing out his cheeks as he kindles the fire with his breath; Silanion cast a metal figure of Apollodorus, who was himself a modeller, and indeed one of quite unrivalled devotion to the art and a severe critic of his own work, who often broke his statues in pieces after he had finished them, his intense passion for his art making him unable to be satisfied, and consequently he was given the surname of the Madman—this quality he brought out in his statue, the Madman, which

hominem ex aere fecit, sed iracundiam—et Achillem
nobilem, item epistaten exercentem athletas; Strongylion Amazonem, quam ab excellentia crurum
eucnemon appellant, ob id in comitatu Neronis
principis circumlatam. idem fecit puerum, quem
amando Brutus Philippiensis cognomine suo inlus-
83 travit. Theodorus, qui labyrinthum fecit Sami,
ipse se ex aere fudit. praeter similitudinis mirabilem famam [1] magna suptilitate celebratur: [2] dextra
limam tenet, laeva tribus digitis quadrigulam
tenuit, tralatam Praeneste parvitatis [3] ut miraculum: [4]
pictam [5] eam currumque et aurigam integeret alis
simul facta musca. Xenocrates, Tisicratis discipulus,
ut alii, Euthycratis, vicit utrosque copia signorum.
et de sua arte composuit volumina.

84 Plures artifices fecere Attali et Eumenis adversus
Gallos proelia, Isigonus, Pyromachus, Stratonicus,
Antigonus, qui volumina condidit de sua arte.
Boëthi, quamquam argento melioris, infans amplex-
ando [6] anserem strangulat. atque ex omnibus, quae
rettuli, clarissima quaeque in urbe iam sunt dicata a

---

[1] similitudinis mirabilem famam B : similitudinem fama
rell.: s. nobilem f. edd. vett.

[2] celebratur J. Müller : celebratus.

[3] parvitatis B : tantae p. rell.

[4] miraculū Mayhoff: ut miraculo B (om. miraculo rell.): ut
mirum dictu Traube : del. ut Urlichs.

[5] pictam B : totam rell. : fictam Stuart Jones : pictam ut
Urlichs.

[6] amplexando Traube : annosum (olim vi annisus)
Buecheler : vi annosum Meister : vi Külb : vi aenum Boisacq :
ex aere H. Stein : sexennis O. Jahn : ex animo Ian : ulnis
Urlichs : eximiū Mayhoff: sex anno B : sex annis B² : eximie
aut eximiae rell.

---

[a] The temple of Hera.

represented in bronze not a human being but anger personified. Silanion also made a famous Achilles, and also a Superintendent Exercising Athletes; Strongylion made an Amazon, which from the remarkable beauty of the legs is called the Eucnemon, and which consequently the emperor Nero caused to be carried in his retinue on his journeys. The same sculptor made the figure rendered famous by Brutus under the name of Brutus's Boy because it represented a favourite of the hero of the battles at 42 B.C. Philippi. Theodorus, who constructed the Labyrinth [a] at Samos, cast a statue of himself in bronze. Besides its remarkable celebrity as a likeness, it is famous for its very minute workmanship; the right hand holds a file, and three fingers of the left hand originally held a little model of a chariot and four, but this has been taken away to Palestrina as a marvel of smallness: if the team were reproduced in a picture with the chariot and the charioteer, the model of a fly, which was made by the artist at the same time, would cover it with its wings. Xenocrates, who was a pupil of Tisicrates, or by other accounts of Euthycrates, surpassed both of the last mentioned in the number of his statues; and he also wrote books about his art.

Several artists have represented the battles of Attalus [b] and Eumenes against the Gauls, Isigonus, Pyromachus, Stratonicus and Antigonus, who wrote books about his art. Boëthus did a Child [c] Strangling a Goose by hugging it, although he is better in silver. And among the list of works I have referred to all the most celebrated have now been dedicated by the

---

[b] Attalus I of Pergamum, who dealt with Gallic invaders of Asia Minor between 240 and 232 B.C.   [c] Copies exist.

Vespasiano principe in templo Pacis aliisque eius operibus, violentia Neronis in urbem convecta et in sellariis domus aureae disposita.

85 Praeterea sunt aequalitate celebrati artifices, sed nullis operum suorum praecipui; Ariston, qui et argentum caelare solitus est, Callides,[1] Ctesias, Cantharus Sicyonius, Dionysius, Diodorus,[2] Critiae discipulus, Deliades, Euphorion, Eunicus et Hecataeus, argenti caelatores, Lesbocles, Prodorus, Pythodicus, Polygnotus, idem pictor e nobilissimis,[3] item e caelatoribus Stratonicus, Scymnus Critiae discipulus.

86 Nunc percensebo eos, qui eiusdem generis opera fecerunt, ut Apollodorus, Androbulus, Asclepiodorus, Aleuas philosophos, Apellas et adornantes [4] se [5] feminas, Antignotus et [luctatores,][6] perixyomenum, tyrannicidasque supra dictos, Antimachus, Athenodorus feminas nobiles, Aristodemus et luctatores bigasque cum auriga, philosophos, anus, Seleucum regem. habet gratiam suam huius quoque dory-

87 phorus. Cephisodoti duo fuere: prioris est Mercurius Liberum patrem in infantia nutriens; fecit et contionantem manu elata—persona in incerto est; sequens philosophos fecit. Colotes, qui cum Phidia Iovem Olympium fecerat, philosophos, item Cleon

---

[1] callases *cd. Par.* 6801 : Callicles *Urlichs* : Calliades *Hardouin.*

[2] Dionysius, Diodorus *Detlefsen* : diodorus *B* : dionysiodorus *aut* dionysodorus *rell.*

[3] *V.l.* idem pictores nobilissimi.   [4] *V.l.* adorantes.

[5] se *cd. Leid. Voss.* : *om. rell.*

[6] et luctatores *cdd.* : *om.* luctatores *B, cd. Par. Lat.* 6797.

emperor Vespasian in the Temple of Peace and his A.D. 75.
other public buildings; they had been looted by
Nero, who conveyed them all to Rome and arranged
them in the sitting-rooms of his Golden Mansion.

Besides these, artists on the same level of merit
but of no outstanding excellence in any of their
works are: Ariston, who often also practised
chasing silver, Callides, Ctesias, Cantharus of Sicyon,
Dionysius, Diodorus the pupil of Critias, Deliades,
Euphorion, Eunicus and Hecatæus the silver chasers,
Lesbocles, Prodorus, Pythodicus, Polygnotus, who
was also one of the most famous among painters,
similarly Stratonicus among chasers, and Critias's
pupil Scymnus.

I will now run through the artists who have
made works of the same class, such as Apollodorus, Androbulus and Asclepiodorus, Aleuas, who
have done philosophers, and Apellas also women
donning their ornaments, and Antignotus also
Man using a Body-scraper and the Men[a] that
Slew the Tyrant, above-mentioned, Antimachus,
Athenodorus who made splendid figures of women,
Aristodemus who also did Wrestlers, and Chariot
and Pair with Driver, figures of philosophers, of old
women, and King Seleucus; Aristodemus's Man
holding Spear is also popular. There were two
artists named Cephisodotus; the Hermes Nursing
Father Liber or Dionysos when an Infant belongs to
the elder, who also did a Man Haranguing with
Hand Uplifted—whom it represents is uncertain.
The later Cephisodotus did philosophers. Colotes
who had co-operated with Pheidias in the Olympian §§ 49, 54.
Zeus made statues of philosophers, as also did Cleon

[a] Harmodius and Aristogeiton. See §§ 70, 72.

et Cenchramis [1] et Callicles et Cepis, Chalcosthenes [2]
et comoedos et athletas, Daippus perixyomenon,
Daiphron et Damocritus et Daemon philosophos.
88 Epigonus omnia fere praedicta imitatus praecessit
in tubicine et matri interfectae infante miserabiliter
blandiente. Eubuli mulier admirans laudatur, Eu-
bulidis digitis computans. Micon athletis spectatur,
Menogenes quadrigis. Nec minus Niceratus omnia,
quae ceteri, adgressus repraesentavit Alcibiaden
lampadumque accensu matrem eius Demaraten
89 sacrificantem. Tisicratis bigae Piston mulierem
inposuit, idem fecit Martem et Mercurium, qui sunt
in Concordiae templo Romae. Perillum nemo
laudet saeviorem Phalaride tyranno, cui taurum
fecit mugitus inclusi [3] hominis pollicitus igni subdito
et [4] primus expertus cruciatum eum iustiore saevitia.
huc a simulacris deorum hominumque devocaverat
humanissimam artem. ideo tot conditores eius
laboraverant, ut ex ea tormenta fierent! itaque una
de causa servantur opera eius, ut quisquis illa videat,
90 oderit manus. Sthennis Cererem, Iovem, Minervam
fecit, qui sunt Romae in Concordiae templo, idem

---

[1] Cenchramus *Overbeck.*
[2] calcostenes *B*: Caecosthenes (=Καϊκοσθένης) *Overbeck.*
[3] inclusi *add. Mayhoff.*
[4] et *B*: *v.ll.* ex, est: exprimere *Detlefsen.*

[a] This should be Dinomache.

and Cenchramis and Callicles and Cepis; Chalco-
sthenes also did actors in comedy and athletes;
Daippus a Man using a Scraper; Daiphron,
Damocritus and Dæmon statues of philosophers.
Epigonus, who copied others in almost all the subjects
already mentioned, took the lead with his Trumpet-
player and his Weeping Infant pitifully caressing
its Murdered Mother. Praise is given to Eubulus's
Woman in Admiration and to Eubulides's Person
Counting on the Fingers. Micon is noticed for his
athletes and Menogenes for his chariots and four.
Niceratus, who likewise attempted all the subjects
employed by any other sculptor, did a statue of
Alcibiades and one of his mother Demarate,[a] represented as performing a sacrifice by torch-light.
Tisicrates did a pair-horse chariot in which Piston
afterwards placed a woman; the latter also made
an Ares and a Hermes now in the Temple of Concord
at Rome. No one should praise Perillus, who was *c.* 570 B.C.
more cruel than the tyrant Phalaris, for whom he
made a bull, guaranteeing that if a man were shut
up inside it and a fire lit underneath the man would
do the bellowing; and he was himself the first to
experience this torture—a cruelty more just than
the one he proposed. Such were the depths to
which the sculptor had diverted this most humane
of arts from images of gods and men! All the
founders of the art had only toiled so that it should
be employed for making implements of torture!
Consequently this sculptor's works are preserved
for one purpose only, so that whoever sees them
may hate the hands that made them. Sthennis did
a Demeter, a Zeus and an Athene that are in the
Temple of Concord at Rome, and also Weeping

193

flentes matronas et adorantes sacrificantesque.
Simon canem et sagittarium fecit, Stratonicus
91 caelator ille philosophos, copas[1] uterque;[2] athletas
autem et armatos et venatores sacrificantesque Baton,
Euchir, Glaucides, Heliodorus, Hicanus, Iophon,[3]
Lyson, Leon, Menodorus, Myagrus, Polycrates,
Polyidus,[4] Pythocritus, Protogenes, idem pictor e
clarissimis, ut dicemus, Patrocles,[5] Pollis, Posidonius,
qui et argentum caelavit nobiliter, natione Ephesius,
Periclymenus, Philon, Symenus, Timotheus, Theomnestus, Timarchides, Timon, Tisias, Thrason.

92   Ex omnibus autem maxime cognomine insignis est
Callimachus, semper calumniator sui nec finem
habentis diligentiae, ob id catatexitechnus appellatus, memorabili[6] exemplo adhibendi et curae
modum.   huius sunt saltantes Lacaenae, emendatum
opus, sed in quo gratiam omnem diligentia abstulerit.
hunc quidem et pictorem fuisse tradunt.   non aere
captus nec arte, unam tantum Zenonis statuam
Cypria expeditione non vendidit Cato, sed quia
philosophi erat, ut obiter hoc quoque noscatur tam
insigne[7] exemplum.

93   In mentione statuarum est et una non praeter-

<hr>

[1] copas *Gerhard* : scopas.

[2] uterque *cdd.* (utrosque B^1, utraque B^2) : utrasque *edd.*
vett.

[3] Iophon *Urlichs*: Leophon *Sillig* : Herophon *Loewy* :
olophon *B* : lophon *rell.*

[4] Polydorus *Hermolaus Barbarus.*

[5] Patroclus *coni. Sillig coll.* § 50.

[6] memorabili *B, cd. Par.* 6801 : memorabilis *rell.*

[7] insigne *Pintianus* : inane.

<hr>

[a] The doubtful text may contain the name Scopas ; see
critical notes.

[b] κατατηξίτεχνος, one who wastes his skill in driblets.

Matrons and Matrons at Prayer and Offering a Sacrifice. Simon made a Dog and an Archer, the famous engraver Stratonicus some philosophers and each of these artists made figures of hostesses of inns.[a] The following have made figures of athletes, armed men, hunters and men offering sacrifice: Baton, Euchir, Glaucides, Heliodorus, Hicanus, Iophon, Lyson, Leon, Menodorus, Myagrus, Polycrates, Polyidus, Pythocritus, Protogenes (who was also, as we shall say later, one of the most famous painters), Patrocles, Pollis and Posidonius (the last also a distinguished silver chaser, native of Ephesus), Periclymenus, Philo, Symenus, Timotheus, Theomnestus, Timarchides, Timon, Tisias, Thraso.

But of all Callimachus is the most remarkable, because of the surname attached to him: he was always unfairly critical of his own work, and was an artist of never-ending assiduity, and consequently he was called the Niggler,[b] and is a notable warning of the duty of observing moderation even in taking pains. To him belongs the Laconian Women Dancing, a very finished work but one in which assiduity has destroyed all charm. Callimachus is reported to have also been a painter. Cato in his expedition to Cyprus sold all the statues found there except one of Zeno; it was not the value of the bronze nor the artistic merit that attracted him, but its being the statue of a philosopher: I mention this by the way, to introduce this distinguished[c] instance also.

In mentioning statues—there is also one we must

xxxv. 101 *sqq.*

flor. c. 400 B.C.

53–56 B.C.

---

[c] The MSS. give 'this empty example,' explained as implying that Cato neglected the example set by his great grandfather, Cato the Censor who disliked the Greeks.

eunda, quamquam auctoris incerti, iuxta rostra,
Herculis tunicati, sola eo habitu Romae, torva facie
sentiensque [1] suprema tunicae.[2] in hac tres sunt
tituli: L. Luculli imperatoris de manubiis, alter:
pupillum Luculli filium ex S. C. dedicasse, tertius:
T. Septimium Sabinum aed. cur. ex privato in
publicum restituisse. tot certaminum tantaeque
dignationis simulacrum id fuit.

94 XX. Nunc praevertemur [3] ad differentias aeris et
mixturas. in Cyprio [coronarium et regulare est
utrumque ductile] [4] coronarium tenuatur in lamnas,
taurorumque felle tinctum speciem auri in coronis
histrionum praebet, idemque in uncias additis auri
scripulis senis praetenui pyropi brattea ignescit.
regulare et in aliis fit metallis, itemque caldarium.
differentia quod caldarium funditur tantum, malleis
fragile, quibus regulare obsequitur ab aliis ductile
appellatum, quale omne Cyprium est. sed et in
ceteris metallis cura distat a caldario; omne enim
diligentius purgatis igni vitiis excoctisque regulare
95 est. In reliquis generibus palma Campano perhi-

---

[1] sentiensque *B* : sentientique *rell.* : sentienteque *edd. vett.*
[2] tunicae *B* : tunica *rell.* : in tunica *edd. vett.*
[3] *V.l.* revertemur.
[4] coronarium . . . . ductile *cd. Vind.* : om. *rell.*

---

[a] *I.e.* the poisoned garment that caused his death.
[b] In campaigns against Mithridates, 74–67 B.C.

not pass over in spite of the sculptor's not being
known—the figure, next to the Beaked Platform, of
Heracles in the Tunic,[a] the only one in Rome that
shows him in that dress; the countenance is stern
and the statue expresses the feeling of the final
agony of the tunic. On this statue there are three
inscriptions, one stating that it had been part of the
booty taken [b] by the general Lucius Lucullus, and
another saying that it was dedicated, in pursuance
of a decree of the Senate, by Lucullus's son while
still a ward, and the third, that Titus Septimius
Sabinus as curule ædile had caused it to be restored
to the public from private ownership. So many
were the rivalries connected with this statue and so
highly was it valued.

XX. But we will now turn our attention particularly to the various forms of copper, and its blends.
In the case of the copper of Cyprus ' chaplet copper '
is made into thin leaves, and when dyed with ox-gall
gives the appearance of gilding on theatrical property
coronets; and the same material mixed with gold
in the proportion of six scruples of gold to the ounce
makes a very thin plate called pyropus, ' fire-coloured '
and acquires the colour of fire. Bar copper also is
produced in other mines, and likewise fused copper.
The difference between them is that the latter can
only be fused, as it breaks under the hammer,
whereas bar copper, otherwise called ductile copper,
is malleable, which is the case with all Cyprus copper.
But also in the other mines, this difference of bar
copper from fused copper is produced by treatment;
for all copper after impurities have been rather
carefully removed by fire and melted out of it
becomes bar copper. Among the remaining kinds

Various forms and blends of copper and bronze.

betur,[1] utensilibus vasis [2] probatissimo. pluribus
fit hoc modis. namque Capuae liquatur non carbonis
ignibus, sed ligni, purgaturque roboreo cribro [3]
profusum in [4] aquam frigidam ac saepius simili modo
coquitur, novissime additis plumbi argentarii Hispa-
niensis denis libris in centenas aeris. ita lentescit
coloremque iucundum trahit, qualem in aliis generibus
96 aeris adfectant oleo ac sale. fit Campano [1] simile in
multis partibus Italiae provinciisque, sed octonas
plumbi libras addunt et carbone recocunt propter
inopiam ligni. quantum ea res differentiae adferat,
in Gallia maxime sentitur, ubi inter lapides cande-
factos funditur; exurente enim coctura nigrum
atque fragile conficitur. praeterea semel recoquunt
quod saepius fecisse bonitati plurimum confert.
id quoque notasse non ab re est, aes omne frigore
magno melius fundi.

97 Sequens temperatura statuaria est eademque
tabularis hoc modo : massa proflatur in primis, mox
in proflatum additur tertia portio aeris collectanei,
hoc est ex usu coempti. peculiare in eo condimen-
tum attritu domiti et consuetudine nitoris veluti
mansuefacti. miscentur et plumbi argentarii pondo
98 duodena ac selibrae centenis proflati. Appellatur

---

[1] perhibetur . . . campano (§ 96 *init.*) *B* : *om. rell.*
[2] vasorum *coni. Warmington coll.* XIII. 72.
[3] ligno *K. C. Bailey.*
[4] in *add. K. C. Bailey* : perfusum aqua frigida *Sillig.*

[a] Tin and lead mixed in equal parts.
[b] Possibly mineral coal.

of copper the palm goes to bronze of Campania, which is most esteemed for utensils. There are several ways of preparing it. At Capua it is smelted in a fire of wood, not of charcoal, and then poured into cold water and cleaned in a sieve made of oak, and this process of smelting is repeated several times, at the last stage Spanish silver lead [a] being added to it in the proportion of ten pounds to one hundred pounds of copper: this treatment renders it pliable and gives it an agreeable colour of a kind imparted to other sorts of copper and bronze by means of oil and salt. Bronze resembling the Campanian is produced in many parts of Italy and the provinces, but there they add only eight pounds of lead, and do additional smelting with charcoal [b] because of their shortage of wood. The difference produced by this is noticed specially in Gaul, where the metal is smelted between stones heated red hot, as this roasting scorches it and renders it black and friable. Moreover they only smelt it again once whereas to repeat this several times contributes a great deal to the quality. It is also not out of place to notice that all copper and bronze fuses better in very cold weather.

The proper blend for making statues is as follows, and the same for tablets: at the outset the ore is melted, and then there is added to the melted metal a third part of scrap copper, that is copper or bronze that has been bought up after use. This contains a peculiar seasoned quality of brilliance that has been subdued by friction and so to speak tamed by habitual use. Silver-lead is also mixed with it in the proportion of twelve and a half pounds to every hundred pounds of the fused metal. There is also

Blends for statues and moulds.

etiamnum et formalis temperatura aeris tenerrimi,
quoniam nigri plumbi decima portio additur et
argentarii vicesima, maximeque ita colorem bibit,
quem Graecanicum vocant. Novissima est quae
vocatur ollaria, vase nomen hoc dante, ternis aut
quaternis libris plumbi argentarii in centenas aeris
additis. Cyprio si addatur plumbum, colos purpurae
fit in statuarum praetextis.

99 XXI. Aera extersa robiginem celerius trahunt
quam neglecta, nisi oleo perunguantur. servari
ea optime in liquida pice tradunt. usus aeris ad
perpetuitatem monimentorum iam pridem tralatus
est tabulis aereis, in quibus publicae constitutiones
inciduntur.

100 XXII. Metalla aeris multis modis instruunt medicinam, utpote cum ulcera omnia ibi ocissime sanen-
tur, maxime tamen prodest[1] cadmea. fit sine dubio
haec et in argenti fornacibus, candidior ac minus
ponderosa, sed nequaquam comparanda aerariae.
plura autem genera sunt. namque ut ipse lapis, ex
quo fit aes, cadmea vocatur, fusuris necessarius,
medicinae inutilis, sic rursus in fornacibus existit
101 alia, quae[2] originis suae nomen[3] recipit. fit autem

[1] prodest *cd. Par.* 6801, *cd. Flor. Ricc.?* prosunt *rell.*
[2] alia quae *aut* aliamque *cdd.*: aliaque aliam *J. Müller.*
[3] originis suae nomen *Mayhoff*: nominis sui originem *cdd.*
item Isid. XVI. 20. 12.

---

[a] A blend for making moulds.
[b] The colour is in fact green. One expects the word
aeruginem here.
[c] See the next two notes.
[d] *Cf.* § 2 of this book (p. 126); mineral calamine and smithsonite = silicate and carbonate of zinc.
[e] Furnace calamine = oxide of zinc. *Cf.* K. C. Bailey, *The
Elder Pliny's Chapters on Chemical Subjects*, II, pp. 166–7.

in addition what is called the mould-blend [a] of bronze of a very delicate consistency, because a tenth part of black lead is added and a twentieth of silver-lead; and this is the best way to give it the colour called Græcanic 'after the Greek.' The last kind is that called pot-bronze, taking its name from the vessels made of it; it is a blend of three or four pounds of silver-lead with every hundred pounds of copper. The addition of lead to Cyprus copper produces the purple colour seen in the bordered robes of statues.

XXI. Things made of copper or bronze get covered with copper-rust [b] more quickly when they are kept rubbed clean than when they are neglected, unless they are well greased with oil. It is said that the best way of preserving them is to give them a coating of liquid vegetable pitch. The employment of bronze was a long time ago applied to securing the perpetuity of monuments, by means of bronze tablets on which records of official enactments are made. *Copper-rust*

XXII. Copper ores and mines supply medicaments in a variety of ways: inasmuch as in their neighbourhood all kinds of ulcers are healed with the greatest rapidity; yet the most beneficial is *cadmea.* [c] This is certainly also produced in furnaces where silver is smelted, this kind being whiter and not so heavy, but it is by no means to be compared with that from copper. There are however several varieties; for while the mineral itself [d] from which the metal is made is called *cadmea*, which is necessary for the fusing process but is of no use for medicine, so again another kind [e] is found in furnaces, which is given a name indicating its origin. It is produced by the thinnest *'Cadmea.'*

egesta flammis atque flatu tenuissima parte materiae
et camaris lateribusque fornacium pro quantitate
levitatis adplicata. tenuissima est in ipso fornacium
ore quam flammae eructarunt,[1] appellata capnitis,
exusta et nimia levitate similis favillae. interior
optuma, camaris dependens et ab eo argumento
botryitis nominata, ponderosior haec priore, levior
102 secuturis—duo eius colores, deterior cinereus, pumicis
melior —, friabilis oculorumque medicamentis utilissima. tertia est in lateribus fornacium, quae propter
gravitatem ad camaras pervenire non potuit. haec
dicitur placitis, et ipsa ab argumento planitiei [2]
crusta verius quam pumex, intus varia, ad psoras
103 utilior et cicatrices trahendas. fiunt [3] ex ea duo
alia genera; onychitis extra paene caerulae, intus
onychis maculis similis, ostracitis tota nigra et e
ceteris sordidissima, volneribus maxime utilis.
omnis autem cadmea, in Cypri [4] fornacibus optima,
iterum a medicis coquitur carbone puro atque,
ubi in cinerem rediit, extinguitur vino Ammineo quae
ad emplastra praeparatur, quae vero ad psoras, aceto.
104 quidam in ollis fictilibus tusam urunt et lavant in

---

[1] *V.ll.* quae *aut* que *aut* qua flamma eructatur *aut* eructantur
aut fluctuantur : eructarunt *Mayhoff.*
[2] planitiei *Salmasius* : planitie.
[3] fiunt *B* : fluunt *rell.*
[4] Cypriis *coni. Mayhoff* : cyprio *aut* cypria *aut* cypri.

part of the substance being separated out by the flames and the blast and becoming attached in proportion to its degree of lightness to the roof-chambers and side-walls of the furnaces, the thinnest being at the very mouth of the furnace, which the flames have belched out; it is called 'smoky *cadmea*' from its burnt appearance and because it resembles hot white ash in its extreme lightness. The part inside is best, hanging from the vaults of the roof-chamber, and this consequently is designated 'grape-cluster *cadmea*': this is heavier than the preceding kind but lighter than those that follow—it is of two colours, the inferior kind being the colour of ash and the better the colour of pumice—and it is friable, and extremely useful for making medicaments for the eyes. A third sort is deposited on the sides of furnaces, not having been able to reach the vaults because of its weight; this is called in Greek 'placitis,' 'caked residue,' in this case by reason of its flatness, as it is more of a crust than pumice, and is mottled inside; it is more useful for itch-scabs and for making wounds draw together into a scar. Of this kind are formed two other varieties, onychitis which is almost blue outside but inside like the spots of an onyx or layered quartz, and ostracitis 'shell-like residue' which is all black and the dirtiest of any of the kinds; this is extremely useful for wounds. All kinds of *cadmea* (the best coming from the furnaces of Cyprus) for use in medicine are heated again on a fire of pure charcoal and, when it has been reduced to ash, if being prepared for plasters it is quenched with Amminean wine, but if intended for itch-scabs with vinegar. Some people pound it and then burn it in earthenware pots, wash it in

mortariis, postea siccant. Nymphodorus lapidem
ipsum quam gravissimum spississimumque urit pruna
et exustum Chio vino restinguit tunditque, mox
linteo cribrat atque in mortario terit, mox aqua
pluvia macerat iterumque terit quod subsedit, donec
cerussae simile fiat, nulla dentium offensa. eadem
Iollae ratio, sed quam purissimum eligit lapidem.
105 XXIII. cadmeae effectus siccare, persanare, sistere
fluctiones, pterygia et sordes oculorum purgare,
scabritiam extenuare et quidquid in plumbi effectu
dicemus.

Et aes ipsum uritur ad omnia eadem, praeterque
albugines oculorum et cicatrices, ulcera quoque
oculorum cum lacte sanat; itaque Aegyptii collyrii
106 id modo terunt in coticulis. facit et vomitiones e
melle sumptum. uritur autem Cyprium in fictilibus
crudis cum sulpuris pari pondere, vasorum [1] circumlito spiramento, in caminis, donec vasa ipsa percoquantur. quidam et salem addunt, aliqui alumen
pro sulpure, alii nihil, sed aceto tantum aspergunt.
ustum teritur in [2] mortario Thebaico, aqua pluvia
lavatur iterumque adiecta largiore teritur et, dum
considat, relinquitur, hoc saepius, donec ad speciem

[1] vasorum *Mayhoff* : vaso *aut* vase.
[2] in *add. Mayhoff.*

---

[a] A medical man of the third century B.C.

mortars and afterwards dry it. Nymphodorus's [a] process is to burn on hot coals the most heavy dense piece of *cadmea* that can be obtained, and when it is thoroughly burnt to quench it with Chian wine, and pound it, and then to sift it through a linen cloth and grind it in a mortar, and then macerate it in rainwater and again grind the sediment that sinks to the bottom till it becomes like white lead and offers no grittiness to the teeth. Iollas' [b] method is the same, but he selects the purest specimens of native *cadmea*. XXIII. The effect of *cadmea* is to dry moisture, to heal lesions, to stop discharges, to cleanse inflamed swellings and foul sores in the eyes, to remove eruptions, and to do everything that we shall specify in dealing with the effect of lead.

Copper itself is roasted to use for all the same purposes and for white-spots and scars in the eyes besides, and mixed with milk it also heals ulcers in the eyes; and consequently people in Egypt make a kind of eye-salve by grinding it in small mortars. Taken with honey it also acts as an emetic, but for this Cyprian copper with an equal weight of sulphur is roasted in pots of unbaked earthenware, the mouth of the vessels being smeared round with oil; and then left in the furnace till the vessels themselves are completely baked. Certain persons also add salt, and some use alum instead of sulphur, while others add nothing at all, but only sprinkle the copper with vinegar. When burnt it is pounded in a mortar of Theban stone, washed with rainwater, and then again pounded with the addition of a larger quantity of water, and left till it settles, and this process is repeated several times, till it is reduced

[b] A Bithynian medical writer of unknown date.

minii redeat. tunc siccatum in sole in aerea pyxide
servatur.

107   XXIV. Et scoria aeris simili modo lavatur, minore [1]
effectu quam ipsum aes. sed et aeris flos medicinae
utilis est. fit aere fuso et in alias fornaces tralato;
ibi flatu crebriore excutiuntur veluti milii squamae,
quas vocant florem; cadunt autem, cum panes aeris
aqua refrigerantur, rubentque similiter squamae
aeris, quam vocant lepida, et sic adulteratur flos, ut
squama veneat pro eo. est autem squama aeris
decussa vi clavis, in quos panes aerei feruminantur,
in Cypri maxime officinis. omnis [2] differentia haec
est, quod squama excutitur ictu isdem panibus, flos
108 cadit sponte. squamae est alterum genus suptilius,
ex summa scilicet lanugine decussum, quod vocant
stomoma.

XXV. Atque haec omnia medici—quod pace
eorum dixisse liceat—ignorant. parent [3] nominibus:
in tantum [4] a conficiendis medicaminibus absunt,
quod esse proprium medicinae solebat. nunc quotiens incidere in libellos, componere ex iis volentes
aliqua, hoc est impendio miserorum experiri [5] com-
mentaria,[6] credunt Seplasiae omnia fraudibus cor-

---

[1] minor *cdd. fere omnes.*
[2] omnis *Mayhoff* (*qui et* summa *coni.*) : omnia.
[3] parent *Urlichs* : paret *B* : pars maior et *rell.* : p.m. paret
Detlefsen. [3-4] parent nominibus hi : tantum *coni. Mayhoff.*
[5] *V.l.* expediri.
[6] commentariaque *B* : *supra post* libellos *trans. Urlichs.*

---

[a] The dross produced when the ore is fused.
[b] Probably in the main red cuprous oxide (not black
cupric oxide) with some metallic copper in it.
[c] Seplasia was the special quarter of Capua where perfumes
were sold.

to the appearance of cinnabar; then it is dried in the
sun and put to keep in a copper box.

XXIV. The slag [a] of copper is also washed in the Slag, scales
same way, but it is less efficacious than copper itself. *and flower of*
The flower [b] of copper also is useful as a medicine. *copper.*
It is made by fusing copper and then transferring
it to other furnaces, where a faster use of the
bellows makes the metal give off layers like scales of
millet, which are called the flower. Also when the
sheets of copper are cooled off in water they shed
off other scales of copper of a similar red hue—this
scale is called by the Greek word meaning 'husk'—
and by this process the flower is adulterated, so that
the scale is sold as a substitute for it—the genuine
flower is a scale of copper forcibly knocked off with
bolts into which are welded cakes of the metal,
specially in the factories of Cyprus. The whole
difference is that the scale is detached from the cakes
by successive hammerings, whereas the flower falls
off of its own accord. There is another finer kind
of scale, the one knocked off from the down-like surface of the metal, the name for which is 'stomoma.'

XXV. But of all these facts the doctors, if they
will permit me to say so, are ignorant—they are
governed by names: so detached they are from
the process of making up drugs, which used to be
the special business of the medical profession. Nowadays whenever they come on books of prescriptions,
wanting to make up some medicines out of them,
which means to make trial of the ingredients in the
prescriptions at the expense of their unhappy
patients, they rely on the fashionable druggists'
shops [c] which spoil everything with fraudulent
adulterations, and for a long time they have been

rumpenti. iam pridem[1] facta emplastra et collyria
mercantur, tabesque mercium aut fraus Seplasiae sic
exhibetur![2]

109 Et squama autem et flos uruntur in patinis fictilibus aut aereis, dein lavantur ut supra ad eosdem usus;
squama[3] et amplius narium carnosa vitia, item sedis,
et gravitates aurium per fistulam in eas flatu inpulsa
et uvas oris farina admota tollit et tonsillas cum
melle. fit ex candido aere squama longe Cypria
inefficacior. nec non urina pueri prius macerant
clavos panesque quidam excussuri squamam, teruntque et aqua pluvia lavant. dant et hydropicis eam
drachmis ii in mulsi hemina et inlinunt cum polline.

110 XXVI. Aeruginis quoque magnus usus est. pluribus fit modis. namque et lapidi, ex quo coquitur
aes, deraditur, et aere candido perforato atque in
cadis suspenso super acetum acre opturatumque
operculo. multo probatior est, si hoc idem squamis
fiat. quidam visa ipsa candidi aeris fictilibus con-

111 dunt in acetum raduntque decumo die. alii vinaceis
contegunt totidemque post dies radunt, alii delima-
tam aeris scobem aceto spargunt versantque spathis
saepius die, donec absumatur. eandem scobem

---

[1] pridem *edd. vett.*: quidem.
[2] sic exhibetur *Warmington*: sic excitetur *Mayhoff* (*qui et*
excitatur *coni.*): exsiccatur *coni. Sillig*: sicce taxetur *Ian*:
alii alia: sic cexatetur B^1: sicce sane duret B^2: sic exteritur
rell. recte?
[3] squama *Mayhoff*: que *cd. Leid. Voss., cd. Flor. Ricc.*:
om. *rell.*

---

[a] Or, if we read *exteritur* (is ground out), 'finds its way into
the mortar' (thus K. C. Bailey).
[b] Brass.   [c] Basic copper carbonate.
[d] Basic copper acetate or true verdigris, which does not
occur in a natural state.

buying plasters and eye-salves ready made; and
thus is deteriorated rubbish of commodities and the
fraud of the druggists' trade put on show.[a]

Both scale however and flower of copper are burnt
in earthenware or copper pans and then washed, as
described above, to be applied to the same purposes; §106.
the scale also in addition removes fleshy troubles in
the nostrils and also in the anus and dullness of
hearing if forcibly blown into the ears through a tube,
and, when applied in the form of powder, removes
swellings of the uvula, and, mixed with honey,
swellings of the tonsils. There is a scale from white
copper [b] that is far less efficacious than the scale from
Cyprus; and moreover some people steep the bolts
and cakes of copper beforehand in a boy's urine
when they are going to detach the scale, and pound
them and wash them with rainwater. It is also
given to dropsical patients in doses of two drams in
half a sextarius of honey-wine; and mixed with
fine flour it is applied as a liniment.

XXVI. Great use is also made of verdigris. *Verdigris.*
There are several ways of making it; it [c] is scraped
off the stone from which copper is smelted, or by [d]
drilling holes in white copper [b] and hanging it up in
casks over strong vinegar which is stopped with a
lid; the verdigris is of much better quality if the
same process is performed with scales of copper.
Some people put the actual vessels, made of white
copper, into vinegar in earthenware jars, and nine
days later scrape them. Others cover the vessels
with grape-skins and scrape them after the same
interval, others sprinkle copper filings with vinegar
and several times a day turn them over with spattles
till the copper is completely dissolved. Others

alii terere in mortariis aereis ex aceto malunt.
ocissime vero contingit coronariorum recisamentis in
112 acetum id [1] additis. adulterant marmore trito
maxime Rhodiam aeruginem, alii pumice aut
cummi. praecipue autem fallit atramento sutorio
adulterata; cetera enim dente deprehenduntur
stridentia in frendendo. experimentum in vatillo
ferreo, nam quae sincera est, suum colorem retinet,
quae mixta atramento, rubescit. deprehenditur et
papyro galla prius macerato, nigrescit enim statim
aerugine inlita. deprehenditur et visu maligne
113 virens. sed sive sinceram sive adulteram [2] aptissi-
mum est elui siccatamque in patina nova uri ac
versari, donec favilla fiat; postea teritur ac re-
conditur. aliqui in crudis fictilibus urunt, donec
figlinum percoquatur. nonnulli et tus masculum
admiscent. lavatur autem aerugo sicut cadmea.
vis eius collyriis oculorum aptissima et delacrima-
tionibus mordendo proficiens, sed ablui necessarium
penicillis calidis, donec rodere desinat.
114 XXVII. Hieracium vocatur collyrium, quod ea [3]
maxime constat. temperatur autem id ham-

---

[1] *seclud.* id *K. C. Bailey.*
[2] *V.l.* adulteratam.
[2] ea *Mayhoff* : illa *quidam apud Dalecamp* : ita.

---

[a] Used for colouring leather. The term probably includes both green vitriol or ferrous sulphate, which is our copperas, and blue vitriol, or cupric sulphate.

[b] This is not true.

[c] Or sponges.

[d] As K. C. Bailey rightly says, not *sal Hammoniacus* (a

prefer to grind copper filings mixed with vinegar in copper mortars. But the quickest result is obtained by adding to the vinegar shavings of coronet copper. Rhodian verdigris is adulterated chiefly with pounded marble, though others use pumicestone or gum. But the adulteration of verdigris that is the most difficult to detect is done with shoemakers' black,[a] the other adulterations being detected by the teeth as they crackle when chewed. Verdigris can be tested on a hot fire-shovel, as a specimen that is pure keeps [b] its colour, but what is mixed with shoemakers' black turns red. It is also detected by means of papyrus previously steeped in an infusion of plantgall, as this when smeared with genuine verdigris at once turns black. It can also be detected by the eye, as it has an evil green colour. But whether pure or adulterated, the best way is to wash it and when it is dry to burn it on a new pan and keep turning it over till it becomes glowing ashes; and afterwards it is crushed and put away in store. Some people burn it in raw earthenware vessels till the earthenware is baked through; some mix in also some male frankincense. Verdigris is washed in the same way as *cadmea*. §106. Its powerfulness is very well suited for eye-salves and its mordant action makes it able to produce watering at the eyes; but it is essential to wash it off with swabs [c] and hot water till its bite ceases to be felt.

XXVII. Hierax's Salve is the name given to an eye-salve chiefly composed of verdigris. It is made by mixing together four ounces of gum of Hammon,[d]

variety of common salt, which itself is not sal ammoniac = ammonium chloride).

moniaci unciis IIII, aeruginis Cypriae II, atramenti
sutorii, quod chalcanthum vocant, totidem, misyos
una, croci VI. haec omnia trita aceto Thasio
colliguntur [1] in pilulas, excellentis remedii contra
initia glaucomatum et suffusionum, contra caligines
aut scabritias et albugines et genarum vitia. cruda
115 autem aerugo volnerariis emplastris miscetur. oris
etiam gingivarumque exulcerationes mirifice emendat et labrorum ulcera cum oleo. quod si et cera
addatur, purgat et ad cicatricem perducit. aerugo
et callum fistularum erodit vitiorumque circa sedem
sive per se sive cum hammoniaco inlita vel collyrii
modo in fistulas adacta. eadem cum resinae tere-
binthinae tertia parte subacta lepras tollit.

16 XXVIII. Est et alterum genus aeruginis, quam vocant scoleca, in Cyprio ⟨mortario Cyprio⟩ [2] aere trito [3]
alumine et sale aut nitro pari pondere cum aceto albo
quam acerrimo. non fit hoc nisi aestuosissimis [4]
diebus circa canis ortum. teritur autem, donec viride
fiat contrahatque se vermiculorum specie, unde
et nomen. quod vitiatum [5] ut emendetur, II partes
quam fuere aceti miscentur urinae pueri inpubis.
idem autem in medicamentis et santerna efficit,
qua diximus aurum feruminari. usus utriusque qui

---

[1] collinuntur *B*.
[2] ⟨mortario Cyprio⟩ *coni. Mayhoff*.
[3] trito *B* : intrito *cd. Flor. Ricc.* : hic trito *rell.* : hoc t.
edd. vett.
[4] aestivosissimis *B*.
[5] vitiatum *Mayhoff* : vitium.

---

[a] Copper pyrites.
[b] See pp. 210–1, note [d].
[c] This sentence is probably defective.

two of Cyprian verdigris, two of the copperas called flower of copper, one of *misy* [a] and six of saffron; all these ingredients are pounded in Thasian vinegar and made up into pills, that are an outstanding specific against incipient glaucoma and cataract, and also against films on the eyes or roughnesses and white ulcerations in the eye and affections of the eyelids. Verdigris in a crude state is used as an ingredient in plasters for wounds also. In combination with oil it is a marvellous cure for ulcerations of the mouth and gums and for sore lips, and if wax is also added to the mixture it cleanses them and makes them form a cicatrix. Verdigris also eats away the callosity of fistulas and of sores round the anus, either applied by itself or with gum of Hammon,[b] or inserted into the fistula in the manner of a salve. Verdigris kneaded up with a third part of turpentine also removes leprosy.

XXVIII. There is also another kind of verdigris called from the Greek worm-like verdigris, made by grinding up in a mortar of true cyprian copper with a pestle of the same metal equal weights of alum and salt or soda with the very strongest white vinegar. This preparation is only made on the very hottest days of the year, about the rising of the Dogstar. The mixture is ground up until it becomes of a green colour and shrivels into what looks like a cluster of small worms, whence its name. To remedy any that is blemished, the urine of a young boy to twice the quantity of vinegar that was used is added to the mixture.[c] Used as a drug, worm-verdigris has the same effect as santerna which we spoke of as used for soldering gold; both of them have the same properties as verdigris. Native worm-verdigris is XXXIII 93.

aeruginis. scolex fit et per se derasus aerario lapidi, de quo nunc dicemus.

117 XXIX. Chalcitim vocant, ex quo et ipso aes coquitur. distat a cadmea, quod illa super terram ex subdialibus petris caeditur, haec ex obrutis, item quod chalcitis friat se statim, mollis natura, ut videatur lanugo concreta. est et alia distinctio, quod chalcitis tria genera continet, aeris et misyos et soreos, de quibus singulis dicemus suis locis.

118 habet autem aeris venas oblongas. probatur mellei coloris, gracili venarum discursu, friabilis nec lapidosa. putant et recentem utiliorem esse, quoniam inveterata sori fiat. vis[1] eius ad excrescentia in ulceribus, sanguinem sistere, gingivas, uvam, tonsillas farina compescere, volvae vitiis in vellere imponi. cum suco vero porri verendorum additur

119 emplastris. maceratur autem in fictili ex aceto circumlito fimo diebus XL, et colorem croci trahit. tum admixto cadmeae pari pondere medicamentum efficit psoricon dictum. quod si II partes chalcitidis tertia cadmeae temperentur, acrius hoc idem fiat; etiamnum vehementius, si aceto quam vino[2] temperetur; tosta vero efficacior fit ad eadem omnia.

120 XXX. Sori Aegyptium maxime laudatur, multum

---

[1] usus *coni. Mayhoff.*   [2] ⟨si⟩ vino *coni. Mayhoff.*

---

[a] Copper pyrites in process of decomposition. For *cadmea* see §§ 2 and 100 and notes.

[b] Copper pyrites.

[c] Probably decomposing marcasite, or sometimes black porous limestone with decomposing pyrites in it (K. C. Bailey).

also obtained by scraping a copper ore of which we shall now speak.

XXIX. Chalcitis, ' copper-stone,' is the name of *Chalcitis.* an ore,[a] that from which copper also, besides *cadmea*, is obtained by smelting. It differs from *cadmea* because the latter is quarried above ground, from rocks exposed to the air, whereas chalcitis is obtained from underground beds, and also because chalcitis becomes immediately friable, being of a soft nature, so as to have the appearance of congealed down. There is also another difference in that chalcitis contains three kinds of mineral, copper, *misy* [b] and *sori*,[c] each of which we shall describe in its place ; §§ 120, 121. and the veins of copper in it are of an oblong shape. The approved variety of chalcitis is honey coloured, and streaked with fine veins, and is friable and not stony. It is also thought to be more useful when fresh, as when old it turns into *sori*. It is used for growths in ulcers, for arresting hæmorrhage and, in the form of a powder, for acting as an astringent on the gums, uvula and tonsils, and, applied in wool, as a pessary for affections of the uterus, while with leek juice it is employed in plasters for the genitals. It is steeped for forty days in vinegar in an earthenware jar, covered with dung, and then assumes the colour of saffron ; then an equal weight of *cadmea* is mixed with it and this produces the drug called psoricon or cure for itch. If two parts of chalcitis are mixed with one of *cadmea* this makes a stronger form of the same drug, and moreover it is more violent if it is mixed in vinegar than if in wine ; and when roasted it becomes more effective for all the same purposes.

XXX. Egyptian *sori* is most highly commended, *Sori.*

superato Cyprio Hispaniensique et Africo, quamquam
oculorum curationi quidam utilius putent Cyprium;
sed in quacumque natione optimum cui maximum
virus olfactu, tritumque pinguiter nigrescens et
spongiosum. stomacho res contraria in tantum, ut
quibusdam olfactum modo vomitiones moveat. et
Aegyptium quidem tale,[1] alterius nationis con-
tritum splendescit ut misy et est lapidosius. prodest
autem et dentium dolori, si contineatur atque
colluat, et ulceribus oris gravibus quaeque serpant.
uritur carbonibus ut chalcitis.

121   XXXI. Misy aliqui tradiderunt fieri exusto lapide
in scrobibus, flore eius luteo miscente se ligni pineae
favillae. re vera autem e supra dicto fit lapide,
concretum natura discretumque vi, optimum in
Cypriorum officinis, cuius notae sunt friati aureae
scintillae et, cum teratur, harenosa natura sine
terra,[2] chalcitidi[3] dissimilis.[4] hoc admiscent qui
aurum purgant. utilitas eius infusi cum rosaceo
auribus purulentis et in lana inpositi capitis ulceribus.
extenuat et scabritias oculorum inveteratas, praecipue
122 utile tonsillis contraque anginas et suppurata. ratio

---

[1] tale est *cd. deperd. Dalecamp.*
[2] terrae *cd. Vind.* : terrea *ed. Lugd.*
[3] chalcitidis *cd. Flor. Ricc.*
[4] dissimilis *K. C. Bailey* : simis *cd. Flor. Ricc.* : sin *B*[1] :
sive *cd. Par. 6801* : similis *rell.*

---

[a] Probably produced in most cases by sulphuretted hydro-
gen.
[b] Of hollow teeth, as is clear from the context and from
Diosc. I, 141, V, 119.
[c] *Chalcitis,* § 117.

being far superior to that of Cyprus and Spain and
Africa, although some people think that Cyprus
sori is more useful for treatment of the eyes; but
whatever its provenance the best is that which has
the most pungent odour,[a] and which when ground up
takes a greasy, black colour and becomes spongy.
It is a substance that goes against the stomach so
violently that with some people the mere smell of
it causes vomiting. This is a description of the *sori*
of Egypt. That from other sources when ground
up turns a bright colour like *misy*, and it is harder;
however, if it is held in the cavities [b] and used
plentifully as a mouth-wash it is good for toothache
and for serious and creeping ulcers of the mouth.
It is burnt on charcoal, like chalcitis.

XXXI. Some people have reported that *misy* Misy.
is made by burning mineral in trenches, its fine
yellow powder mixing itself with the ash of the
pine wood burnt; but as a matter of fact though got
from the mineral [c] above mentioned, it is part of its
substance and separated from it by force, the best
kind being obtained in the copper-factories of Cyprus,
its marks being that when broken it sparkles like
gold and when it is ground it has a sandy appearance, without earth, unlike chalcitis. A mixture of
misy is employed in the magical purification of gold.[d]
Mixed with oil of roses it makes a useful infusion for
suppurating ears and applied on wool a serviceable
plaster for ulcers of the head. It also reduces
chronic roughness of the eyelids, and is especially
useful for the tonsils and against quinsy and suppura-

---

[d] The process of counteracting the supposed evil influence
of gold when held over the head of children, etc. See XXXIII,
84.

ut xvi drachmae in hemina aceti coquantur addito
melle, donec lentescat. sic ad supra dicta utile est.
quotiens opus sit molliri vim eius, mel adspergitur.
erodit et callum fistularum ex aceto foventium et
collyriis additur, sistit et sanguinem ulceraque quae
serpant quaeve putrescant, absumit et excrescentes
carnes. peculiariter virilitatis vitiis utile et feminarum profluvium sistit.

123   XXXII. Graeci cognationem aeris nomine fecerunt
et atramento sutorio; appellant enim chalcanthon.
nec ullius aeque mira natura est. fit in Hispaniae
puteis stagnisve id genus aquae habentibus. deco-
quitur ea admixta dulci pari mensura et in piscinas
ligneas funditur. immobilibus [1] super has transtris
dependent restes lapillis extentae, quibus adhaerescens limus vitreis acinis imaginem quandam uvae
reddit. exemptum ita siccatur diebus xxx. color
est caeruleus perquam spectabili nitore, vitrumque
124 esse creditur; diluendo fit atramentum tinguendis
coriis. fit et pluribus modis : genere terrae eo in
scrobes cavato, quorum e lateribus destillantes
hiberno gelu stirias stalagmian vocant, neque est
purius aliud. sed ex eo, candidum colorem sentiente
125 viola, lonchoton [2] appellant. fit et in saxorum

---

[1] immobilibus *edd. vett.* : immobilis.
[2] *V.l.* locoton : leucoïon *Hermolaus Barbarus* : leucanthon
edd. vett.

---

[a] See n. [a] on § 112.
[b] *I.e.* water holding in solution the substance referred to.
[c] So Diosc. V, 114. But the description suggests ' leu-
coion,' ' violet-white.' The ancient like the modern violas
were of various colours.

tions. The method is to boil 16 drams of it in a twelfth of a pint of vinegar with honey added till it becomes of a viscous consistency: this makes a useful preparation for the purposes above mentioned. When it is necessary to make it softer, honey is sprinkled on it. It also removes the callosity of fistulous ulcers when the patients use it with vinegar as a fomentation; and it is used as an ingredient in eye-salves, arrests hæmorrhage and creeping or putrid ulcers, and reduces fleshy excrescences. It is particularly useful for troubles in the sexual organs in the male, and it checks menstruation.

XXXII. The Greeks by their name for shoe-makers'-black [a] have made out an affinity between it and copper: they call it *chalcanthon*, 'flower of copper'; and there is no substance that has an equally remarkable nature. It occurs in Spain in wells or pools that contain that sort of water.[b] This water is boiled with an equal quantity of pure water and poured into wooden tanks. Over these are firmly fixed cross-beams from which hang cords held taut by stones, and the mud clinging to the cords in a cluster of glassy drops has somewhat the appearance of a bunch of grapes. It is taken off and then left for thirty days to dry. Its colour is an extremely brilliant blue, and it is often taken for glass; when dissolved it makes a black dye used for colouring leather. It is also made in several other ways: earth of the kind indicated is hollowed into trenches, droppings from the sides of which form icicles in a winter frost which are called drop-flower of copper, and this is the purest kind. But some of it, violet with a touch of white, is called *lonchoton*, 'lance-headed.'[c] It is also made in pans hollowed

Shoemakers' black.

catinis pluvia aqua conrivato limo gelante; fit et
salis modo flagrantissimo sole admissas dulces aquas
cogente. ideo quidam duplici differentia fossile
aut facticium appellant, hoc pallidus et quantum
126 colore, tantum bonitate deterius. probant maxime
Cyprium in medicinae usu. sumitur ad pellenda
ventris animalia drachmae pondere cum melle.
purgat et caput dilutum ac naribus instillatum, item
stomachum cum melle aut aqua mulsa sumptum.
medetur et oculorum scabritiae dolorique et caligini
et oris ulceribus. sistit et sanguinem narium, item
haemorroidum. extrahit ossa fracta cum semine
hyoscyami, suspendit epiphoras penicillo fronti
inpositum, efficax et in emplastris ad purganda
127 volnera [1] et excrescentia ulcerum. tollit et uvas,
vel si decocto tangantur, cum lini quoque semine
superponitur emplastris ad dolores tollendos. quod
ex eo candicat, in uno usu praefertur violaceis, ut
gravitati aurium per fistulas inspiretur. volnera
per se inlitum sanat, sed tinguit [2] cicatrices. nuper
inventum ursorum in harena et leonum ora spargere
illo, tantaque est vis in adstringendo, ut non queant
mordere.
128   XXXIII. Etiamnum in aerariis reperiuntur quae

---

[1] *V.l.* ulcera.
[2] *V.l.* tingit : stringit *Caesarius.*

---

[a] Zinc oxide.
[b] Zinc oxide made impure by charcoal-dust and from other
causes (K. C. Bailey).

in the rocks, into which the slime is carried by rain-water and freezes, and it also forms in the same way as salt when very hot sunshine evaporates the fresh water let in with it. Consequently some people distinguish in twofold fashion between the mined flower of copper and the manufactured, the latter paler than the former and as much inferior in quality as in colour. That which comes from Cyprus is most highly approved for medical employment. It is taken to remove intestinal worms, the dose being one dram mixed with honey. Diluted and injected as drops into the nostrils it clears the head, and like-wise taken with honey or honey-water it purges the stomach. It is given as a medicine for roughness of the eyes, pain and mistiness in the eyes, and ulcera-tion of the mouth. It stops bleeding from the nostrils, and also hæmorrhoidal bleeding. Mixed with henbane seed it draws out splinters of broken bones; applied to the forehead with a swab it arrests running of the eyes; also used in plasters it is efficacious for cleansing wounds and gatherings of ulcers. A mere touch of a decoction of it removes swellings of the uvula, and it is laid with linseed on plasters used for relieving pains. The whitish part of it is preferred to the violet kinds for one purpose, that of being blown through tubes into the ears to relieve ear-trouble. Applied by itself as a liniment it heals wounds, but it leaves a discoloration in the scars. There has lately been discovered a plan of sprinkling it on the mouths of bears and lions in the arena, and its astringent action is so powerful that they are unable to bite.

XXXIII. The substances called by Greek names *Zinc oxide.*
meaning ' bubble '[a] and ' ash '[b] are also found in

vocant pompholygem et spodon. differentia, quod
pompholyx lotura separatur, spodos inlota est.
aliqui quod sit candidum levissimumque pompholygem dixere et esse aeris ac cadmeae favillam, spodon
nigriorem esse ponderosioremque, derasam parietibus
fornacium, mixtis scintillis,[1] aliquando et carbonibus.
129 haec aceto accepto odorem aeris praestat et, si
tangatur lingua, saporem horridum. convenit oculorum medicamentis, quibuscumque vitiis occurrens,
et ad omnia quae spodos. hoc solum distat, quod
huius elutior vis est. additur et in emplastra, quibus
lenis refrigeratio quaeritur et siccatio. utilior ad
omnia quae vino lota est.

130   XXXIV. Spodos Cypria optima. fit autem liques-
centibus cadmea et aerario lapide. levissimum hoc
est flaturae totius evolatque e fornacibus et tectis
adhaerescit, a fuligine distans candore. quod minus
candidum ex ea, inmaturae fornacis argumentum est;
hoc quidam pompholygem vocant. quod vero rubi-
cundius ex iis invenitur, acriorem vim habet et
exulcerat adeo, ut, cum lavatur, si attigit oculos,
131 excaecet. est et mellei coloris spodos, in qua plurimum aeris intellegitur. sed quodcumque genus
lavando fit utilius; purgatur ante panno,[2] dein

---

[1] lapillis (*vel* cinere, pilis) *coni. Mayhoff.*
[2] linteo panno *coni. K. C. Bailey*: purgantur ramenta panno
D'Arcy Thompson: ante pinna *aut* ante penna.

the furnaces of copper works. The difference
between them is that bubble is disengaged by washing but ash is not washed out. Some people have
given the name of ' bubble ' to the substance that
is white and very light in weight, and have said that
it is the ashes of copper and *cadmea*, but that ' ash '
is darker and heavier, being scraped off the walls of
furnaces, mixed with sparks from the ore and sometimes also with charcoal. This material when
vinegar is applied to it gives off a smell of copper,
and if touched with the tongue has a horrible taste.
It is a suitable ingredient for eye medicines,
remedying all troubles whatever, and for all the
purposes for which ' ash ' is used; its only difference
is that its action is less violent. It is also used as an
ingredient for plasters employed to produce a gentle
cooling and drying effect. It is more efficacious for
all purposes when it is moistened with wine.

XXXIV. Cyprus ash is the best. It is produced
when *cadmea* and copper ore are melted. The ash
in question is the lightest part of the whole substance produced by blasting, and it flies out of the
furnaces and adheres to the roof, being distinguished
from soot by its white colour. Such part of it as is
less white is an indication of inadequate firing; it is
this that some people call ' bubble.' But the redder
part selected from it has a keener force, and is so
corrosive that if while it is being washed it touches
the eyes it causes blindness. There is also an ash
of the colour of honey, which is understood to indicate
that it contains a large amount of copper. But any
kind is made more serviceable by washing; it is
first purified with a strainer of cloth and then given
a more substantial washing, and the rough portions

crassiore [1] lotura digitis scabritiae excernuntur.[2]
eximia [3] vis est eius, quae vino lavatur. est aliqua
et in genere vini differentia. leni enim lota collyriis
oculorum minus utilis putatur, eademque efficacior
ulceribus, quae manent, vel oris, quae madeant, et
omnibus medicamentis, quae parentur contra gan-
132 graenas. fit et in argenti fornacibus spodos, quam
vocant Lauriotim. utilissima autem oculis adfirma-
tur quae fiat in aurariis, nec in alia parte magis est
vitae ingenia mirari. quippe ne quaerenda essent
metalla, vilissimis rebus utilitates easdem excogitavit.
133 XXXV. Antispodon vocant cinerem fici arboris
vel caprifici vel myrti foliorum cum tenerrimis ramo-
rum partibus vel oleastri vel oleae vel cotonei mali
vel lentisci, item ex moris immaturis, id est candidis,
in sole arefactis vel e buxi coma aut pseudocypiri
aut rubi aut terebinthi vel oenanthes. taurini
quoque glutinis aut linteorum cinerem similiter
pollere inventum est. uruntur omnia ea crudo
fictili in fornacibus, donec figlina percoquantur.
134 XXXVI. In aerariis officinis et smegma fit iam
liquato aere ac percocto additis etiamnum carbonibus
paulatimque accensis, ac repente vehementiore flatu

---

[1] crassiore *aut* crassior *cdd.* (crossiora *cd. Vind.*): crassiora
D'Arcy Thompson: crebriore *coni. Mayhoff.*
[2] scabritiæ (scabritis scabritiae *B*) excernuntur *cdd.*: sca-
britiem exterunt *edd. vett.*: excernit *Caesarius.*
[3] eximia *Mayhoff*: et media *B*: om. et *rell.*

---

[a] So called from Laurium in Attica, where there are still
silver mines.
[b] This word σμῆγμα can, it seems, be used not only for a
detergent or cleansing agent, but also for the stuff removed

are picked out by the fingers. When it is washed with wine it is particularly powerful. There is also some difference in the kind of wine used, as when it is washed with weak wine it is thought to be less serviceable for eye-salves, and at the same time more efficacious for running ulcers or for ulcers of the mouth that are always wet and more useful for all the antidotes for gangrene. An ash called Lauriotis [a] is also produced in furnaces in which silver is smelted; but the kind said to be most serviceable for the eyes is that which is formed in smelting gold. Nor is there any other department in which the ingenuities of life are more to be admired, inasmuch as to avoid the need of searching for metals experience has devised the same utilities by means of the commonest things.

XXXV. The substance called in Greek 'anti-spodos '' substitute ash ' is the ash of the leaves of the figtree or wild fig or myrtle together with the tenderest parts of the branches, or of the wild olive or cultivated olive or quince or mastic and also ash obtained from unripe, that is still pale, mulberries, dried in the sun, or from the foliage of the box or mock-gladiolus, or bramble or turpentine-tree or œnanthe. The same virtues have also been found in the ash of bull-glue or of linen fabrics. All of these are burnt in a pot of raw earth heated in a furnace until the earthenware is thoroughly baked.

XXXVI. Also 'smegma'[b] is made in copper 'Smegma.' forges by adding additional charcoal when the copper has already been melted, and thoroughly fused, and gradually kindling it; and suddenly when a stronger blast is applied a sort of chaff of

by cleansing; so here it means floating impurities containing some copper (K. C. Bailey).

225

exspuitur aeris palea quaedam. solum, quo exci-
piatur, stratum esse debet marilla.[1]

135   XXXVII. Ab ea discernitur quam in isdem officinis
diphrygem vocant Graeci ab eo, quod bis torreatur.
cuius origo triplex. fieri enim traditur ex lapide
pyrite cremato in caminis, donec excoquatur in
rubricam. fit et in Cypro ex luto cuiusdam specus
arefacto [2] prius, mox paulatim circumdatis sarmentis.[3]
tertio fit modo in fornacibus aeris faece subsidente.
differentia est, quod [4] aes ipsum in catinos defluit,
scoria extra fornaces, flos supernatat, diphryges

136   remanent. quidam tradunt in fornacibus globos
lapidis, qui coquatur, feruminari, circa hunc aes
fervere, ipsum vero non percoqui nisi tralatum in
alias fornaces, et esse nodum [5] quendam materiae;
id, quod ex cocto supersit, diphryga vocari. ratio
eius in medicina similis praedictis : siccare et
excrescentia consumere ac repurgare. probatur
lingua, ut eam siccet tactu statim saporemque aeris
reddat.

137   XXXVIII. Unum etiamnum aeris miraculum non
omittemus. Servilia familia inlustris in fastis trientem aereum pascit auro, argento, consumentem
utrumque. origo atque natura eius incomperta mihi

---

[1] marilla *Ian* (*olim* marila) : marili *B* : maxili *rell.* :
maxilla *edd. vett.* : debet. Facile *Hermolaus Barbarus.*

[2] ⟨sole⟩ *vel* ⟨aere⟩ arefacto *coni. K. C. Bailey.*

[3] sarmentis ⟨accensis⟩ *coni. K. C. Bailey* s. ⟨ardentibus⟩
coni. Warmington.

[4] *V.l.* differentiae siquidem : differentia est quidem quod
edd. vett.

[5] *V.l.* nudum : nucleum *coni. Mayhoff.*

copper spirts out. The floor on which it is received ought to be strewn with charcoal-dust.

XXXVII. Distinguished from 'smegma' is the *Other impurities.* substance in the same forges called by the Greeks diphryx, from its being twice roasted. It comes from three different sources. It is said to be obtained from a mineral pyrites which is heated in furnaces till it is smelted into a red earth. It is also made in Cyprus from mud obtained from a certain cavern, which is first dried and then gradually has burning brushwood put round it. A third way of producing it is from the residue that falls to the bottom in copper furnaces; the difference is that the copper itself runs down into crucibles and the slag forms outside the furnace and the flower floats on the top, but the supplies of diphryx remain behind. Some people say that certain globules of stone that is being smelted in the furnaces become soldered together and round this the copper gets red hot, but the stone itself is not fused unless it is transferred into other furnaces, and that it is a sort of kernel of the substance, and that what is called diphryx is the residue left from the smelting. Its use in medicine is similar to that of the substances already described; to dry up moisture and remove excrescent growths and act as a detergent. It can be tested by the tongue—contact with it ought immediately to have a parching effect and impart a flavour of copper.

XXXVIII. We will not omit one further remarkable thing about copper. The Servilian family, famous in our annals, possesses a bronze $\frac{1}{3}$ *as* piece which it feeds with gold and silver and which consumes them both. Its origin and nature are un-

est. verba ipsa de ea re Messallae senis ponam:
Serviliorum familia habet trientem sacrum, cui
summa cum cura magnificentiaque sacra quotannis
faciunt. quem ferunt alias crevisse, alias decrevisse
videri et ex eo aut honorem aut deminutionem
familiae significare.

138 XXXIX. Proxime indicari debent metalla ferri.
optumo pessimoque vitae instrumento est,[1] siquidem
hoc tellurem scindimus, arbores serimus, arbusta
tondemus,[2] vites squalore deciso annis omnibus
cogimus iuvenescere, hoc extruimus tecta, caedimus
saxa, omnesque ad alios usus ferro utimur, sed
eodem ad bella, caedes, latrocinia, non comminus
solum, sed etiam missili volucrique, nunc tormentis
excusso, nunc lacertis, nunc vero pinnato, quam
sceleratissimam humani ingenii fraudem arbitror,
siquidem, ut ocius mors perveniret ad hominem,
alitem illam fecimus pinnasque ferro dedimus.

139 quam ob rem culpa eius non naturae fiat accepta.
aliquot experimentis probatum est posse innocens
esse ferrum. in foedere, quod expulsis regibus
populo Romano dedit Porsina, nominatim compre-
hensum invenimus, ne ferro nisi in agri cultu uteretur.
et tum [3] stilo osseo [4] scribere institutum vetustissimi
auctores prodiderunt. Magni Pompei in tertio

---

[1] est *add. Mayhoff.*
[2] *V.l.* ponemus (p. pomaria *cd. Flor. Ricc. ut videtur*:
ponimus p. *edd. vett.*).
[3] *V.l.* cum.
[4] osseo *add. Mayhoff coll. Isid.* VI. 9. 2.

known to me, but I will put down the actual words of the elder Messala [a] on the subject. 'The family of the Servilii has a holy coin to which every year they perform sacrifices with the greatest devotion and splendour; and they say that this coin seems to have on some occasions grown bigger and on other occasions smaller, and that thereby it portends either the advancement or the decadence of the family.'

XXXIX. Next an account must be given of the mines and ores of iron. Iron serves as the best and the worst part of the apparatus of life, inasmuch as with it we plough the ground, plant trees, trim the trees that prop our vines, force the vines to renew their youth yearly by ridding them of decrepit growth; with it we build houses and quarry rocks, and we employ it for all other useful purposes, but we likewise use it for wars and slaughter and brigandage, and not only in hand-to-hand encounters but as a winged missile, now projected from catapults, now hurled by the arm, and now actually equipped with feathery wings, which I deem the most criminal artifice of man's genius, inasmuch as to enable death to reach human beings more quickly we have taught iron how to fly and have given wings to it. Let us therefore debit the blame not to Nature, but to man. A number of attempts have been made to enable iron to be innocent. We find it an express provision included in the treaty granted by Porsena to the Roman nation after the expulsion of the kings that they should only use iron for purposes of agriculture; and our oldest authors have recorded that in those days it was customary to write with a bone pen. There is extent an edict of Pompey the Great dated

Iron. Its uses and misuses.

508 B.C.
510 B.C.

[a] Consul in 53 B.C.

consulatu extat edictum in tumultu necis Clodianae
prohibentis ullum telum esse in urbe.

140 XL. Et ars antiqua [1] ipsa non defuit honorem
mitiorem habere ferro quoque. Aristonidas artifex,
cum exprimere vellet Athamantis furorem Learcho
filio praecipitato residentem paenitentia, aes ferrumque miscuit, ut robigine eius per nitorem aeris
relucente exprimeretur verecundiae rubor. hoc
141 signum exstat hodie Rhodi.[2] est in eadem urbe et
ferreus Hercules, quem fecit Alcon laborum dei
patientia inductus. videmus et Romae scyphos e
ferro dicatos in templo Martis Ultoris. obstitit
eadem naturae benignitas exigentis ab ferro ipso
poenas robigine eademque providentia nihil in
rebus mortalius [3] facientis [4] quam quod esset
infestissimum mortalitati.

142 XLI. Ferri metalla ubique propemodum reperiuntur, quippe et iam [5] insula Italiae Ilva gignente,
minimaque difficultate adgnoscuntur colore ipso
terrae manifesto. ratio eadem excoquendis venis;
in Cappadocia tantum quaestio est, aquae an terrae
fiat acceptum, quoniam perfusa Ceraso [6] fluvio
terra neque aliter ferrum e fornacibus reddit.

143 differentia ferri numerosa. prima in genere terrae
caelive: aliae molle tantum plumboque vicinum

---

[1] ars antiqua *Mayhoff*: tamen uiquea *B* (t. uique *B*[2]):
tamen vita *rell.*

[2] hodie rhodi *B*: hodierno die *rell.*: h.d. Thebis *Hardouin*:
Thebis hodie *Hermolaus Barbarus.*

[3] mortalius *B*: mortalibus *rell.*

[4] facientis *cd. deperd. Dalecamp*: faciente (facientem *cd.*
Par. 6801, *cd. Leid. Voss.*).

[5] et iam *K. C. Bailey*: etiam.

[6] Ceraso *Urlichs*: certo.

in his third consulship at the time of the disorders 52 b.c.
accompanying the death of Clodius, prohibiting the
possession of any weapon in the city.

XL. Further, the art of former days did not fail to
provide a more humane function even for iron.
When the artist Aristonidas desired to represent the
madness of Athamas subsiding in repentance after
he had hurled his son Learchus from the rock, he
made a blend of copper and iron, in order that the
blush of shame should be represented by rust of the
iron shining through the brilliant surface of the
copper; this statue is still standing at Rhodes.
There is also in the same city an iron figure of
Heracles, which was made by Alcon, prompted by
the endurance displayed by the god in his labours.
We also see at Rome goblets of iron dedicated in
the temple of Mars the Avenger. The same benevo-
lence of nature has limited the power of iron itself
by inflicting on it the penalty of rust, and the same
foresight by making nothing in the world more
mortal than that which is most hostile to mortality.

XLI. Deposits of iron are found almost every- *Iron ores
where, and they are formed even now in the Italian *and smelt-
island of Elba, and there is very little difficulty in *ing. Steel.*
recognizing them as they are indicated by the actual
colour of the earth. The method of melting out the
veins is the same as in the case of copper. In
Cappadocia alone it is merely a question whether the
presence of iron is to be credited to water or to earth,
as that region supplies iron from the furnaces when
the earth has been flooded by the river Cerasus but
not otherwise. There are numerous varieties of
iron; the first difference depending on the kind of
soil or of climate—some lands only yield a soft iron

subministrant, aliae fragile et aerosum rotarumque
usibus et clavis maxime fugiendum, cui prior ratio
convenit; aliud brevitate sola[1] placet clavisque
caligariis, aliud robiginem celerius sentit. stricturae
vocantur hae omnes, quod non in aliis metallis, a[a]
144 stringenda acie vocabulo inposito. et fornacium
magna differentia est, nucleusque quidam ferri
excoquitur in iis ad indurandam aciem, alioque modo
ad densandas incudes malleorumve rostra. summa
autem differentia in aqua, cui subinde candens
inmergitur. haec alibi atque alibi utilior nobilitavit
loca gloria ferri, sicuti Bilbilim in Hispania et Turias-
sonem, Comum in Italia, cum ferraria metalla in iis
145 locis non sint. ex omnibus autem generibus palma
Serico ferro est; Seres hoc cum vestibus suis pelli-
busque mittunt; secunda Parthico. neque alia
genera ferri ex mera[2] acie temperantur, ceteris
enim admiscetur mollior complexus. in nostro orbe
aliubi vena bonitatem hanc praestat, ut in Noricis,
aliubi factura, ut Sulmone, aqua aliubi ut[3] diximus,
quippe cum[4] exacuendo oleariae cotes aquariaeque
146 differant et oleo delicatior fiat acies. tenuiora ferra-

---

[1] bonitate soleis *K. C. Bailey.*
[2] ex mera *B* : ex mira *rell.* (ex nimia *cd. Vind.* ccxxxiv) :
eximia *edd. vett.*
[3] aliubi ut *Warmington* : uti *edd. vett.*: ubi.
[4] *V.l.* cum in.

---

[a] The Chinese; in fact intermediaries are meant.
[b] In the MSS. this sentence comes after the next one.

closely allied to lead, others a brittle and coppery kind that is specially to be avoided for the requirements of wheels and for nails, for which purpose the former quality is suitable; another variety of iron finds favour in short lengths only and in nails for soldiers' boots; another variety experiences rust more quickly. All of these are called ' stricturae,' ' edging ores,' a term not used in the case of other metals; it is, as assigned to these ores, derived from *stringere aciem*, ' to draw out a sharp edge.' There is also a great difference between smelting works, and a certain knurr of iron is smelted in them to give hardness to a blade, and by another process to giving solidity to anvils or the heads of hammers. But the chief difference depends on the water in which at intervals the red hot metal is plunged; the water in some districts is more serviceable than in others, and has made places famous for the celebrity of their iron, for instance Bambola and Tarragona in Spain and Como in Italy, although there are no iron mines in those places. But of all varieties of iron the palm goes to the Seric, sent us by the Seres [a] with their fabrics and skins. The second prize goes to Parthian iron; and indeed no other kinds of iron are forged from pure metal, as all the rest have a softer alloy welded with them. In our part of the world, in some places the lode supplies this good quality, as for instance in the country of the Norici, in other places it is due to the method of working, as at Sulmona, and in others, as we have said, it is due § 144. to the water; inasmuch as for giving an edge there is a great difference between oil whetstones and water whetstones, and a finer edge is produced by oil. It [b] is the custom to quench smaller iron forgings

menta oleo restingui mos est, ne aqua in fragilitatem
durentur.[1] mirumque, cum excoquatur vena, aquae
modo liquari ferrum, postea in spongeas frangi.
a ferro sanguis humanus se ulciscitur, contactum
namque eo celerius robiginem trahit.

147 XLII. De magnete lapide suo loco dicemus
concordiaque, quam cum ferro habet. sola haec
materia virus [2] ab eo lapide accipit retinetque longo
tempore, aliud adprehendens ferrum, ut anulorum
catena spectetur interdum. quod volgus imperitum
appellat ferrum vivum, vulneraque talia asperiora
148 fiunt. hic lapis et in Cantabria nascitur, non ut
ille magnes verus caute continua, sed sparsa bullatione [3]—ita appellant,—nescio an vitro [4] fundendo
perinde utilis, nondum enim expertus est quisquam;
ferrum utique inficit eadem vi. Magnete lapide
architectus Timochares Alexandriae Arsinoes templum concamarare incohaverat, ut in eo simulacrum
e ferro pendere in aëre videretur. intercessit ipsius
mors et Ptolemaei regis, qui id sorori suae iusserat
149 fieri. XLIII. Metallorum omnium vena ferri largis-
sima est. Cantabriae maritima parte, qua oceanus
adluit, mons praealtus—incredibile dictu—totus ex
ea materia est, ut in ambitu oceani diximus.

---

[1] tenuiora . . . durentur *post* acies *transf. Rackham* : *post*
frangi *habent cdd.*
[2] *V.l.* vires (vim *Isid.* XVI. 21. 4).
[3] bulbatione *B* : bullatione *rell.*
[4] vitro *Hermolaus Barbarus* : ultro.

---

[a] As well as in Magnesia.
[b] Wife of Ptolemy II, Philadelphus King of Egypt 286–
247 B.C.
[c] Pliny has not stated this anywhere else. But *cf.* IV. 112.

with oil, for fear that water might harden them and make them brittle. And it is remarkable that when a vein of ore is fused the iron becomes liquid like water and afterwards acquires a spongy and brittle texture. Human blood takes its revenge from iron, as if iron has come into contact with it, it becomes the more quickly liable to rust.

XLII. We will speak in the appropriate place *Lode-stone.* about the lode-stone and the sympathy which it has $\begin{smallmatrix}\text{XXXVI,}\\126\ sqq.\end{smallmatrix}$ with iron. Iron is the only substance that catches the infection of that stone and retains it for a long period, taking hold of other iron, so that we may sometimes see a chain of rings; the ignorant lower classes call this ' live iron,' and wounds inflicted with it are more severe. This sort of stone forms in Biscaya also [a] not in a continuous rocky stratum like the genuine lodestone alluded to but in a scattered pebbly formation or ' bubbling '—that is what they call it. I do not know whether it is equally useful for glass founding, as no one has hitherto tested it, but it certainly imparts the same magnetic property to iron. The architect Timochares had begun to use lodestone for constructing the vaulting in the Temple of Arsinoe [b] at Alexandria, so that the iron statue contained in it might have the appearance of being suspended in mid air; but the project was interrupted by his own death and that of King Ptolemy who had ordered the work to be done in honour of his sister.

XLIII. Iron ore is found in the greatest abundance of all metals. In the coastal part of Biscaya washed by the Atlantic there is a very high mountain which, marvellous to relate, consists entirely of that mineral, as we stated [c] in our account of the lands bordering on the Ocean.

Ferrum accensum igni, nisi duretur ictibus, corrumpitur. rubens non est habile tundendo neque antequam albescere incipiat. aceto aut alumine
150 inlitum fit aeri simile. a robigine vindicatur cerussa et gypso et liquida pice. haec est ferro a Graecis antipathia dicta. ferunt quidem et religione quadam id fieri et exstare ferream catenam apud Euphraten amnem in urbe, quae Zeugma appellatur, qua Alexander Magnus ibi iunxerit pontem, cuius anulos, qui refecti sint, robigine infestari, carentibus ea prioribus.

151 XLIV. Medicina e ferro est et alia quam secandi. namque et circumscribi circulo [1] terve circumlato mucrone et adultis et infantibus prodest contra noxia medicamenta, et praefixisse in limine evulsos sepulchris clavos adversus nocturnas lymphationes, pungique leviter mucrone, quo percussus homo sit, contra dolores laterum pectorumque subitos, qui punctionem adferant. quaedam ustione sanantur, privatim vero canis rabidi morsus, quippe etiam praevalente morbo expaventesque potum usta plaga ilico liberantur. calfit etiam ferro candente potus in multis vitiis, privatim vero dysentericis.

152 XLV. Est et robigo ipsa in remediis, et sic proditur

---

[1] circulo *B* : circulos *rell.*

---

[a] See § 175.
[b] Opposite the modern Birejik.

Iron that has been heated by fire is spoiled unless it is hardened by blows of the hammer. It is not suitable for hammering while it is red hot, nor before it begins to turn pale. If vinegar or alum is sprinkled on it it assumes the appearance of copper. It can be protected from rust by means of lead acetate,[a] gypsum and vegetable pitch; rust is called by the Greeks ‘antipathia,’ ‘natural opposite’ to iron. It is indeed said that the same result may also be produced by a religious ceremony, and that in the city called Zeugma [b] on the river Euphrates there is an iron chain that was used by Alexander the Great in making the bridge at that place, the links of which 331 B.C. that are new replacements are attacked by rust although the original links are free from it.

XLIV. Iron supplies another medicinal service *Medicinal* besides its use in surgery. It is beneficial both for *uses of iron.* adults and infants against noxious drugs for a circle to be drawn round them with iron or for a pointed iron weapon to be carried round them; and to have a fence of nails that have been extracted from tombs driven in in front of the threshold is a protection against attacks of nightmare, and a light prick made with the point of a weapon with which a man has been wounded is beneficial against sudden pains which bring a pricking sensation in the side and chest. Some maladies are cured by cauterization, but particularly the bite of a mad dog, inasmuch as even when the disease is getting the upper hand and when the patients show symptoms of hydrophobia they are relieved at once if the wound is cauterized. In many disorders, but especially in dysenteric cases, drinking water is heated with redhot iron.

XLV. The list of remedies even includes rust

Telephum sanasse Achilles, sive id aerea sive ferrea
cuspide fecit; ita certe pingitur ex ea decutiens
gladio suo.[1] robigo ferri deraditur umido ferro
153 clavis veteribus. potentia eius ligare, siccare, sistere.
emendat alopecias inlita. utuntur et ad scabritias
genarum pusulasque totius corporis cum cera et
myrteo oleo, ad ignes vero sacros ex aceto, item
ad scabiem, paronychia digitorum et pterygia in
linteolis. sistit et feminarum profluvia inposita
in vellere, utilis [2] plagis quoque recentibus vino diluta
et cum murra subacta, condylomatis ex aceto. poda-
gras quoque inlita lenit.

154   XLVI. Squama quoque ferri in usu est ex acie aut
mucronibus, maxime simili, sed acriore vi quam
robigo, quam ob rem et contra epiphoras oculorum
adsumitur. sanguinem sistit, cum volnera ferro
maxime fiant! sistit et feminarum profluvia. in-
ponitur et contra lienium vitia, et haemorroidas
compescit ulcerumque serpentia. et genis prodest
155 farinae modo adspersa paullisper. praecipua tamen
commendatio eius in hygremplastro ad purganda
vulnera fistulasque et omnem callum erodendum et
rasis ossibus carnes recreandas. componitur hoc
modo : propolis [3] oboli VI, Cimoliae cretae drachmae
VI, aeris tusi drachmae II, squamae ferri [4] totidem,

---

[1] suo *K. C. Bailey*: sed.
[2] in vellere, utilis *coni. Mayhoff*: velleribus.
[3] propolis *Mayhoff* (*coll.* XXII. 107, *Scribon.*, 209) *qui et*
galbani *vel* panacis *coni.*: pal. *B* : pari *rell.*: panis *Ian* :
aluminis *coni. Sillig*: picis *Hardouin*.
[4] *V.l.* ferreae.

---

[a] Cf. XXXV, 195 ff.

itself, and this is the way in which Achilles is stated
to have cured Telephus, whether he did it by means
of a copper javelin or an iron one; at all events
Achilles is so represented in painting, knocking the
rust off a javelin with his sword. Rust of iron is
obtained by scraping it off old nails with an iron tool
dipped in water. The effect of rust is to unite
wounds and dry them and staunch them, and applied
as a liniment it relieves fox-mange. They also use
it with wax and oil of myrtle for scabbiness of the
eye-lids and pimples in all parts of the body, but
dipped in vinegar for erysipelas and also for scab,
and, applied on pieces of cloth, for hangnails on the
fingers and whitlows. Applied on wool it arrests
women's discharges and for recent wounds it is useful
diluted with wine and kneaded with myrrh, and for
swellings round the anus dipped in vinegar. Used
as a liniment it also relieves gout.

XLVI. Scale of iron, obtained from a sharp edge
or point, is also employed, and has an effect extremely
like that of rust only more active, for which reason
it is employed even for running at the eyes. It
arrests hæmorrhage, though it is with iron that
wounds are chiefly made! And it also arrests female
discharges. It is also applied against troubles of the
spleen, and it checks hæmorrhoidal swellings and
creeping ulcers. Applied for a brief period in the
form of a powder it is good for the eyelids. But
its chief recommendation is its use in a wet plaster
for cleaning wounds and fistulas and for eating out
every kind of callosity and making new flesh on
bones that have been denuded. The following are
the ingredients: six obols of bee-glue, six drams of
Cimolo earth,[a] two drams of pounded copper, two of

cerae x,[1] olei sextarius. his adicitur, cum sunt
repurganda volnera aut replenda, ceratum.

156   XLVII. Sequitur natura plumbi, cuius duo genera,
nigrum atque candidum. pretiosissimum in [2] hoc
candidum, Graecis appellatum cassiterum fabulo-
seque narratum in insulas Atlantici maris peti vitili-
busque navigiis et circumsutis corio advehi. nunc
certum est in Lusitania gigni et in Gallaecia summa
157 tellure, harenosa et coloris nigri. pondere tantum
ea deprehenditur; interveniunt et minuti calculi,
maxime torrentibus siccatis. lavant eas harenas
metallici et, quod subsedit, cocunt in fornacibus.
invenitur et in aurariis metallis, quae alutias [3] vocant,
aqua immissa eluente calculos nigros paullum candore
variatos, quibus eadem gravitas quae auro, et ideo
in catillis [4] quibus aurum colligitur, cum eo remanent;
postea caminis separantur conflatique in plumbum
158 album resolvuntur. non fit in Gallaecia nigrum, cum
vicina Cantabria nigro tantum abundet, nec ex albo
argentum, cum fiat ex nigro. iungi inter se plumbum nigrum sine albo non potest nec hoc ei sine oleo
ac ne album quidem secum sine nigro. album
habuit auctoritatem et Iliacis temporibus teste
159 Homero, cassiterum ab illo dictum. plumbi nigri

---

[1] x *Mayhoff*: XL *B*[1] : XI *B*[2] : ex *aut* sex *rell.*
[2] in *add. Mayhoff.*
[3] alutias *B* : alutia *aut* aluta *rell.*
[4] catillis *Warmington* : scutulis *Urlichs* : calathis *cd. Par.*
6801 *recte*? : calatis *aut* colatis *aut* cloacis *rell.* (cutalis *B*).

---

  [a] Pliny's ' black lead ' is lead, and his ' white lead ' is tin.
Neither must be confused with the ' black lead ' and ' white
lead ' of modern usage.

  [b] Or, if we read *calathis,* ' baskets.'

scale of iron, ten of wax and a pint of oil. When it is desired to cleanse or fill up wounds, wax plaster is added to these ingredients.

XLVII. The next topic is the nature of lead, of which there are two kinds, black and white.[a] White lead (tin) is the most valuable; the Greeks applied to it the name *cassiteros*, and there was a legendary story of their going to islands of the Atlantic ocean to fetch it and importing it in platted vessels made of osiers and covered with stitched hides. It is now known that it is a product of Lusitania and Gallaecia found in the surface-strata of the ground which is sandy and of a black colour. It is only detected by its weight, and also tiny pebbles of it occasionally appear, especially in dry beds of torrents. The miners wash this sand and heat the deposit in furnaces. It is also found in the goldmines called 'alutiae,' through which a stream of water is passed that washes out black pebbles of tin mottled with small white spots, and of the same weight as gold, and consequently they remain with the gold in the bowls[b] in which it is collected, and afterwards are separated in the furnaces, and fused and melted into white lead. Black lead does not occur in Gallaecia, although the neighbouring country of Biscaya has large quantities of black lead only; and white lead yields no silver, although it is obtained from black lead. Black lead cannot be soldered with black without a layer of white lead, nor can white be soldered to black without oil, nor can even white lead be soldered with white without some black lead. Homer testifies that white lead or tin had a high position even in the Trojan period, he giving it the name of *cassiteros*. There are two different sources of black lead, as it is

Lead and tin.

Il. XI, 25; XVIII, 568, 574, 613; XXIII, 561.

origo duplex est; aut enim sua provenit vena nec
quicquam aliud ex sese parit aut cum argento
nascitur mixtisque venis conflatur. huius qui primus fuit in fornacibus liquor stagnum appellatur;
qui secundus, argentum; quod remansit in fornacibus,
galena, quae fit tertia portio additae venae; haec
rursus conflata dat nigrum plumbum deductis partibus
nonis ii.

160   XLVIII. Stagnum inlitum aereis vasis saporem
facit gratiorem ac compescit virus aeruginis, mirumque, pondus non auget. specula etiam ex eo
laudatissima, ut diximus, Brundisi temperabantur,
donec argenteis uti coepere et ancillae. nunc
adulteratur stagnum addita aeris candidi tertia
portione in plumbum album. fit et alio modo mixtis
albi plumbi nigrique libris; hoc nunc aliqui argentarium appellant. iidem et tertiarium vocant, in
quo duae sunt nigri portiones et tertia albi. pretium eius in libras X xx.[1] hoc fistulae solidantur.
161 inprobiores ad tertiarium additis partibus aequis
albi argentarium vocant et eo quae volunt inco-
quunt. pretium[2] huius faciunt in p. X lxx.[3] albo
per se sincero pretium[4] sunt X lxxx,[5] nigro X vii.[6]

Albi natura plus aridi habet, contraque nigri tota

---

[1] X xx B : xxx rell. (x.x cd. Par. Lat. 6797).
[2] V.l. pretia (pretio cd. Leid. Voss.).
[3] xlxx /// B² : varia cdd.   [4] V.l. pretia.
[5] xlxxx B² : varia cdd.
[6] x.vii cd. Chiffl. apud Dalecamp : xvii rell.

---

[a] The mixture is galena or sulphide of lead, the commonest
lead ore, for which, see XXXIII, 95.

[b] Or stannum, alloy of silver and lead.

[c] Galena is here crude or impure lead, not the modern
galena or sulphide of lead, for which see XXXIII, 95.

[d] Brass.

either found in a vein of its own and produces no other substance mixed with it, or it forms together with silver, and is smelted with the two veins mixed together.[a] Of this substance the liquid that melts first in the furnaces is called *stagnum* [b]; the second liquid is argentiferous lead, and the residue left in the furnaces is impure lead [c] which forms a third part of the vein originally put in; when this is again fused it gives black lead, having lost two-ninths in bulk.

XLVIII. When copper vessels are coated with *stagnum* the contents have a more agreeable taste and the formation of destructive verdigris is prevented, and, what is remarkable, the weight is not increased. Also, as we have said, it used to be employed at Brindisi as a material for making mirrors which were very celebrated, until even servant-maids began to use silver ones. At the present day a counterfeit *stagnum* is made by adding one part of white copper [d] to two parts of white lead; and it is also made in another way by mixing together equal weights of white and black lead: the latter compound some people now call 'silver mixture.' The same people also give the name of 'tertiary' to a compound containing two portions of black lead and one of white; its price is 20 denarii a pound. It is used for soldering pipes. More dishonest makers add to tertiary an equal amount of white lead and call it 'silver mixture,' and use it melted for plating by immersion any articles they wish. They put the price of this last at 70 denarii for 1 lb.: the price of pure white lead without alloy is 80 denarii, and of black lead 7 denarii.

The substance of white lead has more dryness,

XXXIII. 130.

umida est. ideo album nulli rei sine mixtura utile
est. neque argentum ex eo plumbatur, quoniam
162 prius liquescat argentum, confirmantque, si minus
albo nigri, quam satis sit, misceatur, erodi ab eo
argentum. album incoquitur aereis operibus Gal-
liarum invento ita, ut vix discerni possit ab argento,
eaque incoctilia appellant. deinde et argentum
incoquere simili modo coepere equorum maxime
ornamentis iumentorumque ac iugorum in Alesia
163 oppido; reliqua gloria Biturigum fuit. coepere
deinde et esseda sua colisataque ac petorita exornare
simili modo, quae iam luxuria ad aurea quoque, non
modo argentea, staticula[1] pervenit, quaeque in
scyphis cerni prodigum[2] erat, haec in vehiculis adteri
cultus vocatur.

plumbi albi experimentum in charta est, ut liquefactum pondere videatur, non calore, rupisse. India
neque aes neque plumbum habet gemmisque ac
margaritis suis haec permutat.

164 XLIX. Nigro plumbo ad fistulas lamnasque utimur,
laboriosius in Hispania eruto totasque per Gallias,
se din Brittannia summo terrae corio adeo large, ut
lex interdicat ut[3] ne plus certo modo fiat. nigri

[1] vehicula cd. Par. 6801, cd. Par. Lat. 6797, m.2 in ras.
[2] V.l. prodigium.
[3] lex interdicat ut M. Hertz : lex ultro dicatur cdd. : lex
cavere dicatur Urlichs : lex custodiatur Detlefsen : lege
interdicatur Brunn : nec interdicatur coni. Mayhoff.

---

[a] But silver does not melt sooner than ' white lead ' (tin).
Perhaps some alloy is meant here.
[b] At Mont Auxois near Alise.

whereas that of black lead is entirely moist. Consequently white lead cannot be used for anything without an admixture of another metal, nor can it be employed for soldering silver, because the silver melts before the white lead.[a] And it is asserted that if a smaller quantity of black lead than is necessary is mixed with the white, it corrodes the silver. A method discovered in the Gallic provinces is to plate bronze articles with white lead so as to make them almost indistinguishable from silver; articles thus treated are called ' incoctilia.' Later they also proceeded in the town Alesia [b] to plate with silver in a similar manner, particularly ornaments for horses and pack animals and yokes of oxen; the distinction of developing this method belongs to Bordeaux. Then they proceeded to decorate two-wheeled war-chariots, chaises and four-wheeled carriages in a similar manner, a luxurious practice that has now got to using not only silver but even gold statuettes, and it is now called good taste to subject to wear and tear on carriages ornaments that it was once thought extravagant to see on a goblet!

It is a test of white lead when melted and poured on papyrus to seem to have burst the paper by its weight and not by its heat. India possesses neither copper nor lead, and procures them in exchange for her precious stones and pearls.

XLIX. Black lead which we use to make pipes and sheets is excavated with considerable labour in Spain and through the whole of the Gallic provinces, but in Britain it is found in the surface-stratum of the earth in such abundance that there is a law prohibiting the production of more than a certain amount. The various kinds of black lead

generibus haec sunt nomina: Ovetanum,[1] Caprariense,
Oleastrense, nec differentia ulla scoria modo excocta
diligenter. mirum in his solis metallis, quod dere-
165 licta fertilius revivescunt. hoc videtur facere laxatis
spiramentis ad satietatem infusus aër, aeque ut
feminas quasdam fecundiores facere abortus. nuper
id conpertum in Baetica Salutariensi [2] metallo, quod
locari solitum X c̄c̄ [3] annuis, postquam obliteratum
erat, X c̄c̄l̄v̄ [4] locatum est. simili modo Antonianum
in eadem provincia pari locatione pervenit ad HS c̄c̄c̄c̄
vectigalis. mirum et addita aqua non liquescere
vasa e plumbo, eadem, si in aquam [5] addantur calculus
vel aereus quadrans, peruri.
166   L. In medicina per se plumbi usus cicatrices
reprimere adalligatisque lumborum et renium parti
lamnis frigidiore natura inhibere inpetus veneris
visaque in quiete veneria sponte naturae erumpentia
usque in genus morbi. his lamnis Calvus orator
cohibuisse se traditur viresque corporis studiorum
labori custodisse. Nero, quoniam ita placuit diis,
princeps, lamna pectori inposita sub ea cantica
exclamans alendis vocibus demonstravit rationem.
167 coquitur ad medicinae usus patinis fictilibus substrato sulpure minuto, lamnis impositis tenuibus

---

[1] Ovetanum *Hardouin* : iovetanum *B* : iovetantum *rell.*
[2] Salutariensi *coni. Mayhoff* : samariensi *B* : santarensi
aut samiarenci *aut* saremianensi *rell.*
[3] c̄c̄ *Ian* : cc. M. *cd. Leid. Voss., cd. Flor. Ricc.*[2] (ccc
Ricc.[1]) : cc *rell.*
[4] X c̄c̄l̄v̄ *Mayhoff* : varia cdd. et edd.
[5] in aquam *C. F. W. Müller* : sine aqua *K. C. Bailey: coni.
Mayhoff* cum aqua.

---

[a] Not true. K. C. Bailey suggests *si sine aqua* and translates ' a hole is burnt in the same vessels if filled with pebbles
or copper coins instead of water.'

have the following names—Oviedo lead, Capraria lead, Oleastrum lead, though there is no difference between them provided the slag has been carefully smelted away. It is a remarkable fact in the case of these mines only that when they have been abandoned they replenish themselves and become more productive. This seems to be due to the air infusing itself to saturation through the open orifices, just as a miscarriage seems to make some women more prolific. This was recently observed in the Salutariensian mine in Bætica, which used to be let at a rent of 200,000 denarii a year, but which was then abandoned, and subsequently let for 255,000. Likewise the Antonian mine in the same province from the same rent has reached a return of 400,000 sesterces. It is also remarkable that vessels made of lead will not melt if they have water put in them, but if to the water a pebble or quarter-*as* coin is added, the fire burns through *a* the vessel.

L. In medicine lead is used by itself to remove scars, and leaden plates are applied to the region of the loins and kidneys for their comparative chilly nature to check the attacks of venereal passions, and the libidinous dreams that cause spontaneous emissions to the extent of constituting a kind of disease. It is recorded that the pleader Calvus used these plates to control himself and to preserve his bodily strength for laborious study. Nero, whom heaven was pleased to make emperor, used to have a plate of lead on his chest when singing songs *fortissimo*, thus showing a method for preserving the voice. For medical purposes lead is melted in earthen vessels, a layer of finely powdered sulphur being put underneath it; on this thin plates are

Medical uses of lead.

247

opertisque sulpure, veru ferreo [1] mixtis. cum coquatur,[2] munienda in eo opere foramina spiritus convenit; alioqui plumbi fornacium halitus noxius sentitur et pestilens; nocet [3] canibus ocissime, omnium vero metallorum muscis et culicibus, quam ob rem non sunt ea taedia in metallis.

168 Quidam in coquendo scobem plumbi lima quaesitam sulpuri miscent, alii cerussam potius quam sulpur. fit et lotura plumbi usus in medicina. cum se ipso teritur in mortariis plumbeis addita aqua caelesti, donec crassescat; postea supernatans aqua tollitur spongeis; quod crassissimum fuit, siccatum dividitur in pastillos. quidam limatum plumbum sic terunt, quidam et plumbaginem admiscent, alii vero acetum,

169 alii vinum, alii adipem, alii rosam. quidam in lapideo mortario et maxime Thebaico plumbum pistillo lapideo [4] terere malunt, candidiusque ita fit medicamentum. id autem quod ustum est plumbum lavatur ut stibis et cadmea. potest adstringere, sistere, contrahere cicatrices; usu est [5] eodem et in oculorum medicamentis, maxime contra procidentiam eorum et inanitates ulcerum excrescentiave rimasque sedis aut haemorroidas et con

170 dylomata. ad haec maxime lotura plumbi facit,

[1] veru ferreo *coni.* Mayhoff : et ferro *aut* est et foro.
[2] coquuntur *coni.* Mayhoff.
[3] nocet *Warmington* : est *Mayhoff* : et.
[4] plumbum (*Urlichs*) pistillo (*e cd. Par. Lat.* 6797) lapideo (*coll. Diosc.*) *K. C. Bailey* : plumbeum pistillum *aut* plumbeo pistillum *cdd.* (pistillo *cd. Par. Lat.* 6797).
[5] usu est *Mayhoff* : usus et *cdd.* : usus enim ex *ed. vett.* : usus ex *Sillig*.

---

[a] See XXXIII, 98.   [b] See § 175.
[c] Not it seems the plant lead-wort. Cf. Dioscorides, V, 95, 100.

laid and covered with sulphur and stirred up with an iron spit. While it is being melted, the breathing passages should be protected during the operation, otherwise the noxious and deadly vapour of the lead furnace is inhaled: it is hurtful to dogs with special rapidity,[a] but the vapour of all metals is so to flies and gnats, owing to which those annoyances are not found in mines.

Some people during the process of smelting mix lead-filings with the sulphur, and others use lead acetate[b] in preference to sulphur. Another use of lead is to make a wash—it is employed in medicine—pieces of lead with rainwater added being ground against themselves in leaden mortars till the whole assumes a thick consistency, and then water floating on the top is removed with sponges and the very thick sediment left when dry is divided into tablets. Some people grind up lead filings in this way and some also mix in some lead ore,[c] but others use vinegar, others wine, others grease, others oil of roses. Some prefer to grind the lead with a stone pestle in a stone mortar, and especially one made of Thebes stone,[d] and this process produces a drug of a whiter colour. Calcined lead is washed like antimony and *cadmea*.[e] It has the property of acting as an astringent and arresting hæmorrhage and of promoting cicatrization. It is of the same utility also in medicines for the eyes, especially as preventing their procidence, and for the cavities or excrescences left by ulcers and for fissures of the anus or haemorrhoids and swellings of the anus. For these purposes lead lotion is ex-

---

[d] Cf. XXXIII, 68; XXXIV, 106; XXXVI, 63, 157.
[e] See XXXIII, 103; XXXIV, 100-104.

cinis autem usti ad serpentia ulcera aut sordida, eademque quae chartis [1] ratio profectus. uritur autem in patinis per lamnas minutas cum sulpure, versatum rudibus ferreis aut ferulaceis, donec liquor mutetur in cinerem; dein refrigeratum teritur in farinam. alii elimatam scobem in fictili crudo cocunt in caminis, donec percoquatur figlinum. aliqui cerussam admiscent pari mensura aut hordeum teruntque ut in crudo [2] dictum est, et praeferunt sic plumbum spodio Cyprio.

171 LI. Scoria quoque plumbi in usu est. optima quae maxime ad luteum colorem accedit, sine plumbi reliquiis aut sulpuris, specie terrae [3] carens. lavatur haec in mortariis minutim fracta, donec aqua luteum colorem trahat, et transfunditur in vas purum, idque saepius, usque dum subsidat quod utilissimum est. effectus habet eosdem quos plumbum, sed acriores. mirarique succurrit experientiam [4] vitae, ne faece quidem rerum excrementorumque foeditate intemptata tot modis.

172 LII. Fit et spodium ex plumbo eodem modo quo ex Cyprio aere; lavatur in linteis raris aqua caelesti separaturque terrenum transfusione; cribratum teritur. quidam pulverem eum pinnis digerere malunt ac terere in vino odorato.

---

[1] chalcitidis *coni. Mayhoff.*
[2] in crudo *cdd.* : supra *coni. K. C. Bailey.*
[3] specie terrae *B* : et terra *rell.*
[4] experientia *B.*

---

[a] For the medicinal use of burnt papyrus see XXIV, 88.
[b] See § 175.
[c] Some kind of litharge.

tremely efficient, while for creeping or foul ulcers ash
of calcined lead is useful; and the benefit they
produce is on the same lines as in the case of sheets
of papyrus.[a] The lead is burnt in small sheets
mixed with sulphur, in shallow vessels, being stirred
with iron rods or fennel stalks till the molten metal
is reduced to ashes; then after being cooled off it is
ground into powder. Another process is to boil
lead filings in a vessel of raw earth in furnaces till the
earthenware is completely baked. Some mix with
it an equal amount of lead acetate [b] or of barley and
grind this mixture, in the way stated in the case of §168.
raw lead, and prefer the lead treated in this way to
the Cyprus slag.

LI. The dross of lead is also utilized. The best
is that which approximates in colour most closely to
yellow, containing no remnants of lead or sulphur,
and does not look earthy. This is broken up into
small fragments and washed in mortars till the water
assumes a yellow colour, and poured off into a clean
vessel, and the process is repeated several times till
the most valuable part settles as a sediment at the
bottom. Lead dross has the same effects as lead,
but to a more active degree. This suggests a remark
on the marvellous efficacy of human experiment,
which has not left even the dregs of substances and
the foulest refuse untested in such numerous ways!

LII. Slag [c] is also made from lead in the same
way as from Cyprus copper; it is washed with rain §128 sqq.
water in linen sheets of fine texture and the earthy
particles are got rid of by rinsing, and the residue
is sifted and then ground. Some prefer to separate
the powder with a feather, and to grind it up with
aromatic wine.

173   LIII. Est et molybdaena, quam alio loco galenam
appellavimus, vena argenti plumbique communis.
melior haec, quanto magis aurei coloris quantoque
minus plumbosa, friabilis et modice gravis. cocta
cum oleo iocineris colorem trahit. adhaerescit et
auri argentique fornacibus; hanc metallicam vocant.
laudatissima quae in Zephyrio fiat; probantur minime
174 terrenae minimeque lapidosae. coquuntur lavantur-
que scoriae modo. usus in lipara ad lenienda ac
refrigeranda ulcera et emplastris, quae non inligantur,
sed inlita ad cicatricem perducunt in teneris corporibus mollissimisque partibus. compositio eius e
libris III et cerae libra, olei III heminis, quod in senili
corpore cum fracibus additur. temperatur cum
spuma argenti et scoria plumbi ad dysenteriam et
tenesmum fovenda calida.

175   LIV. Psimithium quoque, hoc est cerussam, plum-
bariae dant officinae, laudatissimam in Rhodo. fit
autem ramentis plumbi tenuissimis super vas aceti
asperrimi inpositis atque ita destillantibus. quod ex
eo cecidit in ipsum acetum, arefactum molitur et
cribratur iterumque aceto admixto in pastillos
dividitur et in sole siccatur aestate. fit et alio modo,
addito in urceos aceti plumbo opturatos per dies X
derasoque ceu situ ac rursus reiecto, donec deficiat

---

^a In this word Pliny includes two things: (i) a mineral
which is lead sulphide (still called galena) or perhaps lead
oxide (massicot); (ii) an artificial product which is litharge.
Both Pliny and Dioscorides call the mineral yellow, which is
true of litharge, not of lead sulphide. So the mineral may be
the yellow or yellowish red massicot (K. C. Bailey).
^b Sugar of lead, not the modern basic lead carbonate or
' white lead.'

LIII. There is also *molybdaena* [a] (which in another '*Molybdaena*'.
place we have called *galena*); it is a mineral com- XXXIII,
pound of silver and lead. It is better the more 95;
XXXIV,
golden its colour and the less leaden: it is friable 159.
and of moderate weight. When boiled with oil it
acquires the colour of liver. It is also found adhering
to furnaces in which gold and silver are smelted; in
this case it is called metallic sulphide of lead. The
kind most highly esteemed is produced at Zephyrium.
Varieties with the smallest admixture of earth and of
stone are approved of; they are melted and washed
like dross. It is used in preparing a particular
emollient plaster for soothing and cooling ulcers
and in plasters which are not applied with bandages
but which they use as a liniment to promote cicatrization on the bodies of delicate persons and on the
more tender parts. It is a composition of three
pounds of sulphide of lead and one of wax with half
a pint of oil, which is added with solid lees of olives
in the case of an elderly patient. Also combined
with scum of silver and dross of lead it is applied
warm for fomenting dysentery and constipation.

LIV. ' *Psimithium* ' also, that is cerussa or lead *Sugar of*
acetate,[b] is produced at lead-works. The most *lead.*
highly spoken of is in Rhodes. It is made from
very fine shavings of lead placed over a vessel of
very sour vinegar and so made to drip down. What
falls from the lead into the actual vinegar is dried
and then ground and sifted, and then again mixed
with vinegar and divided into tablets and dried in
the sun, in summertime. There is also another way
of making it, by putting the lead into jars of vinegar
kept sealed up for ten days and then scraping off the
sort of decayed metal on it and putting it back in

176 materia. quod derasum est, teritur et cribratur et
coquitur in patinis misceturque rudiculis, donec
rufescat et simile sandaracae fiat. dein lavatur
dulci aqua, donec nubeculae omnes eluantur. siccatur postea similiter et in pastillos dividitur. vis
eius eadem quae supra dictis, lenissima tantum ex
omnibus, praeterque ad candorem feminarum. est
autem letalis potu sicut spuma argenti. postea
cerussa ipsa, si coquatur, rufescit.

177   LV. Sandaracae quoque propemodum dicta natura
est. invenitur autem et in aurariis et in argentariis
metallis, melior quo magis rufa quoque magis virus
sulpuris [1] redolens ac pura friabilisque. valet purgare, sistere, excalfacere, erodere, summa eius
dote septica. explet alopecias ex aceto inlita;
additur oculorum medicamentis; fauces purgat
cum melle sumpta vocemque limpidam et canoram
facit; suspiriosis et tussientibus iucunde medetur
cum resina terebinthina in cibo sumpta, suffita
quoque cum cedro ipso nidore isdem medetur.

178   LVI. Et arrhenicum ex eadem est materia. quod
optimum, coloris etiam in auro excellentis; quod
vero pallidius aut sandaracae simile est, deterius
iudicatur. est et tertium genus, quo miscetur
aureus color sandaracae. utraque haec squamosa,

---

[1] sulpuris *add. Mayhoff.*

---

[a] Red sulphide of arsenic.
[b] Yellow sulphide of arsenic.

the vinegar, till the whole of it is used up. The stuff scraped off is ground up and sifted and heated in shallow vessels and stirred with small rods till it turns red and becomes like *sandarach*, realgar.[a] Then it is washed with fresh water till all the cloudy impurities have been removed. Afterwards it is dried in a similar way and divided into tablets. Its properties are the same as those of the substances mentioned above, only it is the mildest of them all, and beside that, it is useful for giving women a fair complexion; but like scum of silver, it is a deadly poison. The lead acetate itself if afterwards melted becomes red.

LV. Of realgar also the properties have been *Realgar.* almost completely described. It is found both in goldmines and silvermines; the redder it is and the more it gives off a poisonous scent of sulphur and the purer and more friable it is, the better it is. It acts as a cleanser, as a check to bleeding, as a calorific and a caustic, being most remarkable for its corrosive property; used as a liniment with vinegar it removes fox-mange; it forms an ingredient in eye-washes, and taken with honey it cleans out the throat. It also produces a clear and melodious voice, and mixed with turpentine and taken in the food, is an agreeable remedy for asthma and cough; its vapour also remedies the same complaints if merely used as a fumigation with cedar wood.

LVI. Orpiment [b] also is obtained from the same *Orpiment.* substance. The best is of a colour of even the finest-coloured gold, but the paler sort or what resembles sandarach is judged inferior. There is also a third class which combines the colours of gold and of sandarach. Both of the latter are scaly, but

illud vero siccum purumque, gracili venarum discursu
fissile. vis eadem quae supra, sed acrior. itaque
et causticis additur et psilotris. tollit et pterygia
digitorum carnesque narium et condylomata et
quidquid excrescit. torretur, ut valdius [1] prosit, in
nova testa, donec mutet colorem.

[1] valdius *B* : validius *rell.*

Note on XXXIV. 17 and 70.

The group of two figures (representing Harmodius and
Aristogeiton) made by Antenor in bronze, set up at Athens in
510–9 B.C., was carried off by Xerxes in 480; and a new
bronze pair was made by Critius and Nesiotes and set up at
Athens in 477. Antenor's group was found by Alexander in
Persia, and on his orders, it seems, one of his successors
c. 293–2 restored it to Athens; part of the original base of
Critius' and Nesiotes' group, it seems, has now been found;
the marble group now at Naples is thought to be a Roman
copy of the same group and made in the 2nd cent. A.D. The
bearded head of Aristogeiton can be restored from a head in
the Vatican. Copies of this group can be seen on four Attic
vases of the first half of the fifth century B.C. (Beazley,
Journ. of Hellen. Stud. LXVIII (1948), 28), and one of about
400 B.C. (W. Hahland, *Vasen um Meidias*, p. 6 and pl. 6a).

the other is dry and pure, and divided in a delicate tracery of veins. Its properties are the same as mentioned above, but more active. Accordingly it is used as an ingredient in cauteries and depilatories. It also removes overgrowths of flesh on to the nails, and pimples in the nostrils and swellings of the anus and all excrescences. To increase its efficacy it is heated in a new earthenware pot till it changes its colour.

---

also in the sketch in low relief on a marble throne now at Broomhall. Other copies of Critius' and Nesiotes' work such as that on a coin of Cyzicus *c.* 420 B.C., those on Athena's shield depicted on three Attic amphorae of *c.* 400 B.C. (or a later date), and that on an Athenian tetradrachm of *c.* 400 B.C., are probably imitations made from memory. C. Seltman's opinion that the Broomhall relief suggests a copy of Antenor's group is doubtful (Seltman, *Journ. of Hellen. Stud.*, LXVII (1947), 22–27). The group made by Praxiteles (Pliny, XXXIV. 70) was no doubt a fresh creation of his own, unless there is some blunder on Pliny's part.

BOOK XXXV

LIBER XXXV

1. Metallorum, quibus opes constant, adgnascentiumque iis natura indicata propemodum est, ita conexis rebus, ut immensa medicinae silva officinarumque tenebrae et morosa caelandi fingendique ac tinguendi subtilitas simul dicerentur. restant terrae ipsius genera lapidumque vel numerosiore[1] serie,[2] plurimis singula a[3] Graecis praecipue voluminibus tractata. nos in iis brevitatem sequemur utilem instituto, modo nihil necessarium 2 aut naturale omittentes, primumque dicemus quae restant de pictura, arte quondam nobili—tunc cum expeteretur regibus populisque—et alios nobilitante, quos esset dignata posteris tradere, nunc vero in totum marmoribus pulsa, iam quidem et auro, nec tantum ut parietes toti operiantur, verum et interraso marmore vermiculatisque ad effigies rerum et 3 animalium crustis. non placent iam abaci nec spatia montes in cubiculo dilatantia:[4] coepimus et lapidem[5] pingere. hoc Claudii principatu inventum, Neronis

---

[1] numerosiore *Gelen* : numerosiores.
[2] serie *Gelen* : seriae *B* : erie *aut* aeriae *aut sim. rell.*
[3] a *fortasse delendum (Mayhoff).*
[4] dilatata *coni. Ian* : dilatant iam *Detlefsen.*
[5] lapidem *cd. Par. Lat. 6797, cd. Par. 6801* : lapide *rell.*

---

[a] This translates *lapidem*. If we read *lapide* 'with stone,' Pliny would mean a kind of mosaic. But see §§ 116, 118.

BOOK XXXV

I. We have now practically indicated the nature of metals, in which wealth consists, and of the substances related to them, connecting the facts in such a way as to indicate at the same time the enormous topic of medicine and the mysteries of the manufactories and the fastidious subtlety of the processes of carving and modelling and dyeing. There remain the various kinds of earth and of stones, forming an even more extensive series, each of which has been treated in many whole volumes, especially by Greeks. For our part in these topics we shall adhere to the brevity suitable to our plan, yet omitting nothing that is necessary or follows a law of Nature. And *Painting.* first we shall say what remains to be said about painting, an art that was formerly illustrious, at the time when it was in high demand with kings and nations and when it ennobled others whom it deigned to transmit to posterity. But at the present time it has been entirely ousted by marbles, and indeed finally also by gold, and not only to the point that whole party-walls are covered—we have also marble engraved with designs and embossed marble slabs carved in wriggling lines to represent objects and animals. We are no longer content with panels nor with surfaces displaying broadly a range of mountains in a bedchamber; we have begun even to paint on the masonry.[a] This was invented in the

vero maculas, quae non essent in crustis, inserendo unitatem variare, ut ovatus esset Numidicus, ut purpura distingueretur Synnadicus, qualiter illos nasci optassent deliciae. montium haec subsidia deficientium, nec cessat luxuria id agere, ut quam plurimum incendiis perdat.

4 II. Imaginum quidem pictura, qua maxime similes in aevum propagabantur figurae, in totum exolevit. aerei ponuntur clipei argentea facie,[1] surdo figurarum discrimine; statuarum capita permutantur, volgatis iam pridem salibus etiam carminum. adeo materiam conspici malunt omnes quam se nosci, et inter haec pinacothecas veteribus tabulis consuunt[2] alienasque effigies colunt, ipsi honorem non nisi in pretio

5 ducentes, ut frangat heres forasque[3] detrahat laqueo.[4] itaque nullius effigie vivente imagines pecuniae, non suas, relinquunt. iidem palaestrae[5] athletarum imagi nibus et ceromata sua exornant, Epicuri voltus per cubicula gestant ac circumferunt secum. natali eius sacrificant, feriasque omni mense vicesima luna[6]

---

[1] argentea facie *Mayhoff*: argenteae facies.

[2] conferciunt *quidam apud Dalecamp*: complent *coni. Mayhoff.*

[3] forasque *cd. Par.* 6801: furisque *rell.*: forisque *Detlefsen*: furisve *coni. Brotier.*

[4] laqueo *Detlefsen*: laqueis *Urlichs*: laqueū *aut* laqueūs.

[5] palaestrae *coni. Mayhoff*: palaestras.

[6] vicesima luna *hic Mayhoff*: *post* custodiunt *Ian*: *post* eius *cdd.*

---

[a] Of Synnada in Phrygia in Asia Minor.

[b] So that heads were put on bodies which did not belong to them.

[c] *Ceromata*, Greek for wax ointments used by athletes, and also denoting the rooms where these were applied before or after a match.

[d] Greek εἰκάς, 20th day.

principate of Claudius, while in the time of Nero a
plan was discovered to give variety to uniformity by
inserting markings that were not present in the
embossed marble surface, so that Numidian stone
might show oval lines and Synnadic *a* marble be
picked out with purple, just as fastidious luxury
would have liked them to be by nature. These are
our resources to supplement the mountains when
they fail us, and luxury is always busy in the effort
to secure that if a fire occurs it may lose as much as
possible.

II. The painting of portraits, used to transmit
through the ages extremely correct likenesses of
persons, has entirely gone out. Bronze shields are
now set up as monuments with a design in silver,
with a dim outline of men's figures; heads of statues
are exchanged for others,*b* about which before now
actually sarcastic epigrams have been current: so
universally is a display of material preferred to a
recognizable likeness of one's own self. And in the
midst of all this, people tapestry the walls of their
picture-galleries with old pictures, and they prize
likenesses of strangers, while as for themselves they
imagine that the honour only consists in the price,
for their heir to break up the statue and haul it out
of the house with a noose. Consequently nobody's
likeness lives and they leave behind them portraits
that represent their money, not themselves. The
same people decorate even their own anointing-
rooms *c* with portraits of athletes of the wrestling-
ring, and display all round their bedrooms and carry
about with them likenesses of Epicurus; they offer
sacrifices on his birthday, and keep his festival,
which they call the *eikas* *d* on the 20th day of every

custodiunt, quas icadas vocant, ii maxime, qui se ne
viventes quidem nosci volunt. ita est profecto:
artes desidia perdidit, et quoniam animorum imagines
6 non sunt, negleguntur etiam corporum. aliter apud
maiores in atriis haec erant, quae spectarentur; non
signa externorum artificum nec aera aut marmora:
expressi cera vultus singulis disponebantur ar-
mariis, ut essent imagines, quae comitarentur
gentilicia funera, semperque defuncto aliquo totus
aderat familiae eius qui umquam fuerat populus.
stemmata vero lineis [1] discurrebant ad imagines
7 pictas. tabulina codicibus implebantur et monimentis
rerum in magistratu gestarum. aliae foris et circa
limina animorum ingentium imagines erant adfixis
hostium spoliis, quae nec emptori refigere liceret,
triumphabantque etiam dominis mutatis aeternae [2]
domus. erat haec stimulatio [3] ingens, exprobrantibus
tectis cotidie inbellem dominum intrare in alienum
8 triumphum. exstat Messalae oratoris indignatio,
quae prohibuit inseri genti suae Laevinorum alienam
imaginem. similis causa Messalae seni expressit
volumina illa quae de familiis condidit, cum Scipionis
Pomponiani transisset atrium vidissetque adoptione
testamentaria Salvittones—hoc enim fuerat cogno-

---

[1] linteis *Fröhner coll.* 51.
[2] aeternae *Mayhoff*: ipsae *Gelen*: emptae *Ian, Urlichs*:
tamen *coni. Sillig*: et me *cdd.* (etiã *B²*).
[3] stimulatio *Gelen*: stimmatio *B¹*: estimatio (*deleto* haec)
B²: stima ratio *rell.* (summa r. *cd. Par. Lat.* 6797).

---

[a] A maxim of Epicurus was Λάθε βιώσας 'live unnoticed.'
[b] In private houses.
[c] A branch of the Gens Valeria, to which the Messalae also
belonged.

month—these of all people, whose desire it is not to be known even when alive [a]! That is exactly how things are: indolence has destroyed the arts, and since our minds cannot be portrayed, our bodily features are also neglected. In the halls of our ancestors it was otherwise; portraits were the objects displayed to be looked at, not statues by foreign artists, nor bronzes nor marbles, but wax models of faces were set out each on a separate side-board, to furnish likenesses to be carried in procession at a funeral in the clan, and always when some member of it passed away the entire company of his house that had ever existed was present. The pedigrees too were traced in a spread of lines running near the several painted portraits. The archive-rooms [b] were kept filled with books of records and with written memorials of official careers. Outside the houses and round the doorways there were other presentations of those mighty spirits, with spoils taken from the enemy fastened to them, which even one who bought the house was not permitted to unfasten, and the mansions eternally celebrated a triumph even though they changed their masters. This acted as a mighty incentive, when every day the very walls reproached an unwarlike owner with intruding on the triumphs of another! There is extant an indignant speech by the pleader Messala protesting against the insertion among the likenesses of his family of a bust not belonging to them but to the family of the Laevini.[c] A similar reason extracted from old Messala the volumes he composed 'On Families,' because when passing through the hall of Scipio Pomponianus he had observed the Salvittones [d]—that was their

[d] Probably, like the Scipios, a branch of the Gens Cornelia.

men—Africanorum dedecori inrepentes Scipionum
nomini. sed—pace Messalarum dixisse liceat—
etiam mentiri clarorum imagines erat aliquis vir-
tutum amor multoque honestius quam mereri, ne
quis suas expeteret.

9 Non est praetereundum et novicium inventum,
siquidem non [1] ex auro argentove, at [2] certe ex aere
in bibliothecis dicantur illis, quorum immortales
animae in locis iisdem loquuntur, quin immo etiam
quae non sunt finguntur, pariuntque desideria non

10 traditos vultus, sicut in Homero evenit.[3] utique [4]
maius, ut equidem arbitror, nullum est felicitatis
specimen quam semper omnes scire cupere, qualis
fuerit aliquis. Asini Pollionis hoc Romae inventum,
qui primus bibliothecam dicando ingenia hominum
rem publicam fecit. an priores coeperint Alexandreae et Pergami reges, qui bibliothecas magno

11 certamine instituere, non facile dixerim. imaginum
amorem [5] flagrasse quondam [6] testes sunt Atticus
ille Ciceronis edito de iis volumine, M. Varro benig-
nissimo invento insertis voluminum suorum fecundi-

---

[1] nunc *Ian, Urlichs* : icones *Detlefsen* : non solum *edd. vett.*
[2] at *Mayhoff* : aut.
[3] devenit *B* : id evenit *Ian.*
[4] utique *coni. Mayhoff* : quo.
[5] amorem *B* : amore *rell.*
[6] (amore . . .) quosdam *coni. Brotier.*

---

[a] Scipio Africanus, victor over Hannibal in 202 B.C., and
Scipio Aemilianus Africanus, who destroyed Carthage in
146 B.C.

former surname—in consequence of an act of adoption by will creeping into others' preserves, to the discredit of the Scipios called Africanus.[a] But the Messala family must excuse me if I say that even to lay a false claim to the portraits of famous men showed some love for their virtues, and was much more honourable than to entail by one's conduct that nobody should seek to obtain one's own portraits!

We must not pass over a novelty that has also been invented, in that likenesses made, if not of gold or silver, yet at all events of bronze are set up in the libraries in honour of those whose immortal spirits speak to us in the same places, nay more, even imaginary likenesses are modelled and our affection gives birth to countenances that have not been handed down to us, as occurs in the case of Homer. At any rate in my view at all events there is no greater kind of happiness than that all people for all time should desire to know what kind of a man a person was. At Rome this practice originated with Asinius Pollio, who first by founding a library made works of genius the property of the public. Whether this practice began earlier, with the Kings of Alexandria and of Pergamum,[b] between whom there had been such a keen competition in founding libraries, I cannot readily say. The existence of a strong passion for portraits in former days is evidenced by Atticus the friend of Cicero in the volume he published on the subject and by the most benevolent invention of Marcus Varro, who actually by some means inserted in a prolific output of

Portrait-statues in libraries.

After 39 B.C.

[b] Ptolemy I of Egypt (died 283 B.C.) and Attalus I of Pergamum (241–197 B.C.) both founded libraries. Two at Alexandria became famous under Ptolemies II and III.

tati etiam [1] septingentorum inlustrium aliquo modo imaginibus, non passus intercidere figuras aut vetustatem aevi contra homines valere, inventor muneris etiam dis invidiosi, quando immortalitatem non solum dedit, verum etiam in omnes terras misit, ut praesentes esse ubique ceu [2] di possent. et hoc quidem alienis ille praestitit.

12 III. Verum clupeos in sacro vel publico dicare privatim primus instituit, ut reperio, Appius Claudius qui consul cum P. Servilio fuit anno urbis CCLVIIII. posuit enim in Bellonae aede maiores suos, placuitque in excelso spectari et [3] titulos honorum legi, decora res, utique si liberum turba parvulis imaginibus ceu nidum aliquem subolis pariter ostendat, quales clupeos nemo non gaudens favensque aspicit.

13 IV. post eum M. Aemilius collega in consulatu Quinti Lutatii non in basilica modo Aemilia, verum et domi suae posuit, id quoque Martio exemplo. scutis enim, qualibus apud Troiam pugnatum est, continebantur imagines, unde et nomen habuere clupeorum, non, ut perversa grammaticorum suptilitas voluit, a cluendo. origo plena virtutis,

---

[1] fecunditati etiam *Mayhoff* : f. hominum *Detlefsen* : fecunditatium B^1, *cd. Leid. Voss., cd. Flor. Ricc.* : fecunditati B^2 : fecunditantium non nominibus tantum *cd. Par. Lat.* 6797.
[2] ceu di *M. Hertz, Urlichs* : cludi.
[3] et *coni. Warmington* : in (*recte?*).

---

[a] *E.g.* the shield of Achilles, *Iliad* XVIII, 478 ff., and the shield of Aeneas, *Aeneid* VIII, 625 ff.

volumes portraits of seven hundred famous people, not allowing their likenesses to disappear or the lapse of ages to prevail against immortality in men. Herein Varro was the inventor of a benefit that even the gods might envy, since he not only bestowed immortality but despatched it all over the world, enabling his subjects to be ubiquitous, like the gods. This was a service Varro rendered to strangers.

III. But the first person to institute the custom of privately dedicating the shields with portraits in a temple or public place, I find, was Appius Claudius, the consul with Publius Servilius in the 259th year of the city. He set up his ancestors in 495 B.C. the shrine of the Goddess of War, and desired them to be in full view on an elevated spot, and the inscriptions stating their honours to be read. This is a seemly device, especially if miniature likenesses of a swarm of children at the sides display a sort of brood of nestlings; shields of this description everybody views with pleasure and approval. IV. After him Marcus Aemilius, Quintus Lutatius's colleague 78. B.C. in the consulship, set up portrait-shields not only in the Basilica Aemilia but also in his own home, and in doing this he was following a truly warlike example; for the shields which contained the likenesses resembled those[a] employed in the fighting at Troy; and this indeed gave them their name of *clupei*,[b] which is not derived from the word meaning 'to be celebrated,' as the misguided ingenuity of scholars has made out. It is a copious inspiration of valour for there to be a representation on a shield of the

---

[b] Pliny means that *clupeus* is derived from γλύφω, to carve or emboss, not from the old Latin *cluo* or *clueo*, to be reputed famous.

faciem reddi in scuto eius [1] qui fuerit usus illo.
14 Poeni ex auro factitavere et clupeos et imagines
secumque vexere. in castris [2] certe captis talem
Hasdrubalis invenit Marcius, Scipionum in Hispania
ultor, isque clupeus supra fores Capitolinae aedis
usque ad incendium primum fuit. maiorum quidem
nostrorum tanta securitas in ea re adnotatur, ut
L. Manlio Q. Fulvio cos. anno urbis DLXXV M.
Aufidius tutelae Capitolii [3] redemptor docuerit
patres argenteos esse clupeos, qui pro aereis [4] per
aliquot iam lustra adsignabantur.
15 V. De picturae initiis incerta nec instituti operis
quaestio est. Aegyptii sex milibus annorum aput
ipsos inventam, priusquam in Graeciam transiret,
adfirmant, vana praedicatione, ut palam est; Graeci
autem alii Sicyone, alii aput Corinthios repertam,
omnes umbra hominis lineis circumducta, itaque
primam talem, secundam singulis coloribus et
monochromaton [5] dictam, postquam operosior in-
16 venta erat, duratque talis etiam nunc. inventam
liniarem a Philocle Aegyptio vel Cleanthe Corinthio

[1] eius *Detlefsen* : eiusque *B*[2] : cuiusque.
[2] vexere. in castris *Mayhoff* : in castris vexere *B* : i.c.
tulere *cd. Par.* 6801 : i.c. venere *rell.*
[3] Capitolio *B* : capitoli *cd. Leid. Voss.*
[4] aureis *Detlefsen.*
[5] e monochromato *B* : et monochromaton *rell.*

[a] Publius and Gnaeus Scipio were destroyed in Spain by the
Carthaginians, 212–211 B.C. L. Marcius and T. Fonteius
prevented further disasters.
[b] A conjectural alteration gives ' gold.'

countenance of him who once used it. The Carthaginians habitually made both shields and statues of gold, and carried these with them: at all events Marcius, who took vengeance for the Scipios in Spain,[a] found a shield of this kind that belonged to Hasdrubal, in that general's camp when he captured it, and this shield was hung above the portals of the temple on the Capitol till the first fire. Indeed it is 83 B.C. noticed that our ancestors felt so little anxiety about this matter that in the 575th year of the city, 179 B.C. when the consuls were Lucius Manlius and Quintus Fulvius, the person who contracted for the safety of the Capitol, Marcus Aufidius, informed the Senate that the shields which for a good many censorship periods past had been scheduled as made of bronze [b] were really silver.

V. The question as to the origin of the art of *Origins of painting is uncertain[c] and it does not belong to the *painting.* plan of this work. The Egyptians declare that it was invented among themselves six thousand years ago before it passed over into Greece—which is clearly an idle assertion. As to the Greeks, some of them say it was discovered at Sicyon, others in Corinth, but all agree that it began with tracing an outline round a man's shadow [d] and consequently that pictures were originally done in this way, but the second stage when a more elaborate method had been invented was done in a single colour and called monochrome,[e] a method still in use at the present day. Line-drawing was invented by the Egyptian Philocles or by the Corinthian Cleanthes,

---

[c] Cf. VII, 205. A. Rumpf, *Journ. of Hellenic St.* LXVII, 10 ff.

[d] But study of extant ancient art refutes this idea.

[e] See §§ 29, 56.

primi exercuere Aridices [1] Corinthius et Telephanes
Sicyonius, sine ullo etiamnum hi colore, iam tamen
spargentes linias intus. ideo et quos pinxere [2]
adscribere institutum. primus inlevit [3] eas colore [4]
testae, ut ferunt, tritae [5] Ecphantus [6] Corinthius.
hunc eodem nomine alium [7] fuisse quam [8] tradit
Cornelius Nepos secutum in Italiam Damaratum,
Tarquinii Prisci regis Romani patrem, fugientem a
Corintho tyranni iniurias Cypseli, mox docebimus.

17 VI. Iam enim absoluta erat pictura etiam in
Italia. exstant certe hodieque antiquiores urbe
picturae Ardeae in aedibus sacris, quibus equidem
nullas aeque miror, tam longo aevo durantes [9]
in orbitate tecti veluti recentes.[9] similiter Lanuvi,
ubi Atalante et Helena comminus pictae sunt
nudae ab eodem artifice, utraque excellentissima
forma, sed altera ut virgo, ne ruinis quidem templi
18 concussae. Gaius princeps tollere eas conatus est
libidine accensus, si tectorii natura permisisset.
durant et Caere antiquiores et ipsae, fatebiturque
quisquis eas diligenter aestimaverit nullam artium
celerius consummatam, cum Iliacis temporibus non
fuisse eam appareat.

[1] Aridices *Sillig, Keil* : aradices *B* : ardices *rell.*
[2] pinxere *Mayhoff* : pingerent *edd. vett.* : pingere.
[3] inlevit *Urlichs* : invenit.
[4] colore *B* : colores *rell.* : colorare *Gelen* : colorare colore
Ian.
[5] tritae *Sillig* : triste *B¹* : triste *B²* : ita *aut* it *rell.*
[6] Ecphantus *Sillig, O. Jahn* : ephantus *B* : elephantus *rell.*
[7] eundem nomine alio *Schultz.*
[8] *V.l.* quem : quam quem *Sillig, Ian.*
[9] *V.l.* durantis . . . recentis : durantibus . . . recentibus
coni. Sillig.

but it was first practised by the Corinthian Aridices
and the Sicyonian Telephanes—these were at that
stage not using any colour, yet already adding
lines here and there to the interior of the outlines;
hence it became their custom to write on the pictures
the names of the persons represented. Ecphantus
of Corinth is said to have been the first to daub these
drawings with a pigment made of powdered earthenware. We shall show below that this was another §152.
person, bearing the same name, not the one recorded
by Cornelius Nepos to have followed into Italy
Demaratus the father of the Roman king Tarquinius *trad.* 616–
Priscus when he fled from Corinth to escape the 578 B.C.
violence of the tyrant Cypselus.

VI. For the art of painting had already been *Early*
brought to perfection even in Italy. At all events *Italian*
there survive even to-day in the temples at Ardea *painting.*
paintings that are older than the city of Rome,
which to me at all events are incomparably remarkable, surviving for so long a period as though freshly
painted, although unprotected by a roof. Similarly
at Lanuvium, where there are an Atalanta and a
Helena close together, nude figures, painted by the
same artist, each of outstanding beauty (the former
shown as a virgin), and not damaged even by the
collapse of the temple. The Emperor Caligula from A.D. 37.
lustful motives attempted to remove them, but the
consistency of the plaster would not allow this to be
done. There are pictures surviving at Caere that
are even older. And whoever carefully judges
these works will admit that none of the arts reached
full perfection more quickly, inasmuch as it is
clear that painting did not exist in the Trojan
period.

19   VII. Apud Romanos quoque honos mature huic arti contigit, siquidem cognomina ex ea Pictorum traxerunt Fabii clarissimae gentis, princepsque eius cognominis ipse aedem Salutis pinxit anno urbis conditae CCCCL, quae pictura duravit ad nostram memoriam aede ea Claudi principatu exusta. proxime celebrata est in foro boario aede Herculis Pacui poetae pictura. Enni sorore genitus hic fuit clarioremque artem eam Romae fecit gloria scaenae.

20 postea non est spectata honestis manibus, nisi forte quis Turpilium equitem Romanum e Venetia nostrae aetatis velit referre, pulchris eius operibus hodieque Veronae exstantibus. laeva is manu pinxit, quod de nullo ante memoratur. parvis [1] gloriabatur tabellis extinctus nuper in longa senecta Titedius [2] Labeo praetorius, etiam proconsulatu provinciae Narbonensis functus, sed ea re inrisa [3] etiam contumeliae erat. fuit et principum virorum non

21 omittendum de pictura celebre consilium, cum Q. Pedius, nepos Q. Pedii consularis triumphalisque et a Caesare dictatore coheredis Augusto dati, natura mutus esset. in eo Messala orator, ex cuius familia pueri avia fuerat, picturam docendum

---

[1] paruisse *B* : parvis ipse *coni. Mayhoff.*

[2] Titedius *B* : sit edius *rell.* : Titidius *Sillig* : Antistius *Hardouin* : Aterius *edd. vett.*

[3] inrisa *Mayhoff* : inlisa *B* : in risu *rell.* (irrisu *cd. Par, Lat.* 6797).

---

[a] Roman writer of tragedies, *c.* 220–*c.* 130 B.C.

[b] Famous Roman epic and dramatic poet, 239–169 B.C.

VII. In Rome also honour was fully attained by *Early Roman painting and painters.* this art at an early date, inasmuch as a very distinguished clan of the Fabii derived from it their surname of Pictor, 'Painter,' and the first holder of the name himself painted the Temple of Health in the year 450 from the foundation of the City : the work 304 B.C. survived down to our own period, when the temple was destroyed by fire in the principate of Claudius. Next in celebrity was a painting by the poet Pacuvius *a* in the temple of Hercules in the Cattle Market. Pacuvius was the son of a sister of Ennius,*b* and he added distinction to the art of painting at Rome by reason of his fame as a playwright. After Pacuvius, painting was not esteemed as handiwork for persons of station, unless one chooses to recall a knight of Rome named Turpilius, from Venetia, in our own generation, because of his beautiful works still surviving at Verona. Turpilius painted with his left hand, a thing recorded of no preceding artist. Titedius Labeo, a man of praetorian rank who had actually held the office of Proconsul of the Province of Narbonne, and who died lately in extreme old age, used to be proud of his miniatures, but this was laughed at and actually damaged his reputation. There was also a celebrated debate on the subject of painting held between some men of eminence which must not be omitted, when the former consul and winner of a triumph Quintus Pedius, who was appointed by the Dictator 49–44 .B.C. Caesar as his joint heir with Augustus, had a grandson Quintus Pedius who was born dumb; in this debate the orator Messala, of whose family the boy's grandmother had been a member, gave the advice that the boy should have lessons in painting, and

censuit, idque etiam divus Augustus comprobavit;
22 puer magni profectus in ea arte obiit. dignatio
autem praecipua Romae increvit, ut existimo, a
M'. Valerio Maximo Messala, qui princeps tabulam
[picturam] [1] proelii, quo Carthaginienses et Hieronem
in Sicilia vicerat, proposuit in latere curiae Hostiliae
anno ab urbe condita ccccxc. fecit hoc idem et
L. Scipio tabulamque victoriae suae Asiaticae in
Capitolio posuit, idque aegre tulisse fratrem Afri-
canum tradunt, haut inmerito, quando filius eius
23 illo proelio captus fuerat. non dissimilem offensio-
nem et Aemiliani subiit L. Hostilius [2] Mancinus,
qui primus Carthaginem inruperat, situm eius
oppugnationesque depictas proponendo in foro et
ipse adsistens populo spectanti singula enarrando,
qua comitate proximis comitiis consulatum adeptus
est. habuit et scaena ludis Claudii Pulchri magnam
admirationem picturae, cum ad tegularum simil-
tudinem corvi decepti imagine [3] advolarent.
24 VIII. Tabulis autem externis auctoritatem
Romae publice fecit primus omnium L. Mummius,
cui cognomen Achiaci victoria dedit. namque cum
in praeda vendenda rex Attalus [4] X⌐VI⌐ [5] emisset
tabulam Aristidis, Liberum patrem, pretium miratus

---

[1] *Mayhoff*: picturā *B*: pictam *rell.*
[2] hostili·m. S. *B*: hostilius M.f. *coni. Ian.*
[3] *V.l.* imaginem: imagini *Ian.*
[4] rex attalus distraxisset et *cd. Par.* 6801.
[5] X⌐VI⌐ *Detlefsen*: x.V̄Ī *Hardouin*: xvi *aut* xiii *cdd.*

---

[a] Over Antiochus III in 190 B.C.
[b] Mancinus commanded the Roman fleet in the Third
Punic War when Carthage was taken and destroyed by
Scipio Aemilianus in 146 B.C.
[c] Over the Greeks in 146 B.C., when Mummius destroyed
Corinth.

his late lamented Majesty Augustus also approved
of the plan. The child made great progress in the
art, but died before he grew up. But painting
chiefly derived its rise to esteem at Rome, in my
judgement, from Manius Valerius Maximus Messala,
who in the year 490 after the foundation of the 264 B.C.
city first showed a picture in public on a side wall
of the Curia Hostilia: the subject being the battle
in Sicily in which he had defeated the Carthaginians
and Hiero. The same thing was also done by
Lucius Scipio, who put up in the Capitol a picture of
his Asiatic victory [a]: this is said to have annoyed
his brother Africanus, not without reason, as his son
had been taken prisoner in that battle. Also Lucius
Hostilius Mancinus [b] who had been the first to
force an entrance into Carthage incurred a very
similar offence with Aemilianus by displaying in the
forum a picture of the plan of the city and of the
attacks upon it and by himself standing by it and
describing to the public looking on the details of the
siege, a piece of popularity-hunting which won him
the consulship at the next election. Also the stage
erected for the shows given by Claudius Pulcher 99 B.C.
won great admiration for its painting, as crows were
seen trying to alight on the roof tiles represented
on the scenery, quite taken in by its realism.

VIII. The high esteem attached officially to *Foreign*
foreign paintings at Rome originated from Lucius *paintings in*
Mummius who from his victory [c] received the *Rome.*
surname of Achaicus. At the sale of booty captured *L. Mum-*
King Attalus [d] bought for 600,000 denarii a picture *mius.*
of Father Liber or Dionysus by Aristides, but the

---

[d] Attalus II of Pergamum, 159–138 B.C.

suspicatusque aliquid in ea virtutis, quod ipse
nesciret, revocavit tabulam, Attalo multum que-
rente, et in Cereris delubro posuit, quam primam
arbitror picturam externam Romae publicatam.
25 deinde video et in foro positas volgo. hinc enim
ille Crassi oratoris lepos agentis sub Veteribus;
cum testis compellatus instaret : dic ergo, Crasse,
qualem me noris ? talem, inquit, ostendens in
tabula inficetissime Gallum exerentem linguam.
in foro fuit et illa pastoris senis cum baculo, de qua
Teutonorum legatus respondit interrogatus, quan-
tine[1] eum aestimaret, donari sibi nolle talem vivum
verumque.

26   IX. Sed praecipuam auctoritatem publice tabulis
fecit Caesar dictator Aiace et Media ante Veneris
Genetricis aedem dicatis, post eum M. Agrippa, vir
rusticitati propior quam deliciis. exstat certe eius
oratio magnifica et maximo civium digna de tabulis
omnibus signisque publicandis, quod fieri satius
fuisset quam in villarum exilia pelli. verum eadem
illa torvitas tabulas duas Aiacis et Veneris mercata
est a Cyzicenis HS[2] |XII| ;[3] in thermarum quoque

[1] quanti *cd. Par.* 6801.
[2] HS *Gelen :* h 7777 *B[1] :* hīs *B[2] : om. rell.*
[3] |XII| *Ian : alii alia :* XII *B :* XIII *rell.*

---

a With regard to this story : (i) there was no auction of
pictures; Mummius took to Rome the most valuable and
handed over the rest to Philopoemen. (ii) Attalus was not
present at Corinth (where this scene occurred). When the
Roman soldiers were using the pictures as dice-boards, Philopoemen offered M. 100 talents if he should assign Aristides'
picture to Attalus' share (Paus. VII, 16, 1 ; 8 ; Strabo VIII,
4. 23 = 381).

price surprised Mummius, who suspecting there
must be some merit in the picture of which he was
himself unaware had the picture called back, in
spite of Attalus's strong protests, and placed it in
the Shrine of Ceres: the first instance, I believe, of a
foreign picture becoming state-property at Rome.[a]
After this I see that they were commonly placed
even in the forum: to this is due the famous witticism
of the pleader Crassus, when appearing in a case
Below The Old Shops; a witness called kept asking
him: ' Now tell me, Crassus, what sort of a person
do you take me to be?' 'That sort of a person,'
said Crassus, pointing to a picture of a Gaul putting
out his tongue [b] in a very unbecoming fashion. It
was also in the forum that there was the picture of
the Old Shepherd with his Staff, about which the
Teuton envoy when asked what he thought was the
value of it said that he would rather not have even
the living original as a gift!

IX. But it was the Dictator Caesar who gave *Caesar.*
outstanding public importance to pictures by dedicating paintings of Ajax and Medea in front of the 46 B.C.
temple of Venus Genetrix; and after him Marcus 63–12 B.C.
Agrippa, a man who stood nearer to rustic simplicity
than to refinements. At all events there is preserved
a speech of Agrippa, lofty in tone and worthy of the
greatest of the citizens, on the question of making
all pictures and statues national property, a procedure
which would have been preferable to banishing
them to country houses. However, that same
severe spirit paid the city of Cyzicus 1,200,000
sesterces for two pictures, an Ajax and an Aphrodite;
he had also had small paintings let into the marble

[b] Not apparently as in insult but as an averting act.

calidissima parte marmoribus incluserat parvas tabellas, paulo ante, cum reficerentur, sublatas.

27 X. Super omnes divus Augustus in foro suo cele-berrima in parte posuit tabulas duas, quae Belli faciem pictam habent et Triumphum, item Castores ac Victoriam. posuit et quas dicemus sub artificum mentione in templo Caesaris patris. idem in curia quoque, quam in comitio consecrabat, duas tabulas inpressit parieti. Nemean sedentem supra leonem, palmigeram ipsam, adstante[1] cum baculo sene, cuius supra caput tabella bigae dependet, Nicias scripsit[2] se inussisse; tali enim usus est verbo.

28 alterius tabulae admiratio est puberem filium seni patri similem esse aetatis salva differentia, super-volante aquila draconem complexa; Philochares hoc suum opus esse testatus est, inmensa, vel unam si tantum hanc tabulam aliquis aestimet, potentia artis, cum propter Philocharen ignobilissimos alioqui Glaucionem filiumque eius Aristippum senatus populi Romani tot saeculis spectet! posuit et Tiberius Caesar, minime comis imperator, in templo ipsius Augusti quas mox indicabimus. hactenus dictum sit de dignitate artis morientis.

29 XI. Quibus coloribus singulis primi pinxissent diximus, cum de iis pigmentis traderemus in metallis,

---

[1] adstante *edd. vett.*: adstantem.
[2] asscripsit *coni. Mayhoff.*

---

[a] Castor and Pollux (Polydeuces).
[b] Julius Caesar who had adopted Augustus.
[c] The Nemean forest (personified) where Heracles killed the Nemean lion.
[d] See pp. 356–9.

even in the warmest part of his Hot Baths; which were removed a short time ago when the Baths were being repaired.

X. His late lamented Majesty Augustus went *Augustus* beyond all others, in placing two pictures in the most *and Tiberius.* frequented part of his forum, one with a likeness of War and Triumph, and one with the Castors [a] and Victory. He also erected in the Temple of his father Caesar [b] pictures we shall specify in giving the names of artists. § 91. He likewise let into a wall in the curia which he was dedicating in the comitium: a Nemea [c] seated on a 29 B.C. lion, holding a palm-branch in her hand, and standing at her side an old man leaning on a stick and with a picture of a two-horse chariot hung up over his head, on which there was an inscription saying that it was an ' encaustic ' design—such is the term which he employed—by Nicias. [d] The second picture is remarkable for displaying the close family likeness between a son in the prime of life and an elderly father, allowing for the difference of age: above them soars an eagle with a snake in its claws; Philochares has stated this work to be by him showing the immeasurable power exercised by art if one merely considers this picture alone, inasmuch as thanks to Philochares two otherwise quite obscure persons Glaucio and his son Aristippus after all these centuries have passed still stand in the view of the senate of the Roman nation! The most ungracious emperor Tiberius also placed pictures in the temple A.D. 14–37. of Augustus himself which we shall soon mention. § 131. Thus much for the dignity of this now expiring art.

XI. We stated what were the various single *Painters'* colours used by the first painters when we were *colours.* discussing while on the subject of metals the pig- XXXIII, 117.

quae[1] monochromata a[2] genere[3] picturae vocantur.[4] qui deinde et quae invenerint et quibus temporibus, dicemus in mentione artificum, quoniam indicare naturas colorum prior causa operis instituti est. tandem se ars ipsa distinxit et invenit lumen atque umbras, differentia colorum alterna vice sese excitante. postea deinde adiectus est splendor, alius hic quam lumen. quod inter haec et umbras esset, appellarunt tonon, commissuras vero colorum et transitus harmogen.

30 XII. Sunt autem colores austeri aut floridi. utrumque natura aut mixtura evenit. floridi sunt —quos dominus pingenti[5] praestat—minium, Armenium, cinnabaris, chrysocolla, Indicum, purpurissum; ceteri austeri. ex omnibus alii nascuntur, alii fiunt. nascuntur Sinopis, rubrica, Paraetonium, Melinum, Eretria, auripigmentum; ceteri finguntur, primumque quos in metallis diximus, praeterea e vilioribus ochra, cerussa usta, sandaraca, sandyx, Syricum, atramentum.

---

[1] quae *Sillig* : qui *cdd.* : *del. Littré.*
[2] monochromata a *Mayhoff* : m. ea *Littré* : monochromatea *cd. Par.* 6801 *ut videtur* : mox negrammatae a *B* : mox neogrammatea *rell.*
[3] genere *Mayhoff* : genera.
[4] vocantur *B* : vocaverunt *rell.*
[5] pingenti *ed. Basil.* : fingenti.

---

[a] Study of ancient art does *not* show that painting started with the use of single colours.

[b] The Greek term ἁρμογή means adjustment of parts.

[c] *Minium.* See § 33 (note) and XXXIII, 111–123.

[d] A rich blue colour (from Armenia), the modern azurite. See also § 47.

[e] *Cinnabaris* here in Pliny. See XXXIII, 115–116.

[f] Our ' malachite.'

[g] Earth stained with Tyrian purple.

ments called monochromes from the class of painting
for which they are used. Subsequent [a] inventions
and their authors and dates we shall specify in §§ 353 *sqq*
enumerating the artists, because a prior motive
for the work now in hand is to indicate the nature of
colours. Eventually art differentiated itself, and
discovered light and shade, contrast of colours
heightening their effect reciprocally. Then came
the final adjunct of shine, quite a different thing from
light. The opposition between shine and light on the
one hand and shade on the other was called contrast,
while the juxtaposition of colours and their passage
one into another was termed attunement.[b]

XII. Some colours are sombre and some brilliant,
the difference being due to the nature of the substances or to their mixture. The brilliant colours,
which the patron supplies at his own expense to the
painter, are cinnabar,[c] Armenium,[d] dragon's blood,[e]
gold-solder,[f] indigo, bright purple[g]; the rest are
sombre. Of the whole list some are natural colours
and some artificial. Natural colours are sinopis,[h]
ruddle, Paraetonium,[i] Melinum,[j] Eretrian earth[k] and
orpiment; all the rest are artificial, and first of all
those which we specified among minerals, and XXX, 111,
moreover among the commoner kinds yellow ochre, 158, etc.
burnt lead acetate, realgar, sandyx,[l] Syrian colour[m]
and black.[n]

[h] A brown-red ochre or red oxide of iron from Sinope.
[i] From a white chalk or calcium carbonate, and perhaps
also steatite, of Paraetonium in N. Africa; see note [a] on § 36.
[j] A white marl from Melos.
[k] From Eretria in Euboea; perhaps magnesite.
[l] Mixed oxide of lead and oxide of iron.
[m] See § 40.
[n] See XXXIV, 112, 123.

31 XIII. Sinopis inventa primum in Ponto est; inde nomen a Sinope urbe. nascitur et in Aegypto, Baliaribus, Africa, sed optima in Lemno et in Cappadocia, effossa e speluncis. pars, quae saxis adhaesit, excellit. glaebis suus colos, extra maculosus. hac usi sunt veteres ad splendorem. species Sinopidis tres: rubra et minus rubens atque inter has media. pretium optimae X ii, usus ad penicillum aut si
32 lignum colorare libeat; eius, quae ex Africa venit, octoni asses—cicerculum appellant; magis ceteris rubet, utilior abacis. idem pretium et eius, quae pressior vocatur, et est maxime fusca. usus ad bases abacorum, in medicina vero blandus ⟨. . . pastillis . . .⟩[1] emplastrisque et malagmatis, sive sicca compositione sive liquida facilis, contra ulcera in umore sita, velut oris, sedis. alvum sistit infusa, feminarum profluvia pota denarii pondere. eadem adusta siccat scabritias oculorum, e vino maxime.
33 XIV. Rubricae genus in ea voluere intellegi quidam secundae auctoritatis, palmam enim Lemniae dabant. minio proxima haec est, multum antiquis celebrata cum insula, in qua nascitur. nec nisi signata venumdabatur, unde et sphragidem appel-

---

[1] *Mayhoff.*

---

[a] See note [h], p. 283.
[b] Dark brownish.
[c] This generally is the proper meaning of *minium* except when it is called *m. secundarium* = red lead. See XXXIII, 111–123.

XIII. Sinopis [a] was first discovered in Pontus, *Ochre of Sinope.* and hence takes its name from the city of Sinope. It is also produced in Egypt, the Balearic Islands and Africa, but the best is what is extracted from the caverns of Lemnos and Cappadocia, the part found adhering to the rock being rated highest. The lumps of it are self-coloured, but speckled on the outside. It was employed in old times to give a glow. There are three kinds of Sinopis, the red, the faintly red and the intermediate. The price of the best is 2 denarii a pound : this is used for painting with a brush or else for colouring wood ; the kind imported from Africa costs 8 *as*-pieces a pound, and is called chickpea colour [b] ; it is of a deeper red than the other kinds, and more useful for panels. The same price is charged for the kind called 'low toned' which is of a very dusky colour. It is employed for the lower parts of panelling ; but used as a drug it has a soothing effect in ⟨lozenges and⟩ plasters and poultices, mixing easily either dry or moistened, as a remedy for ulcers in the humid parts of the body such as the mouth and the anus. Used in an enema it arrests diarrhoea, and taken through the mouth in doses of one denarius weight it checks menstruation. Applied in a burnt state, particularly with wine, it dries roughnesses of the eyes.

XIV. Some persons have wished to make out that Sinopis only consists in a kind of red-ochre of inferior quality, as they gave the palm to the red *Ochre of Lemnos.* ochre of Lemnos. This last approximates very closely to cinnabar,[c] and it was very famous in old days, together with the island that produces it ; it used only to be sold in sealed packages, from which it got the name of 'seal red-ochre.' It is used to

34 lavere. hac minium sublinunt adulterantque. in
medicina praeclara res habetur. epiphoras enim
oculorum mitigat ac dolores circumlita et aegilopia
manare prohibet, sanguinem reicientibus ex aceto
datur bibenda. bibitur et contra lienem reniumque
vitia et purgationes feminarum, item et contra
venena et serpentium ictus terrestrium marinorum-
que, omnibus ideo antidotis familiaris.

35 XV. E reliquis rubricae generibus fabris utilissima Aegyptia et Africana, quoniam maxime sor-
bentur tectoriis.[1] rubrica[2] autem nascitur et in
ferrariis metallis. XVI. Ea et fit ochra[3] exusta in
ollis novis luto circumlitis. quo magis arsit in
caminis, hoc melior. omnis autem rubrica siccat
ideoque ex[4] emplastris conveniet[5] igni etiam sacro.

36 XVII. Sinopidis Ponticae selibrae silis lucidi libris x
et Melini Graecensis ii mixtis tritisque una per dies
duodenos[6] leucophorum fit. hoc est glutinum auri,
cum inducitur ligno.

XVIII. Paraetonium loci nomen habet ex Aegypto. spumam maris esse dicunt solidatam cum
limo, et ideo conchae minutae inveniuntur in eo.
fit et in Creta insula atque Cyrenis. adulteratur
Romae creta Cimolia decocta conspissataque. pretium optimo in pondo vi X l. e candidis coloribus

---

[1] tectoriis *Mayhoff* : picturis.
[2] rubrica *hic Mayhoff* : *infra post* exusta.
[3] ea et fit ochra *Mayhoff* : ex ea fit ochra *aut* ochra ex ea fit.
[4] ex *aut* et cdd. (*om. B*).
[5] conveniet *vel* conveniat *Mayhoff* : convenit et.
[6] duodenos *Mayhoff* : duodenis *B* : xii *rell.*

---

[a] Marsa Labeit in N. Africa, between Egypt and the Syrtes.
Cf. n. on § 30.
[b] Cf. XXXV, 195 ff.

supply an undercoating to cinnabar and also for adulterating cinnabar. In medicine it is a substance ranked very highly. Used as a liniment round the eyes it relieves defluxions and pains, and checks the discharge from eye-tumours; it is given in vinegar as a draught in cases of vomiting or spitting blood. It is also taken as a draught for troubles of the spleen and kidneys and for excessive menstruation; and likewise as a remedy for poisons and snake bites and the sting of sea serpents; hence it is in common use for all antidotes.

XV. Among the remaining kinds of red ochre the *Other ochres.* most useful for builders are the Egyptian and the African varieties, as they are most thoroughly absorbed by plaster. Red ochre is also found in a native state in iron mines. XVI. It is also manufactured by burning ochre in new earthen pots lined with clay. The more completely it is calcined in the furnaces the better its quality. All kinds of red ochre have a drying property, and consequently will be found suitable in plasters even for erysipelas.

XVII. Half a pound of sinopis from Pontus, ten pounds of bright yellow ochre and two pounds of Greek earth of Melos mixed together and pounded up for twelve successive days make 'leucophorum,' *Cf.* XXXIII, a cement used in applying gold-leaf to wood. 64.

XVIII. Paraetonium is called after the place [a] *White pigments.* of that name in Egypt. It is said to be sea-foam hardened with mud, and this is why tiny shells are found in it. It also occurs in the island of Crete and in Cyrene. At Rome it is adulterated with Cimolian clay [b] which has been boiled and thickened. The price of the best quality is 50 denarii per 6 lbs. It is the most greasy of all the white colours and makes

287

pinguissimum et tectoriis [1] tenacissimum propter
levorem.

37 XIX. Melinum candidum et ipsum est, optimum
in Melo insula. in Samo quoque [2] nascitur; eo
non utuntur pictores propter nimiam pinguitudinem;
accubantes effodiunt ibi inter saxa venam scrutantes.
in medicina eundem usum habet quem Eretria
creta; praeterea linguam tactu siccat, pilos detrahit
smectica vi.[3] pretium in libras sestertii singuli.

Tertius e candidis colos est cerussa, cuius rationem
in plumbi metallis diximus. fuit et terra per se in
Theodoti fundo inventa Zmyrnae, qua veteres ad
navium picturas utebantur. nunc omnis ex plumbo
et aceto fit, ut diximus.

38 XX. Usta casu reperta est in incendio Piraeei [4]
cerussa in urceis [5] cremata. hac primum usus est
Nicias supra dictus. optima nunc Asiatica habetur,
quae et purpurea appellatur. pretium eius in
libras X vi. fit et Romae cremato sile marmoroso
et restincto aceto. sine usta non fiunt umbrae.

XXI. Eretria terrae suae habet nomen. hac
Nicomachus et Parrhasius usi. refrigerat, emollit,
explet volnera; si coquatur, ad siccanda praecipitur,
utilis [6] et capitis doloribus et ad deprehendenda

---

[1] tectoriis *edd. vett.*: tectorii *Mayhoff* (*recte* ?): tectori *cdd.*
[2] quod *Mayhoff*.
[3] smectica vi *Urlichs*: metcica ut *cd. Flor. Ricc.*: meccica
ut *cd. Leid. Voss., cd. Par. Lat.* 6797 : metica ut *B*.
[4] Piraeei *Gelen*: pyrae *edd. vett*: pira et.
[5] urceis *B²*: urcis *B¹*: orcis *rell.*: hortis *edd. vett*.
[6] praecipitur, utilis *Mayhoff*: praecoquitur utilis *B*: utilis
praecipua *rell.*

---

[a] See note *j* on § 30.
[b] Perhaps lead carbonate, cerussite. From Vitruv. VII,
7, 4 we learn that it was green, perhaps because tinted with
copper salts.

the most tenacious for plasters because of its smoothness.

XIX. Melinum[a] also is a white colour, the best occurring in the island of Melos. It is found in Samos also, but the Samian is not used by painters, because it is excessively greasy. It is dug up in Samos by people lying on the ground and searching for a vein among the rocks. It has the same use in medicine as earth of Eretria; it also dries the tongue by contact, and acts as a depilatory, with a cleansing effect. It costs a sesterce a pound.

The third of the white pigments is ceruse or lead acetate, the nature of which we have stated in speaking of the ores of lead. There was also once a native ceruse earth[b] found on the estate of Theodotus at Smyrna, which was employed in old days for painting ships. At the present time all ceruse is manufactured from lead and vinegar, as we said.

XX. Burnt ceruse was discovered by accident, when some was burnt up in jars in a fire at Piraeus. It was first employed by Nicias above mentioned. Asiatic ceruse is now thought the best; it is also called purple ceruse and it costs 6 denarii per lb. It is also made at Rome by calcining yellow ochre which is as hard as marble and quenching it with vinegar. Burnt ceruse is indispensable for representing shadows.

XXI. Eretrian earth[c] is named from the country that produces it. It was employed by Nicomachus and Parrhasius. It has cooling and emollient effects and fills up wounds; if boiled it is prescribed as a desiccative, and is useful for pains in the head and for detecting internal suppurations, as these are

XXXIV, 175.

Ceruse, etc.

§ 327.

------

[c] See note [k] on § 30.

289

pura; subesse enim ea intellegunt, si ex aqua inlita
continuo [1] arescat.

39 XXII. Sandaracam et ochram Iuba tradidit in insula
Rubri maris Topazo nasci, sed inde non pervehuntur
ad nos. sandaraca quomodo fieret diximus. fit et
adulterina ex cerussa in fornace cocta. color esse
debet flammeus. pretium in libras asses quini.

40 XXIII. Haec si torreatur aequa parte rubrica
admixta, sandycem facit, quamquam animadverto
Vergilium existimasse herbam id esse illo versu:

Sponte sua sandyx pascentis vestiet [2] agnos.

pretium in libras dimidium eius quod sandaracae.
nec sunt alii colores maioris ponderis.

XXIV. Inter facticios est et Syricum, quo minium
sublini diximus. fit autem Sinopide et sandyce
mixtis.

41 XXV. Atramentum quoque inter facticios erit,
quamquam est et terrae,[3] geminae originis. aut
enim salsuginis modo emanat, aut terra ipsa sulpurei
coloris ad hoc probatur. inventi sunt pictores, qui
carbones infestatis [4] sepulchris effoderent.[5] inpor-
tuna haec omnia ac novicia. fit enim e fuligine
pluribus modis, resina vel pice exustis, propter

---

[1] inlita continuo *Mayhoff* : inlita non *cdd.* (inlinunt non *B*).
[2] vestiet *cd. Par.* 6801, *item Verg.* : vestiat *rell.*
[3] terrae *aut* terra *cdd.* : e terra *Madvig.*
[4] infestatis *Mayhoff qui et* infestantes sepulchra *coni.* :
infectant *aut* infectos *cdd.* : iniectos *coni. Sillig* : adfectarent
Detlefsen (sepulchris carbones infectos *cd. Par.* 6801).
[5] effoderent *cd. Tolet.*, *cd. Par.* 6801 : effodere *B* : infoderet
rell. (infoderent *cd. Par. Lat.* 6797).

---

[a] Zeboiget.
[b] Virg. *Ecl.* IV, 45 (*vestiet* Virg.). There is no proof that
Virgil did take sandyx to be a plant.

shown to be present if when it is applied with water it immediately dries up.

XXII. According to Juba sandarach or realgar and ochre are products of the island of Topazus [a] in the Red Sea, but they are not imported from those parts to us. We have stated the method of making sandarach. An adulterated sandarach is also made from ceruse boiled in a furnace. It ought to be flame-coloured. Its price is 5 asses per lb.

XXIII. If ceruse is mixed with red ochre in equal quantities and burnt, it produces sandyx or vermilion—though it is true that I observe Virgil held the view that sandyx is a plant, from the line:

> Sandyx self-grown shall clothe the pasturing lambs.[b]

Its cost per lb. is half that of sandarach. No other colours weigh heavier than these.

XXIV. Among the artificial colours is also Syrian colour, which as we said is used as an undercoating for cinnabar and red lead. It is made by mixing sinopis and sandyx together.

XXV. Black pigment will also be classed among the artificial colours, although it [c] is also derived from earth in two ways; it either exudes from the earth like the brine in salt pits, or actual earth of a sulphur colour is approved for the purpose. Painters have been known to dig up charred remains from graves thus violated to supply it. All these plans are troublesome and new-fangled; for black paint can be made in a variety of ways from the soot produced by burning resin or pitch, owing to which

(marginal notes: XXXIV, 177. / Black pigments.)

[c] For this mineral shoemaker's black, see XXXIV, 112, 123. The other blacks which follow are mostly composed of carbon.

quod etiam officinas aedificavere fumum eum non
emittentes. laudatissimum eodem modo fit e taedis.
adulteratur fornacium balinearumque fuligine quo
42 ad volumina scribenda utuntur. sunt qui et vini
faecem siccatam excoquant adfirmentque, si ex
bono vino faex ea¹ fuerit, Indici speciem id atramentum praebere. Polygnotus et Micon, celeberrimi pictores, Athenis e vinaceis fecere, tryginon
appellantes. Apelles commentus est ex ebore
43 combusto facere, quod elephantinum vocatur. ad-
portatur et Indicum ex India inexploratae adhuc
inventionis mihi. fit etiam aput infectores ex flore
nigro, qui adhaerescit aereis cortinis. fit et ligno e
taedis combusto tritisque in mortario carbonibus.
mira in hoc saepiarum natura, sed ex iis non fit.
omne autem atramentum sole perficitur, librarium
cumme,² tectorium glutino admixto. quod aceto
liquefactum est, aegre eluitur.
44 XXVI. E reliquis coloribus, quos a dominis dari
diximus propter magnitudinem pretii, ante omnes
est purpurissum. creta argentaria cum purpuris
pariter tinguitur bibitque eum colorem celerius
lanis. praecipuum est primum, fervente aheno

¹ faex ea *Mayhoff* : facta *B* : faex *rell.* (fex *cd. Leid. Voss.*).
² cumme *Sillig* : gummi *Gelen* : comme *cdd.* (me *B*¹ et *B*²).

ᵃ Probably the real indigo (§ 46) is meant here.
ᵇ Some unknown carbon pigment, not the indigo of § 46.
ᶜ Or : ' this latter variety is wonderfully like the pigment
of the cuttle-fish, but is never made from these creatures '
(K. C. Bailey).
ᵈ Polishing-powder of pure ground white earth.

factories have actually been built with no exit for
the smoke produced by this process. The most
esteemed black paint is obtained in the same way
from the wood of the pitch-pine. It is adulterated
by mixing it with the soot of furnaces and baths,
which is used as a material for writing. Some people
calcine dried wine-lees, and declare that if the lees
from a good wine are used this ink has the appearance
of Indian ink.[a] The very celebrated painters
Polygnotus and Micon at Athens made black paint
from the skins of grapes, and called it grape-lees ink.
Apelles invented the method of making black from
burnt ivory; the Greek name for this is elephantinon.
There is also an Indian black,[b] imported from India,
the composition of which I have not yet discovered.
A black is also produced with dyes from the black
florescence which adheres to bronze pans. One is
also made by burning logs of pitch-pine and pounding
the charcoal in a mortar. The cuttle-fish has a
remarkable property in forming a black secretion,
but no colour is made from this.[c] The preparation
of all black is completed by exposure to the sun,
black for writing ink receiving an admixture of gum
and black for painting walls an admixture of glue.
Black pigment that has been dissolved in vinegar
is difficult to wash out.

XXVI. Among the remaining colours which be- *Purples.*
cause of their high cost, as we said, are supplied § 30.
by patrons, dark purple holds the first place. It is
produced by dipping silversmiths' earth [d] along with
purple cloth and in like manner, the earth absorbing
the colour more quickly than the wool. The best
is that which being the first formed in the boiling
cauldron becomes saturated with the dyes in their

rudibus medicamentis inebriatum, proximum egesto
eo addita creta in ius idem et, quotiens id factum
est, elevatur bonitas pro numero dilutiore sanie.
45 quare Puteolanum potius laudetur quam Tyrium aut
Gaetulicum vel Laconicum, unde pretiosissimae
purpurae, causa est quod hysgino [1] maxime inficitur
rubiaque, quae [2] cogitur sorbere. vilissimum a
Canusio. pretium a singulis denariis in libras
ad xxx. pingentes sandyce sublita, mox ex [3] ovo
inducentes purpurissum, fulgorem minii faciunt. si
purpurae [4] facere malunt, caeruleum sublinunt,
mox purpurissum ex ovo inducunt.

46   XXVII. Ab hoc maxima auctoritas Indico. ex
India venit harundinum spumae adhaerescente limo.
cum cernatur, nigrum, at in diluendo mixturam
purpurae caeruleique mirabilem reddit. alterum
genus eius est in purpurariis officinis innatans cortinis, et est purpurae spuma. qui adulterant, vero
Indico tingunt stercora columbina aut cretam
Selinusiam vel anulariam vitro inficiunt. probatur
carbone; reddit enim quod sincerum est flammam
excellentis purpurae et, dum fumat, odorem maris.
ob id quidam e scopulis id colligi putant. pretium

---

[1] hysgino *Hermolaus Barbarus* : hygino *B* : yyg- *aut* yog-
cdd. : id genus *cd. Par.* 6801 : iscino *Isid.*
[2] rubiaque quae *Ian* : rubia quae *aut* rubiaque.
[3] ex *add. Mayhoff.*
[4] purpurae *Mayhoff* : purpura *aut* purpuram.

---

[a] A purplish red colour got from the unidentified plant
' hysge.'
[b] From several species of *Indigofera.*

primary state, and the next best produced when white earth is added to the same liquor after the first has been removed; and every time this is done the quality deteriorates, the liquid becoming more diluted at each stage. The reason why the dark purple of Pozzuoli is more highly praised than that of Tyre or Gaetulia or Laconia, places which produce the most costly purples, is that it combines most easily with hysginum [a] and madder which cannot help absorbing it. The cheapest comes from Canosa. The price is from one to thirty denarii per lb. Painters using it put a coat of sandyx underneath and then add a coat of dark purple mixed with egg, and so produce the brilliance of cinnabar; if they wish instead to produce the glow of purple, they lay a coat of blue underneath, and then cover this with dark purple mixed with egg.

XXVII. Of next greatest importance after this is indigo, [b] a product of India, being a slime that adheres to the scum upon reeds. When it is sifted out it is black, but in dilution it yields a marvellous mixture of purple and blue. There is another kind of it that floats on the surface of the pans [c] in the purple dye-shops, and this is the 'scum of purple.' People who adulterate it stain pigeons' droppings with genuine indigo, or else colour earth of Selinus or ring-earth [d] with woad. It can be tested by means of a live coal, as if genuine it gives off a brilliant purple flame and a smell of the sea while it smokes; on this account some people think that it is collected from rocks on the coast. The price of indigo is 20 denarii per

Indigo.

---

[c] Perhaps the vessels containing Tyrian purple.

[d] See § 48. Some white earth; but it is not known whether it came from Selinus in Cilicia or Selinus in Sicily.

Indico ✕ xx [1] in libras.   in medicina Indicum rigores
et impetus sedat siccatque ulcera.

47   XXVIII. Armenia mittit quod eius nomine appellatur. lapis est, hic quoque chrysocollae modo
infectus, optimumque est quod maxime vicinum et
communicato colore cum caeruleo. solebant librae
eius trecenis [2] nummis taxari. inventa per Hispanias harena est similem curam recipiens; itaque
ad denarios senos vilitas rediit. distat a caeruleo
candore modico, qui teneriorem hunc efficit colorem.
usus in medicina ad pilos tantum alendos habet
maximeque in palpebris.

48   XXIX. Sunt etiamnum novicii duo colores e
vilissimis: viride est [3] quod Appianum [4] vocatur et
chrysocollam mentitur, ceu parum multa ficta [5] sint
mendacia eius; fit e creta viridi, aestimatum sestertiis in libras.   XXX. Anulare quod vocant, candidum est, quo muliebres picturae inluminantur;
fit et ipsum e creta admixtis vitreis gemmis e volgi
anulis, inde et anulare dictum.

49   XXXI. Ex omnibus coloribus cretulam amant
udoque inlini recusant purpurissum, Indicum, caeruleum, Melinum, auripigmentum, Appianum, cerussa. cerae tinguntur isdem his coloribus ad eas

[1] ✕ xx *B* : xxx *aut* xx *rell.*
[2] trecenis *B* : tricenis *rell.*
[3] viride est *Mayhoff* : virides.
[4] *fortasse* apianum *vel* apiacum (*item* § 49).
[5] ficta *coni. Mayhoff* : dicta.

---

[a] Azurite.

[b] Probably azurite found mixed with green malachite.

[c] A conjectural emendation *apianum* or *apiacum* gives
'parsley green.'   It was a clay stained by ferrous substances.

pound. Used medicinally it allays cramps and fits and dries up sores.

XXVIII. Armenia sends us the substance [a] *Azurite, etc.* named after it Armenian. This also is a mineral that is dyed like malachite, and the best is that [b] which most closely approximates to that substance, the colour partaking also of dark blue. Its price used to be rated at 300 sesterces per pound. A sand has been found all over the Spanish provinces that admits of similar preparation, and accordingly the price has dropped to as low as six denarii. It differs from dark blue by a light white glow which renders this blue colour thinner in comparison. It is only used in medicine to give nourishment to the hair, and especially the eyelashes.

XXIX. There are also two colours of a very cheap class that have been recently discovered: one is the green called Appian,[c] which counterfeits malachite; just as if there were too few spurious varieties of it already! It is made from a green earth and is valued at a sesterce per pound. XXX. The other colour is that called 'ring-white,' which is used to give brilliance of complexion in paintings of women.[d] This itself also is made from white earth mixed with glass stones from the rings of the lower classes, which accounts for the name 'ring-white.'

XXXI. Of all the colours those which love a dry surface of white clay, and refuse to be applied to a damp plaster, are purple, indigo, blue, Melian, orpiment, Appian [c] and ceruse. Wax is stained with these same colours for encaustic paintings, a

[d] Or 'which shines on the painted faces of women' (K. C. Bailey). Cf. § 46 and note [d] on p. 295.

picturas, quae inuruntur, alieno parietibus genere,
sed classibus familiari, iam vero et onerariis navibus,
quoniam et vehicula[1] expingimus, ne quis miretur
et rogos pingi, iuvatque pugnaturos ad mortem aut
certe caedem speciose vehi. Qua contemplatione
tot colorum tanta varietate subit antiquitatem
mirari.

50 XXXII. Quattuor coloribus solis immortalia illa
opera fecere—ex albis Melino, e silaciis Attico, ex
rubris Sinopide Pontica, ex nigris atramento—
Apelles, Aetion,[2] Melanthius, Nicomachus, claris-
simi pictores, cum tabulae eorum singulae oppidorum
venirent opibus. nunc et purpuris in parietes
migrantibus et India conferente fluminum suorum
limum, draconum elephantorumque saniem nulla
nobilis pictura est. omnia ergo meliora tunc fuere,
cum minor copia. ita est, quoniam, ut supra diximus, rerum, non animi pretiis excubatur.

51 XXXIII. Et nostrae aetatis insaniam in pictura
non omittam. Nero princeps iusserat colosseum se
pingi cxx pedum linteo, incognitum ad hoc tempus.
ea pictura, cum peracta esset in Maianis hortis,
accensa fulmine cum optima hortorum parte con-
52 flagravit. libertus eius, cum daret Anti munus gladia-
torium, publicas porticus occupavit pictura, ut constat,

---

[1] vehicula *coni. Mayhoff*: fericula *Detlefsen*: pericula.
[2] Aetion *Ian*: etion *cdd.* (echion *cd. Par.* 6801).

---

[a] Cicero, *Brutus*, 70 says it was Zeuxis, Polygnotus, Timanthes and others who used four colours only, while in Aetion,
Nicomachus, Protogenes, and Apelles everything had been
brought to perfection. But the Alexander mosaic reproduces
a four-colour original.

[b] Indigo (see § 46) and dragon's blood, which latter is really
a plant-product got chiefly from a species of *Dracaena* or
Pterocarpus in Socotra.

sort of process which cannot be applied to walls but is common for ships of the navy, and indeed nowadays also for cargo vessels, since we even decorate vehicles with paintings, so that no one need be surprised that even logs for funeral pyres are painted; and we like gladiators going into the fray to ride in splendour to the scene of their death or at all events of carnage. Thus to contemplate all these numbers and great variety of colours prompts us to marvel at former generations.

XXXII. Four colours [a] only were used by the illustrious painters Apelles, Aetion, Melanthius and Nicomachus to execute their immortal works—of whites, Melinum; of yellow ochres, Attic; of reds, Pontic Sinopis; of blacks, atramentum—although their pictures each sold for the wealth of a whole town. Nowadays when purple finds its way even on to party-walls and when India contributes [b] the mud of her rivers and the gore of her snakes and elephants, there is no such thing as high-class painting. Everything in fact was superior in the days when resources were scantier. The reason for this is that, as we said before, it is values of § 4. material and not of genius that people are now on the look-out for.

XXXIII. One folly of our generation also in the *Colossal* matter of painting I will not leave out. The Emperor *painting of Nero,* A.D. Nero had ordered his portrait to be painted on a 54–68. colossal scale, on linen 120 ft. high, a thing unknown hitherto; this picture when finished, in the Gardens of Maius, was struck by lightning and destroyed by fire, together with the best part of the Gardens. When a freedman of Nero was giving at Anzio a gladiatorial show, the public porticoes were

gladiatorum ministrorumque omnium veris imaginibus
redditis. hic multis iam saeculis summus animus [1]
in pictura, pingi autem gladiatoria munera atque in
publico exponi coepta a C. Terentio Lucano. is avo
suo, a quo adoptatus fuerat, triginta paria in foro
per triduum dedit tabulamque pictam in nemore
Dianae posuit.

53   XXXIV. Nunc celebres in ea arte quam maxima
brevitate percurram, neque enim instituti operis est
talis [2] executio; itaque quosdam vel [3] in transcursu
et in aliorum mentione obiter nominasse satis erit,
exceptis operum claritatibus quae et ipsa conveniet
attingi, sive exstant sive intercidere.

54   Non constat sibi in hac parte Graecorum diligentia
multas post olumpiadas celebrando pictores quam
statuarios ac toreutas, primumque olympiade LXXXX,
cum et Phidian ipsum initio pictorem fuisse tradatur
clipeumque Athenis ab eo pictum, praeterea in
confesso sit LXXX tertia fuisse fratrem eius Panaenum,
qui clipeum intus pinxit Elide Minervae, quam
fecerat Colotes, discipulus Phidiae et ei in faciendo
55 Iove Olympio adiutor. quid? quod in confesso
perinde est Bularchi pictoris tabulam, in qua erat
Magnetum proelium, a Candaule, rege Lydiae

---

     [1] ambitus *coni. Mayhoff.*
     [2] talis *B, cd. Par. Lat.* 6797: iatis *rell.* (ampla *cd. Par.*
6801): artis *coni. Mayhoff.*
     [3] *V.l.* velut.

---

     [a] Probably not that of Athene Parthenos, which was, on
its inner side, carved in relief.

covered with paintings, so we are told, containing life-like portraits of all the gladiators and assistants. This portraiture of gladiators has been the highest interest in art for many generations now; but it was Gaius Terentius Lucanus who began the practice of having pictures made of gladiatorial shows and exhibited in public; in honour of his grandfather who had adopted him he provided thirty pairs of gladiators in the forum for three consecutive days, and exhibited a picture of their matches in the Grove of Diana.

XXXIV. I will now run through as briefly as possible the artists eminent in painting; and it is not consistent with the plan of this work to go into such detail; and accordingly it will be enough just to give the names of some of them even in passing and in course of mentioning others, with the exception of the famous works of art which whether still extant or now lost it will be proper to particularize. *Famous painters.*

In this department the exactitude of the Greeks is inconsistent, in placing the painters many Olympiads after the sculptors in bronze and chasers in metal, and putting the first in the 90th Olympiad, although it is said that even Phidias himself was a painter to begin with, and that there was a shield [a] at Athens that had been painted by him; and although moreover it is universally admitted that his brother Panaenus came in the 83rd Olympiad, who painted the inner surface of a shield of Athene at Elis made by Colotes, Phidias's pupil and assistant in making the statue of Olympian Zeus. And then, is it not equally admitted that Candaules, the last King of Lydia of the Heraclid line, who was also commonly known by the name of Myrsilus, gave its weight in *Chronology.* 420–417 B.C. *Panaenus and others.* 448–445 B.C.

Heraclidarum novissimo, qui et Myrsilus vocitatus
est, repensam auro? tanta iam dignatio picturae
erat. circa Romuli id aetatem acciderit necesse est,
etenim duodevicensima olympiade interiit Candaules aut, ut quidam tradunt, eodem anno quo
Romulus, nisi fallor, manifesta iam tunc claritate
56 artis, adeo absolutione.[1] quod si recipi necesse est,
simul apparet multo vetustiora principia eosque,
qui monochromatis pinxerint, quorum aetas non
traditur, aliquanto ante fuisse, Hygiaenontem,
Dinian, Charmadan et, qui primus in pictura marem
a femina discreverit, Eumarum Atheniensem, figuras
omnes imitari ausum, quique inventa eius exco-
luerit, Cimonem Cleonaeum. hic catagrapha invenit, hoc est obliquas imagines, et varie formare
voltus, respicientes suspicientesve vel despicientes;
articulis membra distinxit, venas protulit, praeterque
57 in vestibus rugas[2] et sinus invenit. Panaenus
quidem frater Phidiae etiam proelium Atheniensium
adversus Persas apud Marathona factum pinxit.

---

[1] V.ll. absolutioni (B) aut absolutiore: non absolutae
Brotier.
[2] vestibus rugas *Traube*: veste et rugas *Gelen*: veste
brugas B^1: veste rugas B^2, *cd. Tolet.*: verrugas *rell.*

---

[a] An unknown event; it might be the defeat of the Greeks
mentioned in VII, 126; or more likely the great defeat of the
Magnetes by the Treres in 651 B.C. (Strabo XIV, 647).
[b] Candaules was in fact put to death by Gyges about 685 B.C.
[c] See §§ 29, 15.
[d] By painting women's skin paler or white. This is the
stage represented by vase-painting from the seventh century
when women were commonly coloured white, men red or black.

gold for a picture of the painter Bularchus representing a battle *a* with the Magnetes? So high was
the value already set on the art of painting. This
must have occurred at about the time of Romulus,
since Candaules *b* died in the 18th Olympiad, or, 708–705 B.C.
according to some accounts, in the same year as
Romulus, making it clear, if I am not mistaken, trad. 717 B.C.
that the art had already achieved celebrity, and in
fact a perfection. And if we are bound to accept
this conclusion, it becomes clear at the same time
that the first stages were at a much earlier date and
that the painters in monochrome,*c* whose date is not
handed down to us, came considerably earlier—
Hygiaenon, Dinias, Charmadas and Eumarus of
Athens, the last being the earliest artist to distinguish *d* the male from the female sex in painting,
and venturing to reproduce every sort of figure; and
Cimon of Cleonae who improved on the inventions
of Eumarus. It was Cimon who first invented
' catagrapha,' that is, images in ' three-quarter,' *e* and
who varied the aspect of the features, representing
them as looking backward or upward or downward;
he showed the attachments of the limbs, displayed
the veins, and moreover introduced wrinkles and folds
in the drapery. Indeed the brother of Phidias
Panaenus even painted *f* the Battle at Marathon 490 B.C.
between the Athenians and Persians; so widely

e The Greek word meant probably ' foreshortened images,'
but Pliny or his Latin source rightly took it as expressing
' slanting (*obliquus*) images not profile or full-face.' Cf. § 90.
The context may exclude from the word *obliquus* any portraits
where the eyes look back, up, or down.

f On a wooden panel attached to a wall of the στοὰ ποικίλη,
' Painted Portico,' at Athens. The painting was attributed
also to Polygnotus and to Micon; cf. § 59.

adeo iam colorum usus increbruerat adeoque ars
perfecta erat, ut in eo proelio iconicos duces pinxisse
tradatur, Atheniensium Miltiadem, Callimachum,
Cynaegirum, barbarorum Datim, Artaphernen.

58 XXXV. Quin immo certamen etiam picturae
florente eo institutum est Corinthi ac Delphis,
primusque omnium certavit cum Timagora Chal-
cidense, superatus ab eo Pythiis, quod et ipsius
Timagorae carmine vetusto apparet, chronicorum
errore non dubio.

Alii quoque post hos clari fuere ante LXXXX olym-
piadem, sicut Polygnotus Thasius, qui primus
mulieres tralucida [1] veste pinxit, capita earum
mitris versicoloribus operuit plurimumque picturae
primus contulit, siquidem instituit os adaperire,
dentes ostendere, voltum ab antiquo rigore variare.

59 huius est tabula in porticu Pompei, quae ante
curiam eius fuerat, in qua dubitatur ascendentem
cum clupeo pinxerit an descendentem. hic Delphis
aedem pinxit, hic et Athenis porticum, quae Poecile
vocatur, gratuito, cum partem eius Micon mercede
pingeret. vel maior huic auctoritas, siquidem Amphictyones, quod est publicum Graeciae concilium,

---

[1] tralucida *B* : lucida *rell.*

---

[a] Not real portraits if the στοά was built at least thirty
years after 490 B.C.

[b] The Λέσχη, a covered colonnade.

[c] Polygnotus' contribution was a ' Sack of Troy,' Micon's
a ' Battle of the Amazons ' (against Theseus). See also § 57.

established had the employment of colour now become and such perfection of art had been attained that he is said to have introduced actual [a] portraits of the generals who commanded in that battle, Miltiades, Callimachus and Cynaegirus on the Athenian side and Datis and Artaphernes on that of the barbarians. XXXV. Nay more, during the time that Panaenus flourished competitions in painting were actually instituted at Corinth and at Delphi, and on the first occasion of all Panaenus competed against Timagoras of Chalcis, being defeated by him, at the Pythian Games, a fact clearly shown by an ancient poem of Timagoras himself, the chronicles undoubtedly being in error.

After those and before the 90th Olympiad there *420–417 B.C.* were other celebrated painters also, such as Polygnotus of Thasos who first represented women *Polygnotus and Micon.* in transparent draperies and showed their heads covered with a parti-coloured headdress; and he first contributed many improvements to the art of painting, as he introduced showing the mouth wide open and displaying the teeth and giving expression to the countenance in place of the primitive rigidity. There is a picture by this artist in the Portico of Pompeius which formerly hung in front of the Curia which he built, in which it is doubtful whether the figure of a man with a shield is painted as going up or as coming down. Polygnotus painted the temple [b] at Delphi and the colonnade at Athens called the Painted Portico, doing his work gratuitously, although a part of the work was painted by Micon who received a fee.[c] Indeed Polygnotus was held in higher esteem, as the Amphictyones, who are a General Council of Greece, voted him entertainment

hospitia ei gratuita decrevere. Fuit et alius Micon, qui minoris cognomine distinguitur, cuius filia Timarete et ipsa pinxit.

60 XXXVI. lxxxx autem olympiade fuere Aglaophon, Cephisodorus, Erillus,[1] Euenor, pater Parrhasii et praeceptor maximi pictoris, de quo suis annis dicemus, omnes iam inlustres, non tamen in quibus haerere expositio debeat festinans ad lumina artis, in quibus primus refulsit Apollodorus Atheniensis lxxxxiii olympiade. hic primus species exprimere instituit primusque gloriam penicillo iure contulit. eius est sacerdos adorans et Aiax fulmine incensus, quae Pergami spectatur hodie. neque ante eum tabula ullius ostenditur, quae teneat oculos.

61 Ab hoc artis fores apertas Zeuxis Heracleotes intravit olympiadis lxxxxv anno quarto, audentemque iam aliquid penicillum—de hoc enim adhuc loquamur —ad magnam gloriam perduxit, a quibusdam falso in lxxxviiii olympiade positus, cum fuisse [2] necesse est Demophilum Himeraeum et Nesea Thasium, quoniam utrius eorum discipulus fuerit ambigitur.

62 in eum Apollodorus supra scriptus versum fecit, artem ipsis ablatam Zeuxim ferre secum. opes quoque tantas adquisivit, ut in ostentatione [3] earum

[1] Erillus *B*: frilius *rell.* (frillus *cd. Par. Lat.* 6797): Phryllus *Brotier*: Phrylus *edd. vett.*: Herillus *coni. Sillig.*
[2] cum quo f. *Urlichs*: confuisse *Traube.*
[3] ostentationem *Gronov.*

[a] Inventor of shading, and therefore called σκιαγράφος.

at the public expense. There was also another
Micon, distinguished from the first by the surname
of ' the Younger,' whose daughter Timarete also
painted.

XXXVI. In the 90th Olympiad lived Aglaophon, 420–417 B.C.
Cephisodorus, Erillus, and Evenor the father and *Apollodorus and others.*
teacher of Parrhasius, a very great painter (about
Parrhasius we shall have to speak when we come to
his period). All these are now artists of note, yet § 67.
not figures over which our discourse should linger
in its haste to arrive at the luminaries of the art;
first among whom shone out Apollodorus[a] of Athens,
in the 93rd Olympiad. Apollodorus was the first 408–405 B.C.
artist to give realistic presentation of objects, and
the first to confer glory as of right upon the paint
brush. His are the Priest at Prayer and Ajax
struck by Lightning, the latter to be seen at
Pergamum at the present day. There is no painting
now on view by any artist before Apollodorus that
arrests the attention of the eyes.

The gates of art having been now thrown open by *Zeuxis.*
Apollodorus they were entered by Zeuxis of Heraclea
in the 4th year of the 95th Olympiad, who led forward 400–397 B.C.
the already not unadventurous paintbrush—for this
is what we are still speaking of—to great glory.
Some writers erroneously place Zeuxis in the 89th 424–421 B.C.
Olympiad, when Demophilus of Himera and Neseus
of Thasos must have been his contemporaries, as of
one of them, it is uncertain which, he was a pupil.
Of Zeuxis, Apollodorus above recorded wrote an
epigram in a line of poetry to the effect that ' Zeuxis
robbed his masters of their art and carried it off with
him.' Also he acquired such great wealth that he
advertised it at Olympia by displaying his own

Olympiae aureis litteris in palliorum tesseris in-
textum nomen suum ostentaret. postea donare
opera sua instituit, quod nullo pretio satis digno
permutari[1] posse diceret, sicuti Alcmenam Agra-
63 gantinis, Pana Archelao. fecit et Penelopen, in
qua pinxisse mores videtur, et athletam; adeoque
in illo sibi placuit, ut versum subscriberet celebrem
ex eo, invisurum aliquem facilius quam imitaturum.
magnificus est et Iuppiter eius in throno adstantibus
diis et Hercules infans dracones II[2] strangulans
Alcmena matre coram pavente et Amphitryone.
64 reprehenditur tamen ceu grandior in capitibus
articulisque, alioqui tantus diligentia, ut Agra-
gantinis facturus tabulam, quam in templo Iunonis
Laciniae publice dicarent, inspexerit virgines eorum
nudas et quinque elegerit, ut quod in quaque laudatissimum esset pictura redderet. pinxit et monochromata ex albo. aequales eius et aemuli fuere
Timanthes, Androcydes, Eupompus, Parrhasius.
65 descendisse hic in certamen cum Zeuxide traditur
et, cum ille detulisset uvas pictas tanto successu, ut
in scaenam aves advolarent, ipse detulisse linteum
pictum ita veritate repraesentata, ut Zeuxis alitum
iudicio tumens flagitaret tandem remoto linteo

---

[1] permutari *B*? : permutare *rell*.
[2] dracones II *Mayhoff* : draconem *B* : dracones *cd. Par.* 6801 :
dracones in *rell*.

---

[a] King of Macedonia 413–399 B.C.
[b] Μωμήσεταί τις μᾶλλον ἢ μιμήσεται.
[c] Fingers and toes?
[d] Apparently a 'Helen (cf. § 66),' painted in fact for the city
of Croton (Cic. *De Invent.* II, 1, 1; Dionys. Hal., *De Vet. Script.
Cens.* I).
[e] Apparently paintings in pale colours on a dark ground.
[f] The pictures were hung on the front of the stage buildings
in the theatre.

name embroidered in gold lettering on the checked pattern of his robes. Afterwards he set about giving away his works as presents, saying that it was impossible for them to be sold at any price adequate to their value : for instance he presented his Alcmena to the city of Girgenti and his Pan to Archelaus.[a] He also did a Penelope in which the picture seems to portray morality, and an Athlete, in the latter case being so pleased with his own work that he wrote below it a line of verse [b] which has hence become famous, to the effect that it would be easier for someone to carp at him than to copy him. His Zeus seated on a throne with the gods standing by in attendance is also a magnificent work, and so is the Infant Heracles throttling two Snakes in the presence of his mother Alcmena, looking on in alarm, and of Amphitryon. Nevertheless Zeuxis is criticized for making the heads and joints [c] of his figures too large in proportion, albeit he was so scrupulously careful that when he was going to produce a picture [d] for the city of Girgenti to dedicate at the public cost in the temple of Lacinian Hera he held an inspection of maidens of the place paraded naked and chose five, for the purpose of reproducing in the picture the most admirable points in the form of each. He also painted monochromes in white.[e] His contemporaries and rivals were Timanthes, Androcydes, Eupompus and Parrhasius. This last, it is recorded, entered into a competition with Zeuxis, who produced a picture of grapes so successfully represented that birds flew up to the stage-buildings [f]; whereupon Parrhasius himself produced such a realistic picture of a curtain that Zeuxis, proud of the verdict of the birds, requested that the curtain should now

Zeuxis and Parrhasius.

309

ostendi picturam atque intellecto errore concederet
palmam ingenuo pudore, quoniam ipse volucres
66 fefellisset, Parrhasius autem se artificem. fertur
et postea Zeuxis pinxisse puerum uvas ferentem, ad
quas cum advolassent aves,[1] eadem ingenuitate
processit iratus operi et dixit: 'uvas melius pinxi
quam puerum, nam si et hoc consummassem, aves
timere debuerant.' fecit et figlina opera, quae sola
in Ambracia relicta sunt, cum inde Musas Fulvius
Nobilior Romam transferret. Zeuxidis manu Romae
Helena est in Philippi porticibus,[a] et in Concordiae
delubro Marsyas religatus.

67 Parrhasius Ephesi natus et ipse multa contulit.
primus symmetrian picturae dedit, primus argutias
voltus, elegantiam capilli, venustatem[2] oris, con-
fessione artificum in liniis extremis palmam adeptus.
haec est picturae summa subtilitas.[3] corpora enim
pingere et media rerum est quidem magni operis,
sed in quo multi gloriam tulerint; extrema corporum facere et desinentis picturae modum includere
68 rarum in successu artis invenitur. ambire enim se
ipsa debet extremitas et sic desinere, ut promittat
alia et[4] post se[5] ostendatque etiam quae occultat.
hanc ei gloriam concessere Antigonus et Xenocrates,

---

[1] *V.l.* advolarent aves *aut* advolasset avis *aut* advolaret avis.
[2] *V.l.* vetustatem.
[3] suptilitas *B* : sublimitas *rell.*
[4] alia et *Mayhoff* : aliae *cd. Leid. Voss. m.*1 : alia *rell.*
[5] posse *edd. vett.* : pone se *coni. Ian* : alias post se *Traube.*

---

[a] The picture 'Helen' mentioned (not named) in § 64. The
porticoes were built by L. Marcius Philippus in 29 B.C.

be drawn and the picture displayed; and when he
realized his mistake, with a modesty that did him
honour he yielded up the prize, saying that whereas
he had deceived birds Parrhasius had deceived him,
an artist. It is said that Zeuxis also subsequently
painted a Child Carrying Grapes, and when birds
flew to the fruit with the same frankness as before
he strode up to the picture in anger with it and said,
' I have painted the grapes better than the child, as
if I had made a success of that as well, the birds
would inevitably have been afraid of it.' He also
executed works in clay, the only works of art that
were left at Ambracia when Fulvius Nobilior removed 189 B.C.
the statues of the Muses from that place to Rome.
There is at Rome a Helena *a* by Zeuxis in the
Porticoes of Philippus, and a Marsyas Bound, in the
Shrine of Concord.

Parrhasius also, a native of Ephesus, contributed *Parrhasius.*
much to painting. He was the first to give proportions
to painting and the first to give vivacity to the
expression of the countenance, elegance of the hair
and beauty of the mouth; indeed it is admitted by
artists that he won the palm in the drawing of
outlines. This in painting is the high-water mark
of refinement; to paint bulk and the surface within
the outlines, though no doubt a great achievement, is one in which many have won distinction,
but to give the contour of the figures, and make a
satisfactory boundary where the painting within
finishes, is rarely attained in successful artistry.
For the contour ought to round itself off and so
terminate as to suggest the presence of other parts
behind it also, and disclose even what it hides.
This is the distinction conceded to Parrhasius by

qui de pictura scripsere, praedicantes quoque, non
solum confitentes; et alias multa graphidis vestigia
exstant in tabulis ac membranis eius, ex quibus pro-
ficere dicuntur artifices. minor tamen videtur sibi
69 comparatus in mediis corporibus exprimendis. pinxit
demon Atheniensium argumento quoque ingenioso.
ostendebat namque varium iracundum iniustum
inconstantem,[1] eundem exorabilem clementem
misericordem; gloriosum . . . ,[2] excelsum humilem,
ferocem fugacemque et omnia pariter. idem pinxit
et Thesea, quae Romae in Capitolio fuit, et nauar-
chum thoracatum, et in una tabula, quae est Rhodi,
Meleagrum, Herculem, Persea; haec ibi ter fulmine
ambusta neque obliterata hoc ipso miraculum auget.
70 pinxit et archigallum, quam picturam amavit Tiberius princeps atque, ut auctor est Deculo,[3] HS
$\overline{\text{LX}}$ [4] aestimatam cubiculo suo inclusit. pinxit
et Thressam [5] nutricem infantemque in manibus
eius et Philiscum et Liberum patrem adstante
Virtute, et pueros duos, in quibus spectatur securitas
aetatis et simplicitas, item sacerdotem adstante
71 puero cum acerra et corona. sunt et duae picturae
eius nobilissimae, hoplites in certamine ita decur-
rens, ut sudare videatur, alter arma deponens, ut

---

[1] incontinentem *O. Jahn.*
[2] *lac. Mayhoff.*
[3] deculo *B*: depulo, de populo *aut sim. rell.*: Decius
Gelen: Decius Epulo *edd. vett.*: Decius Eculeo *Hermolaus
Barbarus.*
[4] $\overline{\text{LX}}$ *Ian*: LX *B*: LX *rell.*
[5] thressam *B*: cressam *aut* chressam *rell.*

---

[a] Or ' traces of his draughtsmanship.'
[b] Or ' them in various moods.'
[c] Until it perished in the fire of 70 B.C.

Antigonus and Xenocrates who have written on the
art of painting, and they do not merely admit it but
actually advertise it. And there are many other pen-
sketches [a] still extant among his panels and parch-
ments, from which it is said that artists derive profit.
Nevertheless he seems to fall below his own level in
giving expression to the surface of the body inside
the outline. His picture of the People of Athens
also shows ingenuity in treating the subject, since
he displayed them as fickle, [b] choleric, unjust and
variable, but also placable and merciful and compas-
sionate, boastful ⟨and . . . . ⟩, lofty and humble,
fierce and timid—and all these at the same time.
He also painted a Theseus which was once [c] in the
Capitol at Rome, and a Naval Commander in a
Cuirass, and in a single picture now at Rhodes figures
of Meleager, Heracles and Perseus. This last
picture has been three times struck by lightning at
Rhodes without being effaced, a circumstance which
in itself enhances the wonder felt for it. He also
painted a High Priest of Cybele, a picture for which
the Emperor Tiberius conceived an affection and kept A.D.14–37.
it shut up in his bedchamber, the price at which
it was valued according to Deculo being 6,000,000
sesterces. He also painted a Thracian Nurse with
an Infant in her Arms, a Philiscus, and a Father
Liber or Dionysus attended by Virtue, and Two
Children in which the carefree simplicity of childhood
is clearly displayed, and also a Priest attended by
Boy with Incense-box and Chaplet. There are also
two very famous pictures by him, a Runner in the
Race in Full Armour who actually seems to sweat
with his efforts, and the other a Runner in Full
Armour Taking off his Arms, so lifelike that he can

anhelare sentiatur. laudantur et Aeneas Castorque
ac Pollux in eadem tabula, item Telephus, Achilles,
Agamemnon, Ulixes. fecundus artifex, sed quo
nemo insolentius usus sit gloria artis, namque et
cognomina usurpavit habrodiaetum se appellando
aliisque versibus principem artis et eam ab se consummatam, super omnia Apollinis se radice ortum et
Herculem, qui est Lindi, talem a se pictum, qualem
72 saepe in quiete vidisset; et cum [1] magnis suffragiis
superatus a Timanthe esset [2] Sami in Aiace armo-
rumque iudicio, herois nomine se moleste ferre
dicebat, quod iterum ab indigno victus esset—
Pinxit et minoribus tabellis libidines, eo genere
petulantis ioci se reficiens.[3]

73 Nam Timanthis vel plurimum adfuit ingenii.
eius enim est Iphigenia oratorum laudibus celebrata,
qua stante ad aras peritura cum maestos pinxisset
omnes praecipueque patruum et tristitiae omnem
imaginem consumpsisset, patris ipsius voltum velavit,
74 quem digne non poterat ostendere. sunt et alia
ingenii eius exempla, veluti Cyclops dormiens in
parvola tabella, cuius et sic magnitudinem exprimere
cupiens pinxit iuxta Satyros thyrso pollicem eius

---

[1] et cū (= cum) *Mayhoff*: ergo.
[2] Timanthe esset *Mayhoff*: timanthesest B^1: timanthe (*aut*
thimante) est *cdd.*
[3] pinxit . . . reficiens *post* Ulixes 71 *transp. Urlichs.*

---

[a] Showing the healing of Telephus by rust from Achilles'
sword, with Agamemnon and Odysseus looking on.
[b] When the arms of dead Achilles were awarded to Odysseus,
Ajax became mad and at night unknowingly killed sheep in
the belief that he was killing his enemies.
[c] *E.g.* Cicero, *De Oratore* 74.
[d] A picture found at Pompeii may be a copy of this.

314

be perceived to be panting for breath. His Aeneas, Castor and Pollux (Polydeuces), all in the same picture, are also highly praised, and likewise his group *a* of Telephus with Achilles, Agamemnon and Odysseus. Parrhasius was a prolific artist, but one who enjoyed the glory of his art with unparalleled arrogance, for he actually adopted certain surnames, calling himself the ' Bon Viveur,' and in some other verses ' Prince of Painters,' who had brought the art to perfection, and above all saying he was sprung from the lineage of Apollo and that his picture of Heracles at Lindos presented the hero as he had often appeared to him in his dreams. Consequently when *Timanthes.* defeated by Timanthes at Samos by a large majority of votes, the subject of the pictures being Ajax and the Award of the Arms, he used to declare in the name of his hero that he was indignant at having been defeated a second time by an unworthy opponent.*b* He also painted some smaller pictures of an immodest nature, taking his recreation in this sort of wanton amusement.

To return to Timanthes—he had a very high degree of genius. Orators *c* have sung the praises of his Iphigenia,*d* who stands at the altar awaiting her doom ; the artist has shown all present full of sorrow, and especially her uncle,*e* and has exhausted all the indications of grief, yet has veiled the countenance of her father himself,*f* whom he was unable adequately to portray. There are also other examples of his genius, for instance a quite small panel of a Sleeping Cyclops, whose gigantic stature he aimed at representing even on that scale by painting at his side some Satyrs measuring the size of his thumb

e Menelaus.　　　　*f* Agamemnon.

metientes. atque in unius huius operibus intelligitur
plus semper quam pingitur et, cum sit ars summa,
ingenium tamen ultra artem est. pinxit et heroa
absolutissimi operis, artem ipsam complexus viros
pingendi, quod opus nunc Romae in templo Pacis est.

75 Euxinidas hac aetate docuit Aristiden, prae-
clarum artificem, Eupompus Pamphilum, Apellis
praeceptorem. est Eupompi victor certamine gym-
nico palmam tenens. ipsius auctoritas tanta fuit,
ut diviserit picturam[1] : genera, quae ante eum duo
fuere—Helladicum et Asiaticum[2] appellabant—,
propter hunc, qui erat Sicyonius, diviso Helladico
tria facta sunt, Ionicum, Sicyonium, Atticum.

76 Pamphili cognatio et proelium ad Phliuntem ac
victoria Atheniensium, item Ulixes in rate. ipse
Macedo natione, sed . . .[3] primus in pictura omnibus
litteris eruditus, praecipue arithmetica et geometria,
sine quibus negabat artem perfici posse, docuit
neminem talento minoris—annuis X D[4]—, quam
mercedem at Apelles et Melanthius dedere ei.

77 huius auctoritate effectum est Sicyone primum,
deinde in tota Graecia, ut pueri ingenui omissam

---

[1] picturam *Mayhoff*: picturam in *cdd.* (-ras in *cd. Par.*
6801 : -a in *cd. Leid. Voss.*).
[2] asiaticum *B*: asianum *B*[1]? : asiticum quod asiaticum
rell. : quod asiaticum *Gelen.*
[3] *lac. Mayhoff.*
[4] X D *B, cd. Leid. Voss.*: D *rell.* (*om. cd. Flor. Ricc.*).

---

[a] The elder; cf. §§ 108, 111 and note on pp. 410–411.
[b] Possibly the capture of Phlius by the Spartans in 379 B.C.
and the sea-victory of Athens over the Spartans at Naxos in

with a wand. Indeed Timanthes is the only artist in whose works more is always implied than is depicted, and whose execution, though consummate, is always surpassed by his genius. He painted a hero which is a work of supreme perfection, in which he has included the whole art of painting male figures; this work is now in the Temple of Peace in Rome.

It was at this period that Euxinidas had as his pupil the famous artist Aristides,[a] that Eupompus taught Pamphilus who was the instructor of Apelles. *Eupompus and Pamphilus.* A work of Eupompus is a Winner in a Gymnastic Contest holding a Palm branch. Eupompus's own influence was so powerful that he made a fresh division of painting; it had previously been divided into two schools, called the Helladic or Grecian and the Asiatic, but because of Eupompus, who was a Sicyonian, the Grecian school was sub-divided into three groups, the Ionic, Sicyonian and Attic. To Pamphilus belong Family Group, and a Battle at Phlius and a Victory of the Athenians,[b] and also Odysseus on his Raft. He was himself a Macedonian by birth, but ⟨was brought up at Sicyon, and⟩ was the first painter highly educated in all branches of learning, especially arithmetic and geometry, without the aid of which he maintained art could not attain perfection. He took no pupils at a lower fee than a talent, at the rate of 500 drachmae per annum,[c] and this was paid him by both Apelles and Melanthius. It was brought about by his influence, first at Sicyon and then in the whole of Greece as well, that children

376, or the defeat of Sicyonians by Phliasians and Athenians in 367 B.C. The painting may have represented the last event only.
  [c] So that the course of study could last 12 years.

ante [1] graphicen [hoc est picturam [2]] in buxo, docerentur recipereturque ars ea in primum gradum
liberalium. semper quidem honos ei fuit, ut ingenui
eam exercerent, mox ut honesti, perpetuo interdicto ne servitia docerentur. ideo neque in hac
neque in toreutice ullius, qui servierit, opera
celebrantur.

78 Clari et centesima septima olympiade exstitere
Aetion ac Therimachus. Aetionis sunt nobiles
picturae Liber pater, item Tragoedia et Comoedia,
Semiramis ex ancilla regnum apiscens, anus lampadas
praeferens et nova nupta verecundia notabilis.

79 Verum omnes prius genitos futurosque postea
superavit Apelles Cous olympiade centesima duo-
decima. picturae plura solus prope quam ceteri
omnes contulit, voluminibus etiam editis, quae
doctrinam eam continent. praecipua eius in arte
venustas fuit, cum eadem aetate maximi pictores
essent; quorum opera cum admiraretur, omnibus
conlaudatis deesse illam suam venerem [3] dicebat,
quam Graeci χάριτα vocant; cetera omnia contigisse,
80 sed hac sola sibi neminem parem. et aliam gloriam
usurpavit, cum Protogenis opus inmensi laboris ac

---

[1] omissā (= omissam) ante *coni. Mayhoff* : omnia ante B :
omnia anti *rell.* : ante omnia *edd. vett.* : omnes artem *C. F.
Hermann.*
[2] *seclud. Urlichs.*
[3] venustatem *Fröhner (cp.* gratiam *Quintil.* XII. 10. 6).

---

[a] The whole of statuary as contrasted with painting.
[b] Sammuramat, princess of Assyria *c.* 800 B.C.
[c] Really of Ephesus, but some of his famous works were at
Cos.

of free birth were given lessons in drawing on box-wood, which had not been included hitherto, and that this art was accepted into the front rank of the liberal sciences. And it has always consistently had the honour of being practised by people of free birth, and later on by persons of station, it having always been forbidden that slaves should be instructed in it. Hence it is that neither in painting nor in the art of statuary [a] are there any famous works that were executed by any person who was a slave.

In the 107th Olympiad Aetion and Therimachus also attained outstanding distinction. Famous paintings by Aetion are a Father Liber or Dionysus, Tragedy and Comedy and Semiramis [b] the Slave Girl Rising to a Throne; and the Old Woman carrying Torches, with a Newly Married Bride, remarkable for her air of modesty. *352–349 B.C.* *Aetion and Therimachus.*

But it was Apelles of Cos [c] who surpassed all the painters that preceded and all who were to come after him; he dates in the 112th Olympiad. He singly contributed almost more to painting than all the other artists put together, also publishing volumes containing the principles of painting. His art was unrivalled for graceful charm, although other very great painters were his contemporaries. Although he admired their works and gave high praise to all of them, he used to say that they lacked the glamour that his work possessed, the quality denoted by the Greek word *charis*, and that although they had every other merit, in that alone no one was his rival. He also asserted another claim to distinction when he expressed his admiration for the immensely laborious and infinitely meticulous work *Apelles.* *332–329 B.C.* *Apelles and Protogenes.*

319

curae supra modum anxiae miraretur; dixit enim
omnia sibi cum illo paria esse aut illi meliora, sed
uno se praestare, quod manum de tabula sciret [1]
tollere, memorabili praecepto nocere saepe nimiam
diligentiam. fuit autem non minoris simplicitatis
quam artis. Melanthio dispositione cedebat, Ascle-
piodoro de [2] mensuris, hoc est quanto quid a quoque
distare deberet.

81 Scitum inter Protogenen et eum quod accidit.
ille Rhodi vivebat, quo cum Apelles adnavigasset,
avidus cognoscendi opera eius fama tantum sibi
cogniti, continuo officinam petiit. aberat ipse, sed
tabulam amplae magnitudinis in machina aptatam
una [3] custodiebat anus. haec foris esse Protogenen
respondit interrogavitque, a quo quaesitum diceret.
' ab hoc,' inquit Apelles adreptoque penicillo lineam
ex colore duxit summae tenuitatis per tabulam. et

82 reverso Protogeni quae gesta erant anus indicavit.
ferunt artificem protinus contemplatum subtili-
tatem dixisse Apellen venisse, non cadere in alium
tam absolutum opus; ipsumque alio colore tenuiorem lineam in ipsa illa duxisse abeuntemque
praecepisse, si redisset ille, ostenderet adiceretque

---

[1] sciret *B* : non sciret *rell.*
[2] de *fortasse delendum* (*Mayhoff, qui et* dimensuris *coni.*).
[3] una *B* : picturae una *rell.*

---

[a] The expression ' *manum de tabula*,' ' hand from the picture,'
was a saying which expressed ' That's enough.'

[b] Pliny does not say whether it was straight or wavy, or
an outline of some object.

of Protogenes; for he said that in all respects his achievements and those of Protogenes were on a level, or those of Protogenes were superior, but that in one respect he stood higher, that he knew when to take his hand away from a picture [a]—a noteworthy warning of the frequently evil effects of excessive diligence. The candour of Apelles was however equal to his artistic skill: he used to acknowledge his inferiority to Melanthius in grouping, and to Asclepiodorus in nicety of measurement, that is in the proper space to be left between one object and another.

A clever incident took place between Protogenes and Apelles. Protogenes lived at Rhodes, and Apelles made the voyage there from a desire to make himself acquainted with Protogenes's works, as that artist was hitherto only known to him by reputation. He went at once to his studio. The artist was not there but there was a panel of considerable size on the easel prepared for painting, which was in the charge of a single old woman. In answer to his enquiry, she told him that Protogenes was not at home, and asked who it was she should report as having wished to see him. 'Say it was this person,' said Apelles, and taking up a brush he painted in colour across the panel an extremely fine line [b]; and when Protogenes returned the old woman showed him what had taken place. The story goes that the artist, after looking closely at the finish of this, said that the new arrival was Apelles, as so perfect a piece of work tallied with nobody else; and he himself, using another colour, drew a still finer line exactly on the top of the first one, and leaving the room told the attendant to show it to the

hunc esse quem quaereret. atque ita evenit.
revertit enim Apelles et vinci erubescens tertio
colore lineas secuit nullum relinquens amplius
83 subtilitati locum. at Protogenes victum se confessus in portum devolavit hospitem quaerens,
placuitque sic eam tabulam posteris tradi omnium
quidem, sed artificum praecipuo miraculo. con-
sumptam eam priore incendio Caesaris domus in
Palatio audio, spectatam nobis[1] ante, spatiose[2]
nihil aliud continentem quam[3] lineas visum effu-
gientes, inter egregia multorum opera inani similem
et eo ipso allicientem omnique opere nobiliorem.

84 Apelli fuit alioqui perpetua consuetudo numquam
tam occupatum diem agendi, ut non lineam ducendo
exerceret artem, quod ab eo in proverbium venit.
idem perfecta opera proponebat in pergula tran-
seuntibus atque, ipse post tabulam latens, vitia
quae notarentur auscultabat, vulgum diligentiorem
85 iudicem quam se praeferens; feruntque reprehensum
a sutore, quod in crepidis una pauciores intus fecisset
ansas, eodem postero die superbo emendatione
pristinae admonitionis cavillante circa crus, indig-

---

[1] nobis *cdd.*: Rhodi *Mayhoff*: olim *Gronov.*
[2] *V.l.* spatiore (spatio sed B^2): spatio *Pintianus.*
[3] *V.l.* quam in: quam III *Gronov.*

---

[a] Pliny surely indicates that Apelles drew a yet finer
line on top of the other two down their length.
[b] Probably an outline of some object.
[c] *Nulla dies sine linea,* ' No day without a line.'

visitor if he returned and add that this was the person he was in search of; and so it happened; for Apelles came back, and, ashamed to be beaten, cut [a] the lines with another in a third colour, leaving no room for any further display of minute work. Hereupon Protogenes admitted he was defeated, and flew down to the harbour to look for the visitor; and he decided that the panel should be handed on to posterity as it was, to be admired as a marvel by everybody, but particularly by artists. I am informed that it was burnt in the first fire which occurred in Caesar's palace on the Palatine; it had A.D. 4. been previously much admired by us, on its vast surface containing nothing else than the almost invisible lines, so that among the outstanding works of many artists it looked like a blank space, and by that very fact attracted attention and was more esteemed than every masterpiece there.

Moreover it was a regular custom with Apelles never to let a day of business to be so fully occupied that he did not practise his art by drawing a line,[b] which has passed from him into a proverb.[c] Another habit of his was when he had finished his works to place them in a gallery in the view of passers by, and he himself stood out of sight behind the picture and listened to hear what faults were noticed, rating the public as a more observant critic than himself. And it is said that he was found fault with by a shoe-maker because in drawing a subject's sandals he had represented the loops in them as one too few, and the next day the same critic was so proud of the artist's correcting the fault indicated by his previous objection that he found fault with the leg, but Apelles indignantly looked out from behind the

natum prospexisse denuntiantem, ne supra crepidam
sutor iudicaret, quod et ipsum in proverbium abiit.
fuit enim et comitas illi, propter quam gratior
Alexandro Magno frequenter in officinam venti-
tanti—nam, ut diximus, ab alio se pingi vetuerat
edicto—, sed in officina imperite multa disserenti
silentium comiter suadebat, rideri eum dicens a
86 pueris, qui colores tererent. tantum erat auctori-
tati iuris in regem alioqui iracundum. quamquam
Alexander honorem ei clarissimo perhibuit exemplo.
namque cum dilectam sibi e pallacis suis praecipue,
nomine Pancaspen,[1] nudam pingi ob admirationem
formae ab Apelle iussisset eumque, dum paret,
captum amore sensisset, dono dedit ei,[2] magnus
animo, maior imperio sui nec minor hoc facto quam
87 victoria alia, quia[3] ipse se vicit, nec torum tantum
suum, sed etiam adfectum donavit artifici, ne
dilectae quidem respectu motus, cum modo regis ea
fuisset, modo pictoris esset. sunt qui Venerem
anadyomenen ab illo pictam exemplari putent.
Apelles et in aemulis benignus Protogeni digna-

[1] pancaspen *B* : campaspen *aut* -em *rell.* : Pancasten *Sillig.*
[2] ei *Ian* : et.
[3] alia quia *M. Hertz* : alia *Urlichs* : alia qua *B* : aliqua *rell.*

---

[a] *Ne sutor ultra crepidam.* "Let a shoemaker stick to his
last."

picture and rebuked him, saying that a shoemaker in his criticism must not go beyond the sandal— a remark that has also passed into a proverb.[a] In fact he also possessed great courtesy of manners, which made him more agreeable to Alexander the Great, who frequently visited his studio—for, as we have said, Alexander had published an edict forbidding any other artist to paint his portrait; but in the studio Alexander used to talk a great deal about painting without any real knowledge of it, and Apelles would politely advise him to drop the subject, saying that the boys engaged in grinding the colours were laughing at him: so much power did his authority exercise over a King who was otherwise of an irascible temper. And yet Alexander conferred honour on him in a most conspicuous instance; he had such an admiration for the beauty of his favourite mistress, named Pancaspe, that he gave orders that she should be painted in the nude by Apelles, and then discovering that the artist while executing the commission had fallen in love with the woman, he presented her to him, great-minded as he was and still greater owing to his control of himself, and of a greatness proved by this action as much as by any other victory: because he conquered himself, and presented not only his bedmate but his affection also to the artist, and was not even influenced by regard for the feelings of his favourite in having been recently the mistress of a monarch and now belonged to a painter. Some persons believe that she was the model from which the Aphrodite Anadyomene (Rising from the Sea) was painted. It was Apelles also who, kindly among his rivals, first established the reputation of

Apelles and Alexander.

VII, 125.

88 tionem primus Rhodi constituit. sordebat suis, ut
plerumque domestica, percontantique, quanti liceret opera effecta, parvum nescio quid dixerat,
at ille quinquagenis talentis poposcit famamque
dispersit, se emere, ut pro suis venderet. ea res
concitavit Rhodios ad intellegendum artificem, nec
nisi augentibus pretium cessit.

Imagines [1] adeo similitudinis indiscretae pinxit,
ut—incredibile dictu—Apio grammaticus scriptum
reliquerit, quendam ex facie hominum divinantem,
quos metoposcopos vocant, ex iis dixisse aut futurae
89 mortis annos aut praeteritae vitae.[2] non fuerat ei
gratia in comitatu Alexandri cum Ptolemaeo, quo
regnante Alexandriam vi tempestatis expulsus,
subornato fraude aemulorum plano regio invitatus,
ad cenam venit indignantique Ptolemaeo et vocatores
suos ostendenti, ut diceret, a quo eorum invitatus
esset, arrepto carbone extincto e foculo imaginem
in pariete delineavit, adgnoscente voltum plani rege
90 inchoatum protinus. pinxit et Antigoni regis
imaginem altero lumine orbati [3] primus excogitata
ratione vitia condendi; obliquam namque fecit, ut,

[1] imagines *Gelen* : imaginem.
[2] vitae *add. Brunn.*
[3] orbati *Mayhoff* : orbatam.

---

[a] The word μετωποσκόπος means one who gazes at (examines)
foreheads.

[b] Ptolemy I, who died in 286 B.C.

[c] 382–301 B.C. One of Alexander's generals, and King of
Macedonia 306–301.

Protogenes at Rhodes. Protogenes was held in low esteem by his fellow-countrymen, as is usual with home products, and, when Apelles asked him what price he set on some works he had finished, he had mentioned some small sum, but Apelles made him an offer of fifty talents for them, and spread it about that he was buying them with the intention of selling them as works of his own. This device aroused the people of Rhodes to appreciate the artist, and Apelles only parted with the pictures to them at an enhanced price.

He also painted portraits so absolutely lifelike that, incredible as it sounds, the grammarian Apio has left it on record that one of those persons called ' physiognomists,' [a] who prophesy people's future by their countenance, pronounced from their portraits either the year of the subjects' deaths hereafter or the number of years they had already lived. Apelles had been on bad terms with Ptolemy in Alexander's retinue. When this Ptolemy [b] was King of Egypt, Apelles on a voyage had been driven by a violent storm into Alexandria. His rivals maliciously suborned the King's jester to convey to him an invitation to dinner, to which he came. Ptolemy was very indignant, and paraded his hospitality-stewards for Apelles to say which of them had given him the invitation. Apelles picked up a piece of extinguished charcoal from the hearth and drew a likeness on the wall, the King recognizing the features of the jester as soon as he began the sketch. He also painted a portrait of King Antigonus [c] who was blind in one eye, and devised an original method of concealing the defect, for he did the likeness in ' three-quarter,' so that the feature that was lacking in the

Apelles and Ptolemy I.

Apelles and Antigonus.

327

quod deerat corpori, picturae deesse [1] potius videretur, tantumque eam partem e facie ostendit, quam
totam poterat ostendere. sunt inter opera eius et
exspirantium imagines. quae autem nobilissima
91 sint, non est facile dictu. Venerem exeuntem e
mari divus Augustus dicavit in delubro patris Caesaris, quae anadyomene vocatur, versibus Graecis
tali opere,[2] dum [3] laudatur, victo [4] sed [5] inlustrato.[6]
cuius inferiorem partem corruptam qui reficeret
non potuit reperiri, verum ipsa iniuria cessit in
gloriam artificis. consenuit haec tabula carie,
aliamque pro ea substituit Nero in principatu suo
92 Dorothei manu. Apelles inchoaverat et aliam
Venerem Coi,[7] superaturus etiam [8] illam suam
priorem. invidit mors peracta parte, nec qui
succederet operi ad praescripta liniamenta inventus
est. pinxit et Alexandrum Magnum fulmen tenentem in templo Ephesiae Dianae viginti talentis auri.
digiti eminere videntur et fulmen extra tabulam
esse—legentes meminerint omnia ea quattuor
coloribus facta; manipretium eius tabulae in
nummo [9] aureo [10] mensura [11] accepit, non numero.

---

[1] adesse *coni. Mayhoff.*
[2] tantopere *Fröhner.*
[3] aevo dum *J. Müller.*
[4] victa *edd. vett.*: invicto *Schneidewin*: vitio *Fröhner*:
⟨aevis victa⟩ *Mayhoff.*
[5] est *Fröhner.*
[6] illustrata *edd. vett.*: versibus Graecis dum laudatur, tali
opere ⟨aevis⟩ victo, sed inlustrato *coni. Mayhoff.*
[7] *V.l.* Cois.
[8] *V.l.* famam : fama *Urlichs.*
[9] in numero *cd. Par.* 6801 : immane *cd. Flor. Ricc.*
[10] auro *olim Gelen* (*del.* nummo): aureos *edd. vett.*
[11] *V.l.* mensuram.

subject might be thought instead to be absent in the picture, and he only showed the part of the face which he was able to display as unmutilated. Among his works there are also pictures of persons at the point of death. But it is not easy to say which of his productions are of the highest rank. His Aphrodite emerging from the Sea was dedicated by his late lamented Majesty Augustus in the Shrine of his father Caesar; it is known as the Anadyomene; this like other works is eclipsed [a] yet made famous by the Greek verses which sing its praises; the lower part of the picture having become damaged nobody could be found to restore it, but the actual injury contributed to the glory of the artist. This picture however suffered from age and rot, and Nero when emperor substituted another for it, a work by Dorotheus. Apelles had also begun on another Aphrodite at Cos, which was to surpass even his famous earlier one; but death grudged him the work when only partly finished, nor could anybody be found to carry on the task, in conformity with the outlines of the sketches prepared. He also painted Alexander the Great holding a Thunderbolt, in the temple of Artemis at Ephesus, for a fee of twenty talents in gold. The fingers have the appearance of projecting from the surface and the thunderbolt seems to stand out from the picture—readers must remember [b] that all these effects were produced by four colours; the artist received the price of this picture in gold coin measured by weight,[c] not

Various Works by Apelles.

[a] 'Overcome' or 'surpassed' by the poet, who can express more than the painter can; for the painter can represent one moment only.　　[b] See § 50.

[c] It is suggested that this means that the price was the equivalent (in gold coins) of the weight of the panel.

93 pinxit et megabyzi, sacerdotis Dianae Ephesiae,
pompam, Clitum cum equo ad bellum festinantem,
galeam poscenti armigerum porrigentem.[1] Alexandrum et Philippum quotiens pinxerit, enumerare
supervacuum est. mirantur eius Habronem Sami;
Menandrum, regem Cariae, Rhodi, item Antaeum;
Alexandreae Gorgosthenen tragoedum; Romae
Castorem et Pollucem cum Victoria et Alexandro
Magno, item Belli imaginem restrictis ad terga
94 manibus, Alexandro in curru triumphante. quas
utrasque tabulas divus Augustus in fori sui cele-
berrimis partibus dicaverat simplicitate moderata;
divus Claudius pluris existimavit utrisque excisa
Alexandri facie divi Augusti imagines addere.
eiusdem arbitrantur manu[2] esse et in Dianae[3]
templo Herculem aversum, ut, quod est difficillimum,
faciem eius ostendat verius pictura quam promittat.
pinxit et heroa nudum eaque pictura naturam ipsam
95 provocavit. est et equus eius, sive fuit, pictus in
certamine, quo iudicium ad mutas quadripedes
provocavit ab hominibus. namque ambitu praeva-
lere aemulos sentiens singulorum picturas inductis
equis ostendit: Apellis tantum equo adhinnivere.
idque et postea semper evenit, ut experimentum
96 artis illud ostentaretur. fecit et Neoptolemum ex[4]

---

[1] armigero porrigente *coni. Mayhoff.*
[2] manum *B.*
[3] Dianae *Preller* : annae *B* : antoniae *rell.*
[4] ⟨pugnantem⟩ ex *coni. Mayhoff* : *lac. post* Persas
Urlichs.

---

[a] Cf. § 27 and Serv. ad *Aen.* I, 294.
[b] *I.e.* he did not appropriate them for himself.

counted. He also painted a Procession of the
Magabyzus, the priest of Artemis of Ephesus, a
Clitus with Horse hastening into battle; and an
armour-bearer handing someone a helmet at his
command. How many times he painted Alexander
and Philip it would be superfluous to recount. His
Habron at Samos is much admired, as is his Menander,
King of Caria, at Rhodes, likewise his Antaeus, and
at Alexandria his Gorgosthenes the Tragic Actor,
and at Rome his Castor and Pollux with Victory
and Alexander the Great, and also his figure of
War *a* with the Hands Tied behind, with Alexander
riding in Triumph in his Chariot. Both of these
pictures his late lamented Majesty Augustus with
restrained good taste *b* had dedicated in the most
frequented parts of his forum; the emperor Claudius
however thought it more advisable to cut out the
face of Alexander from both works and substitute
portraits of Augustus. The Heracles with Face
Averted in the temple of Diana is also believed to be
by his hand—so drawn that the picture more truly
displays Heracles' face than merely suggests it to
the imagination—a very difficult achievement. He
also painted a Nude Hero, a picture with which he
challenged Nature herself. There is, or was, a
picture of a Horse by him, painted in a competition,
by which he carried his appeal for judgement from
mankind to the dumb quadrupeds; for perceiving
that his rivals were getting the better of him by
intrigue, he had some horses brought and showed
them their pictures one by one; and the horses only
began to neigh when they saw the horse painted by
Apelles; and this always happened subsequently,
showing it to be a sound test of artistic skill. He

equo adversus Persas, Archelaum cum uxore et
filia, Antigonum thoracatum cum equo incedentem.
peritiores artis praeferunt omnibus eius operibus
eundem regem sedentem in equo et Dianam sacri-
ficantium virginum choro mixtam, quibus vicisse
Homeri versus videtur id ipsum describentis. pinxit
et quae pingi non possunt, tonitrua, fulgetra ful-
guraque; Bronten, Astrapen et Ceraunobolian
appellant.

97   Inventa eius et ceteris profuere in arte; unum
imitari nemo potuit, quod absoluta opera atramento
inlinebat ita tenui, ut id ipsum, cum [1] repercussum [2]
claritates [3] colorum [4] omnium [5] excitaret custo-
diretque a pulvere et sordibus, ad manum intuenti [6]
demum appareret, sed et luminum [7] ratione magna,
ne claritas colorum aciem offenderet veluti per
lapidem specularem intuentibus et e longinquo
eadem res nimis floridis coloribus austeritatem
occulte daret.

98   Aequalis eius fuit Aristides Thebanus. is omnium
primus animum pinxit et sensus hominis expressit,

---

[1] cum *add. Mayhoff*.

[2] repercussum B^1: repercussu *rell.*

[3] claritatis B, *cd. Par.* 6801 : claritates *rell.*

[4] colorem B : colorum *rell.* : oculorum *edd. vett.*

[5] ōnium (= omnium) *Mayhoff*: aluum B^1: alium B^2:
om. *rell.*: album *Traube*.

[6] intuenti et B.

[7] et luminum *Mayhoff*: etium B^1: etiam B^2: et cum
rell.: et tum *Hermolaus Barbarus.*

---

[a] One of Alexander's generals.

[b] Two soldiers with this name are recorded as serving under
Alexander.

[c] The One-eyed. See § 90 and note.

also did a Neoptolemus [a] on Horseback fighting against the Persians, an Archelaus [b] with his Wife and Daughter, and an Antigonus [c] with a Breast-plate marching with his horse at his side. Connoisseurs put at the head of all his works the portrait of the same king seated on horseback, and his Artemis in the midst of a band of Maidens offering a Sacrifice, a work by which he may be thought to have surpassed Homer's verses [d] describing the same subject. He even painted things that cannot be represented in pictures—thunder, lightning and thunderbolts, the pictures known respectively under the Greek titles of Bronte, Astrape and Ceraunobolia.

His inventions in the art of painting have been useful to all other painters as well, but there was one which nobody was able to imitate : when his works were finished he used to cover them over with a black varnish of such thinness that its very presence, while its reflexion threw up the brilliance of all the colours and preserved them from dust and dirt, was only visible to anyone who looked at it close up, but also employing great calculation of lights, so that the brilliance of the colours should not offend the sight when people looked at them as if through muscovy-glass and so that the same device from a distance might invisibly give sombreness to colours that were too brilliant.

Contemporary with Apelles was Aristides [e] of _Aristides._ Thebes. He was the first of all painters who depicted the mind and expressed the feelings of a human

---

[d] _Odyssey_ VI, 102 ff., which describe Artemis and maidens wildly ranging amongst boars and deer, not sacrificing. The mistake arises from the two verbs θύω.

[e] The younger, grandson of Aristides, cf. § 75 and note on pp. 410 and 411.

quae vocant Graeci ἤθη, item perturbationes, durior paulo in coloribus. huius opera [1] . . . oppido capto ad matris morientis ex volnere mammam adrepens infans, intellegiturque sentire mater et timere, ne emortuo lacte [2] sanguinem lambat. quam tabulam Alexander Magnus transtulerat Pellam in 99 patriam suam. idem pinxit proelium cum Persis, centum homines tabula ea complexus pactusque in singulos mnas denas a tyranno Elatensium Mnasone. pinxit et currentes quadrigas et supplicantem paene cum voce et venatores cum captura et Leontion Epicuri et anapauomenen propter fratris amorem,[3] item Liberum et Ariadnen [4] spectatos Romae in aede Cereris tragoedum et puerum in Apollinis, 100 cuius tabulae gratia interiit pictoris inscitia, cui tergendam eam mandaverat M. Iunius praetor sub die ludorum Apollinarium. spectata est et in aede Fidei in Capitolio senis cum lyra puerum docentis. pinxit et aegrum sine fine laudatum tantumque arte valuit, ut Attalus rex unam tabulam eius centum talentis emisse tradatur.

101 Simul, ut dictum est, et Protogenes floruit. patria ei Caunus, gentis Rhodiis subiectae. summa paupertas initio artisque summa intentio et ideo

[1] opera *B* : pictura *rell.* : lac. *Mayhoff.*

[2] lacte *B²* : flacte *B¹* : facta *rell.* : e lacte *Mayhoff* : emortuae (*aut* emortua) pro lacte *coni. Warmington.*

[3] propter fratris amorem *supra post* voce *transp.* Urlichs, *infra post* Ariadnen *Dilthey.*

[4] artamenen *B* : arianen *cd. Par. Lat.* 6797 : Artomenen *Dilthey.*

---

[a] There appears to be something lost here.

[b] After he had sacked Thebes in 335 B.C.

[c] It would be one of Alexander's great battles with Darius.

[d] Byblis perhaps, who died of love for her brother Caunus.

being, what the Greeks term *ēthē*, and also the emotions; he was a little too hard in his colours. His works include . . .[a] on the capture of a town, showing an infant creeping to the breast of its mother who is dying of a wound; it is felt that the mother is aware of the child and is afraid that as her milk is exhausted by death it may suck blood; this picture had been removed by Alexander the Great[b] to his native place, Pella. The same artist painted a Battle[c] with the Persians, a panel that contains a hundred human figures, which he parted with to Mnason the Tyrant of Elatea on the terms of ten minae per man. He also painted a Four-horse Chariots Racing, a Suppliant, who almost appeared to speak, Huntsmen with Quarry, Leontion Epicurus's mistress, and Woman[d] At Rest through Love of her Brother; and likewise the Dionysus and the Ariadne once on view in the Temple of Ceres at Rome, and the Tragic Actor and Boy in the Temple of Apollo, a picture of which the beauty has perished owing to the lack of skill of a painter commissioned by Marcus Junius as praetor to clean it in readiness for the festival of the Games of Apollo. There has also been on view in the Temple of Faith in the Capitol his picture of an Old Man with a Lyre giving lessons to a Boy. He also painted a Sick Man which has received unlimited praise; and he was so able an artist that King Attalus is said to have bought a single picture of his for a hundred talents.

Protogenes also flourished at the same time, as has *Protogenes.* been said. He was born at Caunus, in a community § 81. that was under the dominion of Rhodes. At the outset he was extremely poor, and extremely devoted to his art and consequently not very productive.

335

minor fertilitas. quis eum docuerit, non putant
constare; quidam et naves pinxisse usque ad quin-
quagensimum annum; argumentum esse, quod cum
Athenis celeberrimo loco Minervae delubri propylon
pingeret, ubi fecit nobilem Paralum et Hammoniada,
quam quidam Nausicaan vocant, adiecerit parvolas
naves longas in iis, quae pictores parergia appellant,
ut appareret, a quibus initiis ad arcem ostentationis [1]
102 opera sua pervenissent. palmam habet tabularum
eius Ialysus, qui est Romae dicatus in templo Pacis.
cum pingeret eum, traditur madidis lupinis vixisse,
quoniam [2] sic [3] simul et famem sustineret [4] et sitim
nec sensus nimia dulcedine obstrueret.[5] huic picturae quater colorem induxit ceu tria subsidia iniuriae
et vetustatis, ut decedente [6] superiore inferior
succederet. est in ea canis mire factus, ut quem
pariter ars et casus [7] pinxerit. non iudicabat se in
eo exprimere spumam anhelantis, cum in reliqua
parte omni, quod difficillimum erat, sibi ipse satis-
103 fecisset. displicebat autem ars ipsa: nec minui
poterat et videbatur nimia ac longius a veritate
discedere, spumaque pingi, non ex ore nasci. anxio
animi cruciatu, cum in pictura verum esse, non
verisimile vellet, absterserat saepius mutaveratque

---

[1] artis ostentationem *Rochette*. [2] quo *Traube*.
[3] sic *add. Mayhoff.* [4] sustinerent *edd. vett.*
[5] obstrueret *B, cd. Par.* 6801 : obstruerent *rell.*
[6] decidente *B recte?*
[7] ars et casus *Weil* : casus et ars *edd. vett.* : et casus *B* :
casus *rell.*

---

[a] Patron-heroes of sacred Athenian triremes used in state-
services. The Hammonias replaced the older ship Salaminia.
[b] Incidental details of any sort.
[c] With reference perhaps to the Acropolis or stronghold
(*arx*) of Athens. [d] Mythical founder of Ialysus in Rhodes.

Who his teacher was is believed to be unrecorded.
Some people say that until the age of fifty he was
also a ship-painter, and that this is proved by the
fact that when he was decorating with paintings, on
a very famous site at Athens, the gateway of the
Temple of Athene, where he depicted his famous
Paralus and Hammonias,[a] which is by some people
called the Nausicaa, he added some small drawings
of battleships in what painters call the ' side-pieces,' [b]
in order to show from what commencement his work
had arrived at the pinnacle [c] of glorious display.
Among his pictures the palm is held by his Ialysus,[d]
which is consecrated in the Temple of Peace in Rome.
It is said that while painting this he lived on soaked
lupins, because he thus at the same time both
sustained his hunger and thirst and avoided blunting
his sensibilities by too luxurious a diet. For this
picture he used four coats of paint, to serve as three
protections against injury and old age, so that when
the upper coat disappeared the one below it would
take its place. In the picture there is a dog marvellously executed, so as to appear to have been painted
by art and good fortune jointly : the artist's own
opinion was that he did not fully show in it the foam
of the panting dog, although in all the remaining
details he had satisfied himself, which was very
difficult. But the actual art displayed displeased
him, nor was he able to diminish it, and he thought
it was excessive and departed too far from reality—
the foam appeared to be painted, not to be the
natural product of the animal's mouth ; vexed and
tormented, as he wanted his picture to contain the
truth and not merely a near-truth, he had
several times rubbed off the paint and used another

penicillum, nullo modo sibi adprobans. postremo
iratus arti, quod intellegeretur, spongeam inpegit
inviso loco tabulae. et illa reposuit ablatos colores
qualiter cura optaverat, fecitque in pictura fortuna
naturam.

104   Hoc exemplo eius similis et Nealcen successus
spumae equi similiter spongea inpacta secutus
dicitur, cum[1] pingeret[2] poppyzonta retinentem
eum.[3] ita Protogenes monstravit et fortunam.[4]

Propter hunc Ialysum, ne cremaret tabulam,
Demetrius rex, cum ab ea parte sola posset Rhodum
capere, non incendit, parcentemque picturae fugit
105 occasio victoriae. erat tunc Protogenes in suburbano
suo hortulo, hoc est Demetrii castris,[5] neque inter-
pellatus proeliis incohata opera intermisit omnino
nisi accitus a rege, interrogatusque, qua fiducia
extra muros ageret, respondit scire se cum Rhodiis
illi bellum esse, non cum artibus. disposuit rex in
tutelam eius stationes, gaudens quod manus ser-
varet, quibus pepercerat, et, ne saepius avocaret,
ultro ad eum venit hostis relictisque victoriae suae

---

[1] *V.l.* dicuntur cum (disceret *B*) : dum celetem *Traube*.
[2] pingitur *B²* : pingatur *B¹* : pingit ac *Traube*.
[3] *V.l.* retinent pane cum : *varia temptant edd.*
[4] ita . . . fortunam *transp. vult Warmington supra post*
naturam § 103 : *idem coni.* ita ⟨et iram.⟩.
[5] hoc . . . castris *delendum ? (Urlichs)*.

---

[a] See § 102, p. 337.

brush, quite unable to satisfy himself. Finally he fell into a rage with his art because it was perceptible, and dashed a sponge against the place in the picture that offended him, and the sponge restored the colours he had removed, in the way that his anxiety had wished them to appear, and chance produced the effect of nature in the picture!

It is said that Nealces also following this example of his achieved a similar success in representing a horse's foam by dashing a sponge on the picture in a similar manner, in a representation of a man clucking in his cheek to soothe a horse he was holding. Thus did Protogenes indicate the possibilities of a stroke of luck also.

It was on account of this Ialysus *a* that King Demetrius, in order to avoid burning a picture, abstained from setting fire to Rhodes when the city could only be taken from the side where the picture was stored, and through consideration for the safety of a picture lost the chance of a victory! Protogenes at the time was in his little garden on the outskirts of the city, that is in the middle of the ' Camp of Demetrius,' and would not be interrupted by the battles going on, or on any account suspend the works he had begun, had he not been summoned by the King, who asked him what gave him the assurance to continue outside the walls. He replied that he knew the King was waging war with the Rhodians, not with the arts. The King, delighted to be able to safeguard the hands which he had spared, placed guardposts to protect him, and, to avoid repeatedly calling him from his work, actually though an enemy came to pay him visits, and quitting his aspirations for his own victory, in the thick of battles and the

Protogenes and King Demetrius. 305–4 B.C.

339

votis inter arma et murorum ictus spectavit artificem; sequiturque tabulam illius temporis haec
fama, quod eam Protogenes sub gladio pinxerit:
106 Satyrus hic est, quem anapauomenon vocant, ne
quid desit temporis eius securitati, tenentem tibias.

Fecit et Cydippen et Tlepolemum, Philiscum
tragoediarum scriptorem meditantem, et athletam [1]
et Antigonum regem, matrem Aristotelis philosophi,
qui ei suadebat, ut Alexandri Magni opera pingeret
propter aeternitatem rerum; impetus animi et quaedam artis libido in haec potius eum tulere; novissime pinxit Alexandrum [2] ac Pana. fecit et signa
ex aere, ut diximus.

107 Eadem aetate fuit Asclepiodorus, quem in
symmetria mirabatur Apelles. huic Mnaso tyrannus
pro duodecim diis dedit in singulos mnas tricenas,
idemque Theomnesto in singulos heroas vicenas.

108 His adnumerari debet et Nicomachus, Aristidis [3]
filius ac discipulus. pinxit raptum Proserpinae, quae
tabula fuit in Capitolio in Minervae delubro supra
aediculam Iuventatis, et in eodem Capitolio, quam
Plancus imperator posuerat, Victoria quadrigam in
sublime rapiens. Ulixi primus addidit pilleum.
109 pinxit et Apollinem ac Dianam, deumque matrem
in leone sedentem, item nobiles Bacchas obrep-

[1] Alcetam *Gronov.*
[2] Alexandream *Fröhner.*
[3] Aristidis *Mayhoff coll.* 111, 122: Aristidi *Urlichs*:
Aristidae illius *Oemichen*: Aristiaei *Sillig*: aristiaci *B*:
ariste(-i-)cheimi *rell.*

---

[a] Phaestis or Phaestias.
[b] The elder; cf. § 75 and note on pp. 410 and 411.
[c] Before the fire of A.D. 64.
[d] Munatius, who triumphed in 43 B.C.
[e] Cybele.

battering down of walls, looked on at the work of an
artist. And even to this day the story is attached to
a picture of that date that Protogenes painted it
with a sword hanging over him. The picture is the
one of a Satyr, called the Satyr Reposing, and to
give a final touch to the sense of security felt at the
time, the figure holds a pair of flutes.

Other works of Protogenes were a Cydippe, a *Other works* *of Protogenes.*
Tlepolemus, a Philiscus the Tragic Poet in Medita-
tion, an Athlete, a portrait of King Antigonus, and
one of the Mother [a] of Aristotle the philosopher.
Aristotle used to advise the artist to paint the
achievements of Alexander the Great, as belonging
to history for all time. The impulse of his mind
however and a certain artistic capriciousness led him
rather to the subjects mentioned. His latest works
were pictures of Alexander and of Pan. He also
made bronze statues, as we have said. XXXIV, 91.

In the same period there was also Asclepiodorus, *Asclepeio-*
who was admired by Apelles for his proportions. For *dorus.*
a picture of the Twelve Gods the tyrant Mnaso paid
him three hundred minae per god. The same patron
paid Theomnestus twenty minae for each of the
heroes in a picture.

To the list of these artists must also be added *Nicomachus*
Nicomachus son and pupil of Aristides.[b] He painted *and others.*
a Rape of Persephone, a picture formerly [c] in the
Shrine of Minerva on the Capitol, just above the
Chapel of Youth; and there was also in the Capitol,
where it was placed by General Plancus,[d] his Victory
hurrying her Chariot aloft. He was the first painter
who represented Odysseus wearing a felt skull-cap.
He also painted an Apollo and Artemis, and the
Mother [e] of the Gods seated on a Lion, and likewise

tantibus Satyris, Scyllamque, quae nunc est Romae
in templo Pacis. nec fuit alius in ea arte velocior.
tradunt namque conduxisse pingendum ab Aristrato,
Sicyoniorum tyranno, quod is faciebat Telesti poetae
monimentum praefinito die, intra quem perageretur,
nec multo ante venisse, tyranno in poenam accenso,
paucisque diebus absolvisse et celeritate et arte
110 mira. Discipulos habuit Aristonem fratrem et
Aristiden [1] filium et Philoxenum Eretrium, cuius
tabula nullis postferenda, Cassandro regi picta,
continuit Alexandri proelium cum Dario. idem
pinxit et lasciviam, in qua tres Sileni comissantur.
hic celeritatem praeceptoris secutus breviores etiamnum quasdam picturae conpendiarias invenit.
111 Adnumeratur his et Nicophanes, elegans ac con-
cinnus ita, ut venustate ei pauci conparentur;
cothurnus et gravitas artis multum a Zeuxide et
Apelle abest. Apellis discipulus Perseus, ad quem
de hac arte scripsit, huius fuerat aetatis. Aristidis
Thebani discipuli fuerunt et filii Niceros et Ariston,
cuius est Satyrus cum scypho coronatus, discipuli
Antorides [2] et Euphranor, de quo mox dicemus.
112 XXXVII. Namque subtexi par est minoris picturae
celebres in penicillo, e quibus fuit Piraeicus [3]

---

[1] *V.l.* Aristoclem.

[2] Antenorides *Letronne.*

[3] Piraeicus *Ian* : pirasicus *B* : praeicus *aut* preicus *rell.*
(peritus *cd. Par.* 6801).

---

[a] The younger. Cf. §98 and note on pp. 410–411.

[b] King of Macedonia 306–297 B.C.

[c] The younger confused with the elder, §§ 75, 98, 108.

[d] Really pupils of the elder Aristides.

[e] Really pupils of the elder Aristides.

a fine picture of Bacchants with Satyrs prowling
towards them, and a Scylla that is now in the Temple
of Peace in Rome. No other painter was ever a more
rapid worker. Indeed it is recorded that he accepted *c.* 355 B.C.
a commission from the tyrant of Sicyon Aristratus
to paint by a given date a monument that he was
erecting to the poet Telestes, and that he only *fl. c.* 398 B.C.
arrived not long before the date; the wrathful
tyrant threatened to punish him, but in a few days
he finished the work with a speed and an artistic
skill that were both remarkable. Among his pupils
were his brother Ariston and his son Aristides,[a]
and Philoxenus of Eretria, who painted for King
Cassander [b] a picture that holds the highest rank,
containing a battle between Alexander and Darius.
He also painted a picture with a wanton subject
showing three Sileni at their revels. Imitating the
rapidity of his master he introduced some shorthand
methods of painting, executed with still more
rapidity of technique.

With these artists is also reckoned Nicophanes,
an elegant and finished painter with whom few can be
compared for gracefulness, but who for tragic feeling
and weight of style is far from Zeuxis and Apelles.
Perseus, the pupil to whom Apelles dedicated his §79.
volumes on the art of painting, had belonged to the
same period. Aristides [c] of Thebes also had as his
pupils his sons Niceros and Ariston,[d] the latter the
painter of a Satyr Crowned with a Wreath and
Holding a Goblet; and other pupils of Aristides were
Antorides and Euphranor [e]; about the latter we
shall speak later on. §128.

XXXVII. For it is proper to append the artists *Piraeicus*
famous with the brush in a minor style of painting. *and others.*

343

arte paucis postferendus: proposito nescio an distinxerit[1] se, quoniam humilia quidem secutus
humilitatis tamen summam adeptus est gloriam.
tonstrinas sutrinasque pinxit et asellos et obsonia
ac similia, ob haec cognominatus rhyparographos,
in iis consummatae voluptatis, quippe eae pluris
113 veniere quam maximae multorum. e diverso
Maeniana, inquit Varro, omnia operiebat Serapionis
tabula sub Veteribus. hic scaenas optime pinxit,
sed hominem pingere non potuit. contra Dionysius
nihil aliud quam homines pinxit, ob id anthropo-
114 graphos cognominatus. parva et Callicles fecit, item
Calates comicis tabellis, utraque Antiphilus. namque
et Hesionam nobilem pinxit et Alexandrum ac
Philippum cum Minerva, qui sunt in schola in
Octaviae porticibus, et in Philippi Liberum patrem,
Alexandrum puerum, Hippolytum tauro emisso
expavescentem, in Pompeia vero Cadmum et
Europen. idem iocoso[2] nomine Gryllum deridiculi
habitus pinxit, unde id genus picturae grylli vocantur. ipse in Aegypto natus didicit a Ctesidemo.

---

[1] distinxerit *Mayhoff*: distrinxerit *Fröhner*: distruxerit
aut destruxerit.   [2] iocoso *edd. vett.*: iocosis *aut* locosis.

---

[a] Balconies on houses in Rome first built by one Maenius.
[b] Large and small pictures.

Among these was Piraeicus, to be ranked below few painters in skill; it is possible that he won distinction by his choice of subjects, inasmuch as although adopting a humble line he attained in that field the height of glory. He painted barbers' shops and cobblers' stalls, asses, viands and the like, consequently receiving a Greek name meaning 'painter of sordid subjects'; in these however he gives exquisite pleasure, and indeed they fetched bigger prices than the largest works of many masters. On the other hand 'a picture by Serapio,' says Varro, 'covered the whole of the Maenian Balconies *a* at the place Beneath the Old Shops.' Serapio was a most successful scene-painter, but he could not paint a human being. On the contrary, Dionysius painted nothing else but people, and consequently has a Greek name meaning 'Painter of Human Beings.' Callicles also made small pictures, and so did Calates of subjects taken from comedy; both classes *b* were painted by Antiphilus, who executed *Antiphilus.* the famous picture of Hesione and an Alexander and a Philip *c* with Athene which are now in the school in Octavia's Porticoes, and in Philippus' *d* Portico a Father Liber or Dionysus, a Young Alexander, a Hippolytus alarmed by the Bull rushing upon him, and in Pompey's Portico a Cadmus and Europa. He also painted a figure in an absurd costume known by the joking name of Gryllus, the name consequently applied to every picture of that sort. He was himself born in Egypt and a pupil of Ctesidemus.

c King of Macedon, father of Alexander.
d Of L. Marcius Philippus; built in 29 B.C.; Octavia's were built after 27 B.C., Pompey's (see below) in 55 B.C.

115   Decet non sileri et Ardeatis templi pictorem,
praesertim civitate donatum ibi et carmine,[1] quod
est in ipsa pictura his versibus:

> Dignis dignu'[2] loco[3] picturis condecoravit
> reginae Iunonis supremi coniugis templum
> Plautius Marcus[4]; cluet Asia lata[5] esse oriundus,
> quem nunc et post semper ob artem hanc Ardea
>   laudat,

116 eaque sunt scripta antiquis litteris Latinis; non
fraudando[6] et S. Tadio[7] divi Augusti aetate, qui
primus instituit amoenissimam parietum picturam,
villas et porticus[8] ac topiaria opera, lucos, nemora,
colles, piscinas, euripos, amnes, litora, qualia quis
optaret, varias ibi obambulantium species aut navi-
gantium terraque villas adeuntium asellis aut
vehiculis, iam piscantes, aucupantes aut venantes

117 aut etiam vindemiantes. sunt in eius exemplaribus
nobiles palustri[9] accessu[10] villae, succollatis[11] spon-
sione[12] mulieribus labantes trepidis quae[13] feruntur,
plurimae praeterea tales argutiae facetissimi salis.

---

    [1] carmen *Schneidewin.*
    [2] dignu' *Hermolaus Barbarus*: digna.
    [3] *fortasse* Lyco : *fortasse* dignu' loco dignis picturis *vel* dignis
digna loco. picturis.
    [4] plaucius marcus *B* : mareus plautis marcus *rell.*
    [5] cluet Asia lata *Bergk* : Cleoetas Alalia *Sillig* : cluetas
alata *B* : cloet (do et *cd. Flor. Ricc.*) asia lata *rell.*
    [6] fraudanda *Mayhoff.*
    [7] S. Tadio *Urlichs, Ian* : studio *B* : ludio *rell.*
    [8] porticus *cd. Par. Lat. 6797, ut videtur* : portus *rell.*
    [9] palustri *B* : paulstri *rell.* : plaustri *edd. vett.*
    [10] accessu *B* : ac censu *rell.*
    [11] succollatis *cdd.* (suae collatis *B*) : subcollantium *Hermo-
laus Barbarus.*
    [12] specie *Hermolaus Barbarus.*
    [13] trepidis quae *B, cd. Leid. Voss.* : trepidisque *rell.*

It is proper also not to pass over the painter of the *Italian painters.* temple at Ardea, especially as he was granted the *Plautius.* citizenship of that place and honoured with an inscription on the picture, consisting in the following verses:

> One Marcus Plautius, a worthy man,
> Adorned, with paintings worthy of this place,[a]
> The shrine of Juno, Queen of Spouse supreme,
> This Marcus Plautius, as men know, was born
> In Asia wide. Now, and hereafter always,
> Ardea applauds him for this work of art.

These lines are written in the antique Latin script. Nor must Spurius Tadius [b] also, of the period of his *Tadius.* late lamented Majesty Augustus, be cheated of his due, who first introduced the most attractive fashion *Before* of painting walls with pictures of country houses and *A.D. 37.* porticoes and landscape gardens, groves, woods, hills, fish-ponds, canals, rivers, coasts, and whatever anybody could desire, together with various sketches of people going for a stroll or sailing in a boat or on land going to country houses riding on asses or in carriages, and also people fishing and fowling or hunting or even gathering the vintage. His works include splendid villas approached by roads across marshes,[c] men tottering and staggering along carrying women on their shoulders for a bargain, and a number of humorous drawings of that sort besides, extremely wittily designed. He

---

[a] But perhaps the right reading is *Dignis digna.* *Lyco...* 'To the worthy, worthy reward; Lycon adorned. . . .' *I.e.* the artist was M. Plautius Lycon, keeping his Greek name when he received a new one on becoming a citizen at Ardea.

[b] Or Studius or Ludius. The reading is uncertain.

[c] Or: 'well known among his works are men approaching a country house across marshes. . . .' The Latin text of much of this sentence is uncertain.

idem subdialibus maritimas urbes pingere instituit, blandissimo aspectu minimoque inpendio.

118 Sed nulla gloria artificum est nisi qui tabulas pinxere. eo venerabilior antiquitatis prudentia apparet. non enim parietes excolebant dominis tantum nec domos uno in loco mansuras, quae ex incendiis rapi non possent. casa Protogenes contentus erat in hortulo suo; nulla in Apellis tectoriis pictura erat. nondum libebat parietes totos tinguere; omnium eorum ars urbibus excubabat, pictorque res communis terrarum erat. [a]

119 Fuit et Arellius Romae celeber paulo ante divum Augustum, ni flagitio insigni corrupisset artem, semper ei lenocinans feminae, cuius[1] amore flagraret, et[2] ob id deas pingens, sed dilectarum imagine.

120 itaque in pictura eius scorta numerabantur. fuit et nuper gravis ac severus idemque floridis tumidus[3] pictor Famulus.[4] huius erat Minerva spectantem spectans, quacumque aspiceretur. paucis diei horis pingebat, id quoque cum gravitate, quod semper togatus, quamquam in machinis. carcer eius artis domus aurea fuit, et ideo non extant exempla alia magnopere. post eum fuere in auctoritate Cornelius Pinus et Attius Priscus, qui Honoris

---

[1] feminae cuius *Mayhoff*: cuius feminae.
[2] flagraret et *Urlichs*: flagrans esset *coni. Sillig*: flagrans et.
[3] floridis tumidus *coni. Ian*: floridis (floridus *B²*) umidus *B*: floridus humilis *cd. Par. Lat.* 6797: f. h. rei *cd. Par.* 6801: f. humidis *cd. Flor. Ricc.*: f. umidis *rell.*: floridus et vividus *Traube*: floridus *Sillig*: floridissimus *Urlichs*: floridis multus *vel* f. nitidus *vel* invictus *coni. Mayhoff*: S. Ummidius *Fröhner*.
[4] Famulus *cdd.*: famulus *Fröhner*: Fabullus *ed. princ.*

---

[a] *I.e.* canvases or panels, not wall-paintings.   [b] Nero's palace.

also introduced using pictures of seaside cities to decorate uncovered terraces, giving a most pleasing effect and at a very small expense.

But among artists great fame has been confined to painters of pictures only,[a] a fact which shows the wisdom of early times to be the more worthy of respect, for they did not decorate walls, merely for owners of property, or houses, which would remain in one place and which could not be rescued from a fire. Protogenes was content with a cottage in his little garden; Apelles had no wall-frescoes in his house; it was not yet the fashion to colour the whole of the walls. With all these artists their art was on the alert for the benefit of cities, and a painter was the common property of the world.

A little before the period of his late lamented *Arellius.* Majesty Augustus, Arellius also was in high esteem at Rome, had he not prostituted his art by a notorious outrage, by always paying court to any woman he happened to fall in love with, and consequently painting goddesses, but in the likeness of his mistresses; and so his pictures included a number of portraits of harlots. Another recent painter was *Famulus.* Famulus, a dignified and severe but also very florid artist; to him belonged a Minerva who faced the spectator at whatever angle she was looked at. Famulus used to spend only a few hours a day in painting, and also took his work very seriously, as he always wore a toga, even when in the midst of his easels. The Golden House [b] was the prison that contained his productions, and this is why other examples of his work are not extant to any considerable extent. After him in esteem were Cornelius Pinus and Attius Priscus, who painted

Virtutis aedes Imperatori Vespasiano Augusto
restituenti pinxerunt, Priscus antiquis similior.

121 XXXVIII. Non est omittenda in picturae mentione celebris circa Lepidum fabula, siquidem in
triumviratu quodam loco deductus a magistratibus
in nemorosum hospitium minaciter cum iis postero
die expostulavit somnum ademptum sibi volucrum
concentu; at illi draconem in longissima membrana
depictum circumdedere luco, eoque terrore aves
tunc siluisse narratur et postea posse compesci.

122 XXXIX. Ceris pingere ac picturam inurere quis
primus excogitaverit, non constat. quidam Aristidis
inventum putant, postea consummatum a Praxitele;
sed aliquanto vetustiores encaustae picturae exstitere, ut Polygnoti et Nicanoris, Mnesilai[1] Pa-
riorum. Elasippus[2] quoque Aeginae picturae suae
inscripsit ἐνέκαεν, quod profecto non fecisset nisi
encaustica inventa.

123 XL. Pamphilus quoque, Apellis praeceptor, non
pinxisse solum encausta, sed etiam docuisse
traditur Pausian Sicyonium, primum in hoc genere
nobilem. Bryetis filius hic fuit eiusdemque primo
discipulus. pinxit et ipse penicillo parietes Thespiis,
cum reficerentur quondam a Polygnoto picti,

[1] Mnesilai *Mayhoff*: mens im *B* : ae *aut* e manesilai *rell.*
ac Mnasilai *Detlefsen*: et Archesilai *Hermolaus Barbarus* (et
arcesilai *cd. Par.* 6801 *ut videtur*).

[2] Elasippus *Schneidewin*: eiasippus *B* : lassippus *rell.*:
Lysippus *edd. vett.*

[a] With Octavian and Antony, formed in 43 B.C.
[b] Apparently the elder; cf. § 75.

the temples of Honour and Virtue for the Emperor
Vespasian's restoration of them; Priscus was nearer
in style to the artists of old days.

XXXVIII. In speaking of painting one must
not omit the famous story about Lepidus. During
his Triumvirate,[a] when entertained by the magistrates of a certain place, he was given lodging in a
house buried in trees; and the next day he
complained to them in threatening language that
he had been robbed of sleep by the singing of the
birds; however the authorities had a picture of a
large snake made on an extremely long strip of
parchment and fixed it up round the wood, and the
story goes that this at once frightened the birds
into silence, and that subsequently it was possible
to keep them in check.

XXXIX. It is not agreed who was the inventor *Painting in wax; en-caustic.*
of painting in wax and of designs in encaustic.
Some people think it was a discovery of Aristides,[b]
subsequently brought to perfection by Praxiteles,
but there were encaustic paintings in existence at a
considerably earlier date, for instance those of
Polygnotus, and Nicanor and Mnasilaus of Paros.
Also Elasippus of Aegina has inscribed on a picture
enekaën (' burnt in '), which he would not have done
if the art of encaustic painting had not been invented.

XL. It is recorded also that Pamphilus, the *Pausias.*
teacher of Apelles, not only painted in encaustic but
also taught it to Pausias of Sicyon, the first artist
who became famous in this style. Pausias was the
son of Bryetes, and started as his father's pupil. He
himself also did some wall-painting with the brush
at Thespiae, when some old paintings by Polygnotus
were being restored, and he was deemed to come

multumque comparatione superatus existimabatur,
124 quoniam non suo genere certasset. idem et lacu-
naria primus pingere instituit, nec camaras ante eum
taliter adornari mos fuit; parvas pingebat tabellas
maximeque pueros. hoc aemuli interpretabantur
facere eum, quoniam tarda picturae ratio esset illi.
quam ob rem daturus ei celeritatis famam absolvit
uno die tabellam quae vocata est hemeresios, puero
125 picto. amavit in iuventa Glyceram municipem suam,
inventricem coronarum, certandoque imitatione eius
ad numerosissimam florum varietatem perduxit
artem illam. postremo pinxit et[1] ipsam sedentem
cum corona, quae e nobilissimis tabula est, appellata
stephanoplocos, ab aliis stephanopolis, quoniam
Glycera venditando coronas sustentaverat paupertatem. huius tabulae exemplar, quod apographon
vocant, L. Lucullus duobus talentis emit. . . .[2]
126 Dionysius[3] Athenis. Pausias autem fecit et grandes
tabulas, sicut spectatam in Pompei porticu boum
immolationem. eam primus invenit picturam, quam
postea imitati sunt multi, aequavit nemo. ante
omnia, cum longitudinem bovis ostendi vellet,
adversum eum pinxit, non traversum, et abunde

---

[1] et B: om. rell.   [2] lac. Mayhoff.
[3] Dionysiis Pintianus: a Dionysio edd. vett.

---

[a] There is no proof that perspective is meant, but somehow
Pausias gave the figure due relief.

off very second best in comparison with the original artist, having entered into competition in what was not really his line. Pausias also first introduced the painting of panelled ceilings, and it was not customary before him to decorate arched roofs in this way. He used to paint miniatures, and especially children. His rivals explained this practice as being due to the slow pace of his work in painting; and consequently to give his work also the reputation of speed he finished a picture in a single day, a picture of a boy which was called in Greek Hemeresios, meaning One-day Boy. In his youth he fell in love with a fellow-townswoman named Glycera, who invented chaplets of flowers; and by imitating her in rivalry he advanced the art of encaustic painting so as to reproduce an extremely numerous variety of flowers. Finally he painted a portrait of the woman herself, seated and wearing a wreath, which is one of the very finest of pictures; it is called in Greek Stephanoplocos, Girl making Wreaths, or by others Stephanopōlis, Girl selling Wreaths, because Glycera had supported her poverty by that trade. A copy (in Greek *apographon*) of this picture was bought by Lucius Lucullus at Athens for two talents; ⟨it had been made by⟩ 88–7 B.C. Dionysius at Athens. But Pausias also did large pictures, for instance the Sacrifice of Oxen which formerly was to be seen in Pompey's Portico. He first invented a method of painting which has afterwards been copied by many people but equalled by no one; the chief point was that although he wanted to show the long body of an ox he painted the animal facing the spectator and not standing sideways, and its great size is fully conveyed.[a]

353

127 intellegitur amplitudo. dein, cum omnes, quae
volunt eminentia videri, candicanti faciant colore,
quae condunt, nigro, hic totum bovem atri coloris
fecit umbraeque corpus ex ipsa [1] dedit, magna prorsus
arte in aequo extantia ostendente et in confracto
solida omnia. Sicyone et hic vitam egit, diuque illa
fuit patria picturae. tabulas inde e publico omnes
propter aes alienum civitatis addictas Scauri aedilitas
Romam transtulit.

128 Post eum eminuit longe ante omnes Euphranor
Isthmius olympiade CIIII, idem qui inter fictores
dictus est nobis. fecit et colossos et marmorea et
typos [2] scalpsit, docilis ac laboriosus ante omnes et in
quocumque genere excellens ac sibi aequalis. hic
primus videtur expressisse dignitates heroum et
usurpasse symmetrian, sed fuit in universitate
corporum exilior et capitibus articulisque grandior.

129 volumina quoque composuit de symmetria et coloribus. opera eius sunt equestre proelium, XII dei,
Theseus, in quod dixit eundem apud Parrhasium
rosa pastum esse, suum vero carne. nobilis eius
tabula Ephesi est, Ulixes simulata insania bovem
cum equo iungens et palliati cogitantes, dux gladium
condens.

---

[1] *V.l.* ipso.   [2] scyphos *cd. Par.* 6801.

---

[a] Pliny perhaps means that in spite of varying ('broken')
tones of black, all the black looks solid.

[b] *I.e.* later than P. But this is wrong. Pliny's mistake
has been traced to his confusion of the two artists named
Aristides. Cf. §§ 75, 108, 111 and note on pp. 410–11.

[c] Fingers and toes?

[d] Which preceded the battle of Mantinea, 362 B.C. This

Next, whereas all painters ordinarily execute in light colour the parts they wish to appear prominent and in dark those they wish to keep less obvious, this artist has made the whole ox of a black colour and has given substance to the shadow from the shadow itself, with quite remarkable skill that shows the shapes standing out on a level surface and a uniform solidity on a broken ground.[a] Pausias also passed his life at Sicyon, which was for a long period a native place of painting. But all the pictures there had to be sold to meet a debt of the community, and were removed from the ownership of the state to Rome by Scaurus as aedile.   56 B.C.

After Pausias,[b] Euphranor the Isthmian distin- *Euphranor.* guished himself far before all others, in the 104th 364–361 B.C. Olympiad; he has also appeared in our account of XXXIV, 50. statuaries. His works included colossal statues, works in marble, and reliefs, as he was exceptionally studious and diligent, excelling in every field and never falling below his own level. This artist seems to have been the first fully to represent the lofty qualities of heroes, and to have achieved good proportions, but he was too slight in his structure of the whole body and too large in his heads and joints.[c] He also wrote books about proportions and about colours. Works of his are a Cavalry Battle,[d] the Twelve Gods, and a Theseus, in respect of which he said that Parrhasius's Theseus had lived on a diet of roses, but his was a beef-eater. There is a celebrated picture by him at Ephesus, Odysseus Feigning Madness and yoking an ox with a horse, with men in cloaks reflecting, and the leader sheathing his sword.

and the next two pictures were both in the Stoa of Zeus Eleutherios at Athens. Paus. I, 3, 3–4.

130 Eodem tempore fuere Cydias,[1] cuius tabulam Argonautas HS cxxxxiiii Hortensius orator mercatus est eique aedem fecit in Tusculano suo. Euphranoris autem discipulus Antidotus. huius est clipeo dimicans Athenis et luctator tubicenque inter pauca laudatus. ipse diligentior quam numerosior et in coloribus severus[2] maxime inclaruit discipulo Nicia Atheniense, qui diligentissime mulieres pinxit.

131 lumen et umbras custodiit atque ut eminerent e tabulis picturae[3] maxime curavit. operum[4] eius Nemea advecta ex Asia Romam a Silano, quam in curia diximus positam, item Liber pater in aede Concordiae, Hyacinthus, quem Caesar Augustus delectatus eo secum deportavit Alexandrea capta, et ob id Tiberius Caesar in templo eius dicavit hanc tabulam, et Danae,[5] Ephesi vero est megabyzi,

132 sacerdotis Ephesiae Dianae, sepulchrum, Athenis necyomantea Homeri. hanc vendere Attalo regi noluit talentis LX potiusque patriae suae donavit abundans opibus. fecit et grandes picturas, in quibus sunt Calypso et Io et Andromeda; Alexander quoque in Pompei porticibus praecellens et Calypso sedens huic eidem[6] adscribuntur.

---

[1] Cydias *Detlefsen* : et Cydias Cythnius *Urlichs, Bergk* : cydi (*aut* cidi) et cydias (*aut* cidias).
[2] *V.ll.* severior.   [3] figurae *coni. Mayhoff.*
[4] operū (= operum) *Mayhoff* : opera.
[5] danae // *cd. Leid. Voss.* : *v.ll.* danaen, danen, diana.
[6] eidem *Sillig, Schultz* : quidem *edd. vett.*: fidem.

---

[a] In the forum at Rome; built by Tiberius, A.D. 14.
[b] ' Place of prophecy of the dead,' *Odyssey* Bk. XI.
[c] Attalus I of Pergamum, 241–197 B.C. But Plutarch is probably right in giving the king's name as Ptolemy I Soter, ruler of Egypt 323–284 B.C.

Contemporaries of Euphranor were Cydias, for *Cydias.* whose picture of the Argonauts the orator Hortensius *114–50 B.C.* paid 144,000 sesterces, and made a shrine for its reception at his villa at Tusculum. Euphranor's pupil was Antidotus. Works by the latter are a *Antidotus.* Combatant with a Shield at Athens and a Wrestler and a Trumpeter which has been exceptionally praised. Antidotus himself was more careful in his work than prolific, and severe in his use of colours; his chief distinction was being the teacher of the Athenian Nicias, who was an extremely careful *Nicias the younger.* painter of female portraits. Nicias kept a strict watch on light and shade, and took the greatest pains to make his paintings stand out from the panels. Works of his are : a Nemea, brought to Rome from *75 B.C.* Asia by Silanus and deposited in the Senate-house as we have said, and also the Father Liber or Dionysus *§ 27.* in the Shrine of Concord, a Hyacinthus with which Caesar Augustus was so delighted that when he took *30 B.C.* Alexandria he brought it back with him—and consequently Tiberius Caesar dedicated this picture in the Temple *a* of Augustus—and a Danaë; while at Ephesus there is the tomb of a megabyzus or priest of Diana of Ephesus, and at Athens there is a Necyomantea *b* of Homer. The last the artist refused to sell to King Attalus *c* for 60 talents, and preferred to present it to his native place, as he was a wealthy man. He also executed some large pictures, among them a Calypso, an Io *d* and an Andromeda; and also the very fine Alexander in Pompey's Porticoes and a Seated Calypso are assigned to him.

d One or two extant later paintings may be copies of this. A. Rumpf, *Journ. Hellen. St.,* LXVII, 21.

133 Quadripedum[1] prosperrime canes expressit. hic
est Nicias, de quo dicebat Praxiteles interrogatus, quae maxime opera sua probaret in marmoribus: quibus Nicias manum admovisset; tantum
circumlitioni eius tribuebat. non satis discernitur,
alium eodem nomine an hunc eundem quidam
faciant olympiade cxii.

134 Niciae comparatur et aliquando praefertur Athenion
Maronites, Glaucionis Corinthii discipulus, austerior
colore et in austeritate iucundior, ut in ipsa pictura
eruditio eluceat. pinxit in templo Eleusine phy-
larchum et Athenis frequentiam, quam vocavere
syngenicon, item Achillem virginis habitu occul-
tatum Ulixe deprendente et in una tabula vi signa,[2]
quaque maxime inclaruit, agasonem cum equo.
quod nisi in iuventa obiisset, nemo compararetur.

135 Est nomen et Heraclidi Macedoni. initio naves
pinxit captoque Perseo rege Athenas commigravit.
ubi eodem tempore erat Metrodorus, pictor idemque
philosophus, in utraque scientia magnae auctoritatis.
itaque cum L. Paulus devicto Perseo petiisset ab
Atheniensibus, ut ii sibi quam probatissimum
philosophum mitterent ad erudiendos liberos, item
pictorem ad triumphum excolendum, Athenienses
Metrodorum elegerunt, professi eundem in utroque
desiderio praestantissimum, quod ita Paulus quoque

---

[1] quadripedum *Madvig*: quadripedē (quadripedes *B*).
[2] vi signa *Gronov*: insigni *Durand*: ut signa.

---

[a] It must be remembered that Greek marbles were painted.

In drawings of animals he was most successful with dogs. It is this Nicias of whom Praxiteles used to say, when asked which of his own works in marble he placed highest, ' The ones to which Nicias has set his hand '—so much value did he assign to his colouring of surfaces.[a] It is not quite clear whether it is another artist of the same name or this Nicias whom some people put in the 112th Olympiad.

Nicias the elder.

332–329 B.C.
Athenion.

With Nicias is compared Athenion of Maronea, and sometimes to the disadvantage of the former. Athenion was a pupil of Glaucion of Corinth; he is more sombre in his colour than Nicias and yet therewithal more pleasing, so that his extensive knowledge shines out in his actual painting. He painted a Cavalry Captain in the temple at Eleusis and at Athens the group of figures which has been called the Family Group, and also an Achilles Disguised in Female Dress detected by Odysseus, a group of six figures in a single picture, and a Groom with a Horse, which has specially contributed to his fame. If he had not died in youth, there would have been nobody to compare with him.

Heraclides of Macedon is also a painter of note. He began by painting ships, and after the capture of King Perseus he migrated to Athens, where at the same period was the painter Metrodorus, who was also a philosopher and a great authority in both fields. Accordingly when Lucius Paulus after conquering Perseus requested the Athenians to send him their most esteemed philosopher to educate his children, and also a painter to embellish his triumphal procession, the Athenians selected Metrodorus, stating that he was most distinguished in both of these requirements alike, as to which Paulus also

Heraclides.

168 B.C.
Metrodorus.

136 iudicavit. Timomachus Byzantius Caesaris dictatoris aetate Aiacem et Mediam pinxit, ab eo in Veneris Genetricis aede positas, LXXX talentis venundatas. talentum Atticum X̄ V̄I̅ [1] taxat M. Varro. Timomachi aeque laudantur Orestes, Iphigenia in Tauris et Lecythion, agilitatis exercitator, cognatio nobilium, palliati, quos dicturos pinxit, alterum stantem, alterum sedentem. praecipue tamen ars ei favisse in Gorgone visa est.

137 Pausiae filius et discipulus Aristolaus e severissimis pictoribus fuit, cuius sunt Epaminondas, Pericles, Media, Virtus, Theseus, imago Atticae plebis, boum immolatio. sunt quibus et Nicophanes, eiusdem Pausiae discipulus, placeat diligentia, quam intellegant soli artifices, alias durus in coloribus et sile multus. nam Socrates iure omnibus placet; tales sunt eius cum Aesculapio filiae Hygia, Aegle,[2] Panacea, Iaso,[3] et piger, qui appellatur Ocnos, spartum torquens, quod asellus adrodit.

138 Hactenus indicatis proceribus in utroque genere non silebuntur et primis proximi: Aristoclides, qui [4] pinxit aedem Apollinis Delphis. Antiphilus puero ignem conflante laudatur ac pulchra alias domo [5]

---

[1] V̄I̅ *Hardouin* : VI.M. *B* : XVI *rell.*

[2] Hygia, Aegle *edd. vett.*: thygiaegle *B* : hygiagle *rell.*

[3] Panacea, Iaso *Hermolaus Barbarus* : panaca iasus *B* : p. lacus *cd. Par. Lat.* 6797 : penaca lacus *aut* locus *rell.*

[4] qui *delendum coni. Mayhoff.*

[5] laudatur ê pictura atra foculo *coni. Mayhoff.*

---

[a] Copies of this picture exist.

[b] Probably a mask of Medusa.

[c] Hence a Latin proverb: *ocnus spartum torquens,* 'sloth twisting a rope.' *I.e.* 'Labour in vain.'

[d] Both large and small pictures.

held the same view. Timomachus of Byzantium *Timoma-*
in the period of Caesar's dictatorship[a] painted an *chus.*
Ajax and a Medea, placed by Caesar in the temple *46 B.C.*
of Venus Genetrix, having been bought at the price
of 80 talents (Marcus Varro rates the Attic talent
at 6000 denarii). Equal praise is given to
Timomachus's Orestes, his Iphigenia among the
Tauri and his Gymnastic-Master Lecythion; also his
Noble Family and his Two Men wearing the Pallium,
whom he has represented as about to converse; one
is a standing figure and the other seated. It is in
his painting of a Gorgon[b] however that his art seems
to have given him most success.

Pausias's son and pupil Aristolaus was one of the *Aristolaus.*
painters of the very severe style; to him belong an
Epaminondas, a Pericles, a Medea, a Virtue, a
Theseus, a figure representing the Athenian People,
and a Sacrifice of Oxen. Some persons also admire
Nicophanes, who was likewise a pupil of Pausias, *Nicophanes.*
for his careful accuracy which only artists can appreciate, though apart from that he is hard in his
colouring and lavish in his use of ochre. As for
Socrates he is justly a universal favourite; popular *Socrates.*
pictures by him are his group of Asclepius with his
daughters Health, Brightness, All-Heal and Remedy,
and his Sluggard, bearing the Greek name of Ocnos,
Laziness, and represented as twisting a rope of
broom which an ass is nibbling.[c]

Having so far pointed out the chief painters in both
branches,[d] we will also mention those of the rank
next to the first: Aristoclides who decorated the *Aristoclides*
Temple of Apollo at Delphi, Antiphilus who is *and others.*
praised for his Boy Blowing a Fire, and for the
apartment, beautiful in itself, lit by the reflection

splendescente ipsiusque pueri ore, item lanificio,
in quo properant omnium mulierum pensa, Ptolemaeo venante, sed nobilissimo Satyro cum pelle
pantherina, quem aposcopeuonta appellant, Aristophon Ancaeo [1] vulnerato [2] ab apro cum socia doloris
Astypale [3] numerosaque tabula, in qua sunt Priamus,
Helena, Credulitas, Ulixes, Deiphobus, Dolus.[4]

139 Androbius pinxit Scyllum ancoras praecidentem
Persicae classis, Artemon Danaen mirantibus eam
praedonibus, reginam Stratonicen, Herculem et
Deianiram, nobilissimas autem, quae sunt in Octaviae operibus, Herculem ab Oeta monte Doridos
exusta mortalitate consensu deorum in caelum
euntem, Laomedontis circa Herculem et Nep-
tunum historiam; Alcimachus Dioxippum, qui
pancratio Olympiae citra pulveris [5] iactum, quod
vocant ἀκονιτί, vicit; Coenus stemmata.

140 Ctesilochus, Apellis discipulus, petulanti pictura
innotuit, Iove Liberum parturiente depicto mitrato
et muliebriter ingemescente inter obstetricia dearum,
Cleon Cadmo, Ctesidemus Oechaliae expugnatione,
Laodamia, Ctesicles [6] reginae Stratonices iniuria.
nullo enim honore exceptus ab ea pinxit volu-

[1] Ancaeo *Gelen*   Ancaeum *Detlefsen* : ancaiu B^1 : angaiu
B^2 : anchalū *rell.*
[2] vulnerato *edd. vett.*: muineratumo *B* : vulneratū *rell.*
[3] Astypalaea *Brunn.*   [4] Dolon *Caesarius.*
[5] pulveris *edd. vett.*: pueris.
[6] Ctesicles *Ian* : etesides *B* : clesides *rell.*

---

[a] ᾿Αποσκοπεύων, shading his eyes with his hand (Athenaeus,
XIV, 629 f.).   The gesture is a common one in satyrs on vases.
[b] At Artemisium, 480 B.C.
[c] Probably S. who was wife of Seleucus I Nicator, King of
Nearer Asia 312–281 B.C.
[d] Heracles saved Hesione from a monster sent by Posidon

from the fire and the light thrown on the boy's face; and likewise for his Spinning-room, in which all the women are busily plying their tasks, and his Ptolemy Hunting, but, most famous of all, his Satyr with Leopard's Skin, called in Greek the Man Shading his Eyes.[a] Aristophon did an Ancaeus Wounded by the Boar, with Astypale sharing his grief, and a picture crowded with figures, among them Priam, Helen, Credulity, Odysseus, Deiphobus, Craft. Androbius painted a Scyllus Cutting the Anchor-ropes of the Persian Fleet,[b] Artemon a Danae admired by the Robbers, a Queen Stratonice,[c] and a Heracles and Deianira; but the finest of all his works, now in Octavia's Buildings, are his Heracles Ascending to Heaven with the consent of the Gods after his mortal remains were burnt on Mount Oeta in Doris, and the story of Laomedon in the matter of Heracles and Posidon.[d] Alcimachus painted Dioxippus, who won the All-round Bout at Olympia 'without raising any dust,'[e] akoniti as the Greek word is. Coenus painted pedigrees.

Ctesilochus a pupil of Apelles became famous for a saucy burlesque painting which showed Zeus in labour[f] with Dionysus, wearing a woman's nightcap and crying like a woman, while goddesses act as midwives; Cleon for his Cadmus, Ctesidemus for his Storming of Oechalia[g] and his Laodamia. Ctesicles won notoriety by the insult he offered to Queen Stratonice,[h] because as she did not give him an honourable reception he painted a picture of her

to ravage the land of Troy after Hesione's father King Lao-medon broke a promise.

[e] *I.e.* without any difficulty.

[f] Dionysus was born from Zeus' thigh.

[g] By Heracles.   [h] See note c.

tantem cum piscatore, quem reginam amare sermo
erat, eamque tabulam in portu Ephesi proposuit,
ipse velis raptus. regina tolli vetuit, utriusque
similitudine mire expressa. Cratinus comoedos[1]
Athenis in pompeo[2] pinxit; Eutychides[3] bigam:
141 regit[4] Victoria. Eudorus scaena spectatur—idem
et ex aere signa fecit—, Hippys[5] Neptuno et
Victoria. Habron Amicitiam[6] et Concordiam pinxit
et deorum simulacra, Leontiscus Aratum victorem
cum tropaeo, psaltriam, Leon Sappho, Nearchus
Venerem inter Gratias et Cupidines, Herculem
tristem insaniae paenitentia, Nealces Venerem,
142 ingeniosus et sollers, . . .[7] ime siquidem, cum proelium navale Persarum et Aegyptiorum pinxisset,
quod in Nilo cuius est aqua maris similis[8] factum
volebat intellegi, argumento declaravit quod arte
non poterat: asellum enim bibentem in litore
143 pinxit et crocodilum insidiantem ei; Oenias
syngenicon, Philiscus officinam pictoris ignem conflante puero, Phalerion Scyllam, Simonides Agathar-
chum et Mnemosynen, Simus iuvenem requiescen-

[1] comoedos *Caesarius* : comoedus.
[2] pompeo *Gelen* : pompeio.
[3] Eutychides *Hermolaus Barbarus* : eutychidis *cd. Flor. Ricc.
ut videtur* : euthycides *B* : euclides *cd. Par.* 6801.
[4] regis *cd. Par.* 6801 : regis cum *Gelen* : quam regit *quid.
apud Dalecamp* : Eutychides biga quam regit Victoria,
Eudorus scaena spectatur *coni. Mayhoff.*
[5] Hippys *Keil* : Hippias *Hardouin* : Iphis *edd. vett.* : hyppis
cd. Par. Lat. 6797 : hyppus *B* : hypis *rell.*
[6] Amicitiam *edd. vett.* : amicam.
[7] *lac. Mayhoff.*
[8] cuius . . . similis *delenda esse putant Urlichs, Mayhoff.*

[a] At the city gates; from it solemn processions started.
[b] I.e. apparently the river is so extensive that in the
picture it might be mistaken for the sea.

romping with a fisherman with whom gossip said she was in love, and put it on exhibition at Ephesus Harbour, himself making a hurried escape on ship-board. The Queen would not allow the picture to be removed, the likeness of the two figures being admirably expressed. Cratinus painted the Comic Actors in the Processional Building [a] at Athens, Eutychides a Chariot and Pair driven by Victory. Eudorus is famous for a scene-painting—he also made bronze statues—and Hippys for his Posidon and his Victory. Habron painted a Friendship and a Harmony and figures of gods, Leontiscus an Aratus with the Trophies of Victory, and a Harpist Girl, Leon a Sappho, Nearchus Aphrodite among the Graces and the Cupids, and a Heracles in Sorrow Repenting his Madness, Nealces an Aphrodite. This Nealces was a talented and clever artist, inas-much as when he painted a picture of a naval battle between the Persians and the Egyptians, which he desired to be understood as taking place on the river Nile, the water of which resembles [b] the sea, he suggested by inference what could not be shown by art: he painted an ass standing on the shore drinking, and a crocodile lying in wait for it. [c] Oenias has done a Family Group, Philiscus a Painter's Studio with a boy blowing the fire, Phalerion a Scylla, Simonides an Agatharchus and a Mnemosyne, Simus a Young Man Reposing, a Fuller's Shop

[c] But it is certain that the picture referred to a battle in the Persian Artaxerxes III Ochus' conquest of Egypt in 350 B.C. The Egyptians called him ' Ass ' (with allusion to the ass-shaped Seth Typhon who represented the wicked foe); and the likeness of ὄνος (ass) to Ὦχος (Ochus) became a joke amongst Greeks who fought on both sides.

tem, officinam fullonis quinquatrus celebrantem,
144 idemque Nemesim egregiam, Theorus [1] se inungentem,[2] idem ab Oreste matrem et Aegisthum
interfici, bellumque Iliacum pluribus tabulis, quod
est Romae in Philippi porticibus, et Cassandram,
quae est in Concordiae delubro, Leontium Epicuri
cogitantem, Demetrium regem, Theon Orestis
insaniam, Thamyram citharoedum, Tauriscus disco-
bolum, Clytaemestram, Paniscon, Polynicen regnum
repetentem et Capanea.

145 Non omittetur inter hos insigne exemplum.
namque Erigonus, tritor colorum Nealcae pictoris,
in tantum ipse profecit, ut celebrem etiam discipulum reliquerit Pasiam, fratrem Aeginetae pictoris.[3] illud vero perquam rarum ac memoria
dignum est, suprema opera artificum inperfectasque
tabulas, sicut Irim Aristidis, Tyndaridas Nicomachi,
Mediam Timomachi et quam diximus Venerem
Apellis, in maiore admiratione esse quam perfecta,
quippe in iis liniamenta reliqua ipsaeque cogitationes
artificum spectantur, atque in lenocinio commendationis dolor est manus, cum id ageret, exstinctae.

146 Sunt etiamnum non ignobiles quidem in transcursu tamen dicendi Aristocydes, Anaxander, Aristobulus Syrus, Arcesilas [4] Tisicratis filius, Coroebus [5]

---

[1] V.l. Theodorus.
[2] se inungentem Sillig, Ian : emungentem B : et inungentem aut et mungentem rell.
[3] pictoris Detlefsen : pictores B : fictoris aut fictores rell.
[4] Arcesilaus cd. Par. 6801 : arcesillas B.
[5] Coroebus Keil : Corybas edd. vett. : corbios aut corbius
aut cordius.

---

[a] Celebrated for five days, March 19th–23rd, by persons
whose trades were under Minerva's patronage. The original
doubtless depicted some festival of Athene.

Celebrating the Quinquatrus,[a] and also a Nemesis of great merit; Theorus a Man Anointing Himself, and also Orestes killing his Mother and Aegisthus, and the Trojan War in a series of pictures now in Philippus' Porticoes at Rome and a Cassandra, in the Shrine of Concord, a Leontion Epicurus's mistress in Contemplation, a King Demetrius; Theon a Madness of Orestes, a Thamyras the Harper; Tauriscus a Man throwing a Quoit, a Clytaemnestra, a Young Pan, a Polynices Claiming the Sovereignty,[b] and a Capaneus.

Among these artists the following remarkable case is not to be left out; the man who ground the colours for the painter Nealces, Erigonus, attained such proficiency on his own account that he actually left behind him a famous pupil, Pasias, the brother of the painter Aeginetas. It is also a very unusual and memorable fact that the last works of artists and their unfinished pictures such as the Iris of Aristides,[c] the Tyndarus' Children [d] of Nicomachus, the Medea of Timomachus and the Aphrodite of Apelles which we have mentioned, are more admired §92. than those which they finished, because in them are seen the preliminary drawings left visible and the artists' actual thoughts, and in the midst of approval's beguilement we feel regret that the artist's hand while engaged in the work was removed by death.

There are still some artists who are not undistinguished but who only need be mentioned in passing—Aristocydes, Anaxander, Aristobulus of Syria, Arcesilas son of Tisicrates, Coroebus the pupil of

[b] Of Thebes, against his brother Eteocles.

[c] See §§ 75, 98, 108, 111.

[d] These were Castor, Polydeuces (Pollux), Helen, and Clytaemnestra.

Nicomachi discipulus, Charmantides [1] Euphranoris, Dionysodorus [2] Colophonius, Dicaeogenes,[3] qui cum Demetrio rege vixit, Euthymides,[4] Heraclides Macedo, Milon Soleus,[5] Pyromachi [6] statuarii discipuli, Mnasitheus [7] Sicyonius, Mnasitimus Aristonidae filius et discipulus, Nessus Habronis filius, Polemon Alexandrinus, Theodorus Samius et Stadius [8] Nicosthenis discipuli, Xenon, Neoclis discipulus, Sicyonius.

147   Pinxere et mulieres: Timarete, Miconis filia, Dianam, quae in tabula Ephesi est antiquissimae [9] picturae; Irene, Cratini pictoris filia et discipula, puellam, quae est Eleusine, Calypso, senem et praestigiatorem Theodorum, Alcisthenen saltatorem; Aristarete, Nearchi filia et discipula, Aesculapium. Iaia [10] Cyzicena, perpetua virgo, M. Varronis iuventa [11] Romae et penicillo pinxit et cestro in ebore imagines mulierum maxime et Neapoli anum in grandi tabula,

148 suam quoque imaginem ad speculum. nec ullius velocior in pictura manus fuit, artis vero tantum, ut multum manipretiis antecederet celeberrimos eadem aetate imaginum pictores Sopolim et Dionysium, quorum tabulae pinacothecas inplent. pinxit et

[1] Charmantides *Keil*: charmanides *B*: carmanides *rell.*
[2] Dionysodorus *Keil*: dionysiodorus.
[3] Dicaeogenes *Keil*: dicaogenes (diogenes *cd. Par.* 6801).
[4] euthymides *cd. Par. Lat.* 6797, *ut videtur*: euthymedes *cd. Par.* 6801: eutymides *B*.
[5] Soleus *Gelen*: solaeuus *B*: solus *rell.*
[6] Philomachi *edd. vett.*   [7] Mnesitheus *Hardouin.*
[8] stadius *cd. Flor. Ricc.*: statius *cd. Leid. Voss.*: stadios *rell.*: Tadius *edd. vett.*
[9] antiquissimae *B*: in antiquissimis *rell.*
[10] iaia *B*: lala *rell.*: Laia *Schneidewin*: Maia *Fröhner.*
[11] inventa *Rochette.*

[a] Of Macedon 249–287 B.C.

Nicomachus, Charmantides, the pupil of Euphranor,
Dionysodorus of Colophon, Dicaeogenes resident
at the court of King Demetrius,[a] Euthymides, the
Macedonian Heraclides and Milon of Soli, pupils of
Pyromachus, the sculptor of the human figure,
Mnasitheus of Sicyon, Mnasitimus the son and pupil
of Aristonides, Nessus son of Habron, Polemo of
Alexandria, Theodorus of Samos and Stadius, both
pupils of Nicosthenes, Xenon of Sicyon, pupil of
Neocles.

There have also been women artists—Timarete the *Women artists.*
daughter of Micon who painted the extremely
archaic panel picture of Artemis at Ephesus, Irene
daughter and pupil of the painter Cratinus who
did the Maiden at Eleusis, a Calypso,[b] an Old
Man and Theodorus the Juggler, and painted also
Alcisthenes the Dancer; Aristarete the daughter
and pupil of Nearchus, who painted an Asclepius.
When Marcus Varro was a young man, Iaia of *116-26 B.C.*
Cyzicus, who never married, painted pictures with
the brush at Rome (and also drew with the *cestrum*
or graver[c] on ivory), chiefly portraits of women, as
well as a large picture on wood of an Old Woman at
Naples, and also a portrait of herself, done with a
looking-glass. No one else had a quicker hand
in painting, while her artistic skill was such that in
the prices she obtained she far outdid the most
celebrated portrait painters of the same period,
Sopolis and Dionysius, whose pictures fill the
galleries. A certain Olympias also painted; the

[b] Or, if Calypso is the name of a woman artist, . . .
'Eleusis, Calypso, who painted an Old Man . . .'.

[c] The *cestrum* was, it seems, a graver, spoon-shaped at one
end (for holding colours over heat), and with the handle-end
thickened or flattened out for levelling the colours.

369

quaedam Olympias, de qua hoc solum memoratur,
discipulum eius fuisse Autobulum.

149   XLI. Encausto pingendi duo fuere antiquitus
genera, cera et in ebore cestro, id est vericulo,[1]
donec classes pingi coepere. hoc tertium accessit
resolutis igni ceris penicillo utendi, quae pictura
navibus nec sole nec sale ventisve corrumpitur.

150   XLII. Pingunt et vestes in Aegypto, inter pauca
mirabili genere, candida vela, postquam attrivere,
inlinentes non coloribus, sed colorem sorbentibus
medicamentis. hoc cum fecere, non apparet in
velis, sed in cortinam pigmenti ferventis mersa post
momentum extrahuntur picta. mirumque, cum sit
unus in cortina colos, ex illo alius atque alius fit in
veste accipientis medicamenti qualitate mutatus,
nec postea ablui potest. ita cortina, non dubie
confusura colores, si pictos acciperet, digerit ex
uno pingitque, dum coquit, et adustae eae vestes
firmiores usibus fiunt quam si non urerentur.

151   XLIII. De pictura satis superque. contexuisse
his et plasticen conveniat. eiusdem opere [2] terrae
fingere ex argilla similitudines Butades Sicyonius

---

[1] vericulo *Sillig* : viriculo. *Verba* id est vericulo *fortasse
delenda*.

[2] operae *B, cd. Leid. Voss. m.*2 : opere *rell.* : operis *coni
Mayhoff.*

---

[a] These words look like a gloss. Pliny has already mentioned
the *cestrum* in § 147. The two kinds of encaustic painting
here mentioned are with wax and graver on wood, and with
wax and graver on ivory.

[b] Far back though it is, Pliny seems to refer to § 1 of this
book. But the right reading is not certain.

only fact recorded about her is that Autobulus was her pupil.

XLI. In early days there were two kinds of *Encaustic.* encaustic painting, with wax and on ivory with a graver or *cestrum* (that is a small pointed graver [a]); but later the practice came in of decorating battle-ships. This added a third method, that of employing a brush, when wax has been melted by fire; this process of painting ships is not spoilt by the action of the sun nor by salt water or winds.

XLII. In Egypt they also colour cloth by an *Egyptian* exceptionally remarkable kind of process. They *dyeing.* first thoroughly rub white fabrics and then smear them not with colours but with chemicals that absorb colour. When this has been done, the fabrics show no sign of the treatment, but after being plunged into a cauldron of boiling dye they are drawn out a moment later dyed. And the remarkable thing is that although the cauldron contains only one colour, it produces a series of different colours in the fabric, the hue changing with the quality of the chemical employed, and it cannot afterwards be washed out. Thus the cauldron which, if dyed fabrics were put into it, would un-doubtedly blend the colours together, produces several colours out of one, and dyes the material in the process of being boiled; and the dress fabrics when submitted to heat become stronger for wear than they would be if not so heated.

XLIII. Enough and more than enough has now *Plastic art.* been said about painting. It may be suitable to *Early* append to these remarks something about the *stages.* plastic art. It was through the service of that *others.* same earth [b] that modelling portraits from clay was *Butades and*

figulus primus invenit Corinthi filiae opera, quae
capta amore iuvenis, abeunte illo peregre, umbram
ex facie eius ad lucernam in pariete lineis circumscripsit, quibus pater eius inpressa argilla typum
fecit et cum ceteris fictilibus induratum igni proposuit, eumque servatum in Nymphaeo, donec
152 Mummius Corinthum everterit, tradunt. sunt qui
in Samo primos omnium plasticen invenisse Rhoecum
et Theodorum tradant multo ante Bacchiadas
Corintho pulsos, Damaratum vero ex eadem urbe
profugum, qui in Etruria Tarquinium regem populi
Romani genuit, comitatos fictores Euchira, Diopum,
Eugrammum; ab iis Italiae traditam plasticen.
Butadis inventum est rubricam addere aut ex rubra
creta fingere, primusque personas tegularum extremis imbricibus inposuit, quae inter initia prostypa
vocavit; postea idem ectypa fecit. hinc et fastigia
templorum orta. propter hunc plastae appellati.
153 XLIV. Hominis autem imaginem gypso e facie
ipsa primus omnium expressit ceraque in eam
formam gypsi infusa emendare instituit Lysistratus
Sicyonius, frater Lysippi, de quo diximus. hic et
similitudines[1] reddere instituit; ante eum quam

---

[1] similitudines *Sillig* : similitudini *coni. Mayhoff* : similitudinis *B* : similitudinem *rell.*

---

[a] Traditionally they invented the art of casting bronze,
not of making casts in clay.

[b] Of these fictitious names *Eucheir* means 'skilful-handed,'
and *Eugrammus* 'skilled drawer.' *Diopus* would be connected
with διόπτρα, an instrument for taking levels.

[c] In low relief.

[d] In high relief.

first invented by Butades, a potter of Sicyon, at Corinth. He did this owing to his daughter, who was in love with a young man; and she, when he was going abroad, drew in outline on the wall the shadow of his face thrown by a lamp. Her father pressed clay on this and made a relief, which he hardened by exposure to fire with the rest of his pottery; and it is said that this likeness was preserved in the Shrine of the Nymphs until the destruction of Corinth by Mummius. Some authorities state that the plastic art was first invented by Rhoecus and Theodorus[a] at Samos, long before the expulsion of the Bacchiadae from Corinth, but that when Damaratus, who in Etruria became the father of Tarquin king of the Roman people, was banished from the same city, he was accompanied by the modellers Euchir, Diopus and Eugrammus,[b] and they introduced modelling to Italy. The method of adding red earth to the material or else modelling out of red chalk, was an invention of Butades, and he first placed masks as fronts to the outer gutter-tiles on roofs; these at the first stage he called *prostypa*,[c] but afterwards he likewise made *ectypa*.[d] It was from these that the ornaments on the pediments of temples originated. Because of Butades modellers get their Greek name of *plastae*.

XLIV. The first person who modelled a likeness in plaster of a human being from the living face itself, and established the method of pouring wax into this plaster mould and then making final corrections on the wax cast, was Lysistratus of Sicyon, the brother of Lysippus of whom we have spoken. Indeed he introduced the practice of giving likenesses, the object aimed at previously having been to

146 B.C.

581–580 B.C.

trad. 616–578 B.C.

Lysistratus.
XXXIV, 61.

pulcherrimas facere studebant. idem et de signis
effigies exprimere invenit, crevitque res in tantum,
ut nulla signa statuaeve sine argilla fierent. quo
apparet antiquiorem hanc fuisse scientiam quam
fundendi aeris.

154 XLV. Plastae laudatissimi fuere Damophilus et
Gorgasus, iidem pictores, qui Cereris aedem Romae
ad circum maximum utroque genere artis suae
excoluerant, versibus inscriptis Graece, quibus sig-
nificarent ab dextra opera Damophili esse, ab laeva [1]
Gorgasi. ante hanc aedem Tuscanica omnia in
aedibus fuisse auctor est Varro, et ex hac, cum
reficeretur, crustas parietum excisas tabulis mar-
ginatis inclusas esse, item signa ex fastigiis dispersa.

155 fecit et Chalcosthenes cruda opera Athenis, qui
locus ab officina eius Ceramicos appellatur. M.
Varro tradit sibi cognitum Romae Possim nomine, a quo facta poma et uvas ut non posses [2]
aspectu discernere a veris.[3] idem magnificat Arcesi-
laum, L. Luculli familiarem, cuius proplasmata [4]
pluris venire solita artificibus ipsis quam aliorum

156 opera; ab hoc factam Venerem Genetricem in foro
Caesaris et, priusquam absolveretur, festinatione

---

[1] ab laeva *Gronov*: a parte laeva *cd. Par.* 6801 : ab imia
B : aplane *aut* aplone *rell.*
[2] ut non posses *Hardouin* (u. n. possis *Gronov*): nemo
posset *Mayhoff*: non possis *Ian*: ita ut non sit *Gelen*:
alitem nescisse *Traube*: item pisces *B* : item piscis *rell.* (poscis
cd. Par. Lat. 6797) : item pisces quos *cd. Poll.*
[3] veris vix posses *cd. Poll.*
[4] proplasmata *Gelen*: propriasmata *B* : proplasticen *edd.*
vett.: plastica *cd. Par.* 6801 : pleurosamta *rell.*

---

[a] Dedicated in 493 B.C.   [b] See XXXIV, 34.
[c] It was restored (after the fire of 31 B.C.) by Augustus in
27 B.C.   [d] Or Caecosthenes (= Καϊκοσθένης) ? cf. XXXIV. 87.

make as handsome a face as possible. The same artist also invented taking casts from statues, and this method advanced to such an extent that no figures or statues were made without a clay model. This shows that the knowledge of modelling in clay was older than that of casting bronze.

XLV. Most highly praised modellers were Damophilus and Gorgasus, who were also painters; they had decorated the Shrine [a] of Ceres in the Circus Maximus at Rome with both kinds of their art, and there is an inscription on the building in Greek verse in which they indicated that the decorations on the right hand side were the work of Damophilus and those on the left were by Gorgasus. Varro states that before this shrine was built everything in the temples was Tuscanic work [b]; and that when this shrine was undergoing restoration,[c] the embossed work of the walls was cut out and enclosed in framed panels; and that the figures also were taken from the pediment and dispersed. Chalcosthenes [d] also executed at Athens some works in unbaked clay, at the place named the Ceramicus, Potters Quarter, after his workshop. Marcus Varro records that he knew at Rome an artist named Possis who made fruit and grapes in such a way that nobody could tell by sight from the real things. Varro also speaks very highly of Arcesilaus, who was on terms of intimacy with Lucius Lucullus, and says that his sketch-models of clay used to sell for more, among artists themselves, than the finished works of others; and that this artist made the statue of Venus Genetrix in Caesar's Forum and that it was erected before it was finished as there was a great haste to dedicate it; and that the same artist had

Damophilus Gorgasus, and others.

Arcesilaus. c. 110–56 B.C.

46 B.C.

dedicandi positam; eidem a Lucullo HS |x̄|[1] signum
Felicitatis locatum, cui mors utriusque viderit;
Octavio equiti Romano cratera facere volenti exemplar e gypso factum talento. laudat et Pasitelen,
qui plasticen matrem caelaturae et statuariae
scalpturaeque[2] dixit et, cum esset in omnibus iis
summus, nihil umquam fecit ante quam finxit.
157 praeterea elaboratam hanc artem Italiae et maxime
Etruriae; Vulcam[3] Veis accitum, cui locaret Tarquinius Priscus Iovis effigiem in Capitolio dicandam;
fictilem eum fuisse et ideo miniari solitum; fictiles
in fastigio templi eius quadrigas, de quibus supra[4]
diximus; ab hoc eodem factum Herculem, qui
hodieque materiae nomen in urbe retinet. hae
enim tum effigies deorum erant lautissimae, nec
paenitet nos illorum, qui tales eos coluere; aurum
enim et argentum ne diis quidem conficiebant.
158 XLVI. durant etiam nunc plerisque in locis talia
simulacra; fastigia quidem templorum etiam in
urbe crebra et municipiis, mira caelatura et arte
suique firmitate, sanctiora auro, certe innocentiora.
in sacris quidem etiam inter has opes hodie non

---

[1] |x̄| *Detlefsen*: L̄X̄ *Sillig*: IXI *B*: LX *rell.*
[2] scalpturae *B*[1]: sculpturae *B*[2]: scalturae *rell.*
[3] uulcam *B*[1]: uulcani *B*[2]: uulgam *rell.* (turianum *cd. Par.*
6801): Volcaniam *coni. Ian.*
[4] supra *coni. Ian.*: saepe.

---

[c] Since Arcesilaus was still doing work for Caesar in 46 B.C.,
it may well be that the Lucullus here mentioned is the one
who was killed at Philippi in 42 B.C.
[b] The Hercules Fictilis, ' Hercules in Clay.'

contracted with Lucullus to make a statue of Happiness for 1,000,000 sesterces, which was prevented by the death of both parties [a]; and that when a Knight of Rome Octavius desired him to make a wine-bowl he made him a model in plaster for the price of a talent. He also praises Pasiteles, who said that *Pasiteles.* modelling was the mother of chasing and of bronze statuary and sculpture, and who, although he was eminent in all these arts, never made anything before he had made a clay model. He also states that this art had already been brought to perfection by Italy and especially by Etruria; that Vulca was summoned from Veii to receive the contract from Tarquinius Priscus for a statue of Jupiter to be *trad. 616-* consecrated in the Capitol, and that this Jupiter *578 B.C.* was made of clay and consequently was regularly painted with cinnabar; and that the four-horse VIII, 161 chariots about which we spoke above on the pediment of the temple were modelled in clay; and that the figure of Hercules, which even to-day retains in the city the name [b] of the material it is made of, was the work of the same artist. For these were the most splendid images of gods at that time; and we are not ashamed of those ancestors of ours for worshipping them in that material. For they used not formerly to work up silver and gold even for gods. XLVI. Statues of this kind are still to be found at various places. In fact even at Rome and in the Municipal Towns there are many pediments of temples, remarkable for their carving and artistic merit and intrinsic durability, more deserving of respect than gold, and certainly less baneful. At the present day indeed, even in the midst of our present rich resources the preliminary libation is

murrinis crystallinisve, sed fictilibus prolibatur
simpulis,[1] inenarrabili Terrae benignitate, si quis
singula aestimet, etiam ut omittantur in frugum,
159 vini, pomorum, herbarum et fruticum, medicamen-
torum, metallorum generibus beneficia eius, quae [2]
adhuc diximus. neque [3] adsiduitate satiant figli-
narum opera, doliis ad vina excogitatis, ad aquas
tubulis,[4] ad balineas mammatis,[5] ad tecta imbricibus,[6] coctilibus laterculis ad parietes fundamentaque,[7] aut quae [8] rota fiunt, propter quae [9] Numa
160 rex septimum collegium figulorum instituit. quin
et defunctos sese multi fictilibus soliis condi maluere,
sicut M. Varro, Pythagorio modo in myrti et oleae
atque populi nigrae foliis. maior pars hominum
terrenis utitur vasis. Samia etiam nunc in escu-
lentis laudantur. retinent hanc nobilitatem et
Arretium in Italia et calicum tantum Surrentum,
Hasta, Pollentia, in Hispania Saguntum, in Asia
161 Pergamum. habent et Trallis ibi opera sua et in
Italia Mutina, quoniam et sic gentes nobilitantur
et haec quoque per maria, terras ultro citro por-
tantur, insignibus rotae officinis.[10] Erythris in templo
hodieque ostenduntur amphorae duae propter tenui-
tatem consecratae discipuli magistrique certamine,

---

[1] simpulis *edd. vett.*: sin puls B^1: sinpulsa B^2: simpuuiis
cd. Par. 6801: simpuis *rell.*

[2] *V.ll.* quaeque, que quae: quaequae *Detlefsen.*

[3] neque *Mayhoff*: vel quae B: vel *rell.*

[4] *V.l.* tabulis.

[5] hamatis *Hermolaus Barbarus.*

[6] imbricibus *hic Mayhoff*: *supra post* opera.

[7] ad parietes fundamentaque *coni. Mayhoff* (ad fundamenta
Detlefsen): fundamentisque.

[8] ob quae *Hardouin*: quae aut *Detlefsen.*

[9] rota fiunt propter quae B: om. *rell. In* § 159 *alia
temptant edd.* [10] officiis *coni. Rackham.*

made at sacrifices not from fluor-spar or crystal vessels but with small ladles of earthenware, thanks to the ineffable kindness of Mother Earth, if one considers her gifts in detail, even though we omit her blessings in the various kinds of corn, wine, fruit, herbs and shrubs, drugs and metals, all the things that we have so far mentioned. Nor do our *Pottery.* products even in pottery satisfy our needs with their unfailing supply, with jars invented for our wine, and pipes for water, conduits for baths, tiles for our roofs, baked bricks for our house-walls and foundations, or things that are made on a wheel, because of which King Numa established a seventh Guild, the *trad.* Potters.[a] Indeed moreover many people have 715–672 B.C. preferred to be buried in earthenware coffins, for instance Marcus Varro who was interred in the Pythagorean style, in leaves of myrtle, olive and 26 B.C. black poplar; the majority of mankind employs earthenware receptacles for this purpose. Among table services Samian pottery is still spoken highly of; this reputation is also retained by Arezzo in Italy, and, merely for cups, by Sorrento, Asti, and Pollenza, and by Saguntum in Spain and Pergamum in Asia Minor. Also Tralles in Asia Minor and Modena in Italy have their respective products, since even this brings nations fame, and their products also, so distinguished are the workshops of the potter's wheel, are carried to and fro across land and sea. In a temple at Erythrae even to-day are on view two wine-jars which were dedicated on account of their fine material, owing to a competition between a master potter and his apprentice as to which

---

[a] The text of part of § 159 is very uncertain.

uter tenuiorem humum duceret. Cois ea laus[1]
maxima, Hadrianis firmitas, nonnullis circa hoc
162 severitatis quoque exemplis. Q. Coponium invenimus ambitus damnatum, quia vini amphoram
dedisset dono ei, cui suffragi latio erat. atque ut
e luxu[2] quoque aliqua contingat auctoritas figlinis :
tripatinium, inquit Fenestella, appellabatur summa
cenarum lautitia ; una erat murenarum, altera
luporum, tertia mixti piscis, inclinatis iam scilicet
moribus, ut tamen eos praeferre Graeciae etiam
philosophis possimus, siquidem in Aristotelis here-
dum auctione septuaginta patinas venisse traditur.
163 nos cum unam Aesopi tragoediarum histrionis in
natura avium diceremus HS $\bar{c}$[3] stetisse, non dubito
indignatos legentes. at, Hercules, Vitellius in
principatu suo $\lceil\overline{x}\rfloor$[4] HS condidit patinam, cui faciendae
fornax in campis exaedificata erat, quoniam eo
pervenit luxuria, ut etiam fictilia pluris constent
164 quam murrina. propter hanc Mucianus altero consulatu suo in conquestione exprobravit patinarum
paludes Vitelli memoriae, non illa foediore, cuius
veneno Asprenati reo Cassius Severus accusator
165 obiciebat interisse convivas cxxx. nobilitantur his
quoque oppida, ut Regium et Cumae. Samia testa

---

[1] ea laus *Mayhoff* : illa laus *Urlichs* : levitas *Fröhner* : laus.
[2] e luxu *coni. Ian* : fruxu *aut* frucu *aut* fruxo *cdd.* (fluxu
cd. Poll.).
[3] $\bar{c}$ *cd. Leid. Voss.* [c] *B* : c' *aut* centum *aut* DC *rell.*
[4] $\lceil\overline{x}\rfloor$ *B* : $\overline{x}$ *aut* x *rell.*

would make thinner earthenware. The pottery
of Cos is most famous for this, but that of Adria is
most substantial; while there are also some instances
of severity also in relation to pottery. We find
that Quintus Coponius was found guilty of bribery
because he made a present of a jar of wine to a
person who had the right to a vote. And so that
luxury also may contribute some importance to
earthenware, the name of a service of three dishes,
we are told by Fenestella, used to denote the most
luxurious possible banquet: one dish was of lamprey,
a second of pike and a third of a mixture of fish.
Clearly manners were already on the decline, though
nevertheless we can still prefer them even to those
of the philosophers of Greece, inasmuch as it is
recorded that at the auction held by the heirs of 322 B.C.
Aristotle seventy earthenware dishes were sold.
We have already stated when on the subject of birds x. 141.
that a single dish cost the tragic actor Aesop 100,000
sesterces, and I have no doubt that readers felt
indignant; but, good heavens, Vitellius when
emperor had a dish made that cost 1,000,000 sesterces, A.D. 69.
and to make which a special furnace was constructed
out in open country, as luxury has reached a point
when even earthenware costs more than vessels
of fluor-spar. It was owing to this dish that
Mucianus in his second consulship, in a protest A.D. 70.
which he delivered, reproached the memory of
Vitellius for dishes as broad as marshes, although
this particular dish was not more disgraceful than
the poisoned one by which Cassius Severus when
prosecuting Asprenas charged him with having caused
the death of 130 guests. Artistic pottery also con-
fers fame on towns, for instance Reggio and Cumae.

Matris deum sacerdotes, qui Galli vocantur, viri-
litatem amputare nec aliter citra perniciem, M.
Caelio credamus, qui linguam sic amputandam
obiecit gravi probro, tamquam et ipse iam tunc
eidem Vitellio malediceret. quid non excogitat
vita[1] fractis etiam testis utendo, sic ut firmius
durent, tunsis calce addita, quae vocant Signina!
quo genere etiam pavimenta excogitavit.

166   XLVII. Verum et ipsius terrae sunt alia com-
menta. quis enim satis miretur pessumam eius
partem ideoque pulverem appellatam in Puteolanis
collibus opponi maris fluctibus, mersumque protinus
fieri lapidem unum inexpugnabilem undis et for-
tiorem cotidie, utique si Cumano misceatur cae-
167 mento? eadem est terrae natura et in Cyzicena
regione, sed ibi non pulvis, verum ipsa terra qua
libeat magnitudine excisa et demersa in mare
lapidea extrahitur. hoc idem circa Cassandream
produnt fieri, et in fonte Cnidio dulci intra octo
menses terram lapidescere. ab Oropo quidem
Aulida usque quidquid attingitur mari terrae mutatur in saxa. non multum a pulvere Puteolano

---

[1] excogitat vita *B* : excogitavit vita *Ian* : excogitavit ars
cd. Par. 6801 : excogitavit a *rell.*

---

[a] Cybele.
[b] Volcanic ash or earth, now called *pozzolana*.
[c] The ancient Potidaea in the Chalcidic peninsula.
[d] Of Boeotia.

The priests of the Mother [a] of the Gods called Galli castrate themselves, if we accept the account of Marcus Caelius, with a piece of Samian pottery, the only way of avoiding dangerous results; and Caelius proposed as a penalty for an abominable offence that the guilty person should have his tongue cut out in the same way, just as if he were already himself inveighing against the same Vitellius in anticipation. What is there that experience cannot devise? For it employs even broken crockery, making it more solid and durable by pounding it up and adding what is called Segni lime, a kind of material used in a method which experience has also invented for making pavements.

XLVII. But there are other inventions also that belong to Earth herself. For who could sufficiently marvel at the fact that the most inferior portion of the earth's substance, which is in consequence designated dust,[b] on the hills of Pozzuoli, encounters the waves of the sea and as soon as it is submerged turns into a single mass of stone that withstands the attacks of the waves and becomes stronger every day, especially if it mixed with broken quarry-stone from Cumae? In the Cyzicus district also the nature of the earth is the same, but there not dust but the earth itself is cut out in blocks of any size wanted and plunged into the sea; and when drawn out, it is of the consistency of stone. The same is said to take place in the neighbourhood of Cassandrea,[c] and it is stated that in a fresh water spring at Cnidus earth becomes petrified in less than eight months. Or the coast [d] from Oropus to Aulis all the earth that the sea touches is turned into rocks. The finest portion of the sand from the Nile is not very different

Pozzolana and other earths.

distat e Nilo harena tenuissima sui parte, non ad
sustinenda maria fluctusque frangendos, sed ad
168 debellanda corpora palaestrae studiis. inde certe
Patrobio, Neronis principis liberto, advehebatur.
quin et Cratero et Leonnato ac Meleagro, Alexandri
Magni ducibus, sabulum [1] hoc portari [2] cum reliquis
militaribus commerciis reperio, plura de hac parte
non dicturus, non, Hercules, magis quam de terrae
usu in ceromatis, quibus exercendo iuventus nostra
169 corporis vires perdit animorum. XLVIII. quid?
non in Africa Hispaniaque e terra parietes, quos
appellant formaceos, quoniam in forma circumdatis
II utrimque tabulis inferciuntur verius quam struun-
tur, aevis durant, incorrupti imbribus, ventis, ignibus
omnique caemento firmiores? spectat etiam nunc
speculas Hannibalis Hispania terrenasque turres
iugis montium inpositas. hinc et caespitum natura
castrorum vallis accommodata contraque fluminum
impetus aggeribus. inlini quidem crates parietum
luto et ut [3] lateribus crudis exstrui quis ignorat?
170   XLIX. Lateres non sunt ex sabuloso neque
harenoso multoque minus calculoso ducendi solo, sed
e cretoso et albicante aut ex rubrica vel etiam e

---

[1] sabulum *cd. Par.* 6801, *cd. Par. Lat.* 6797 : sabium *B* :
sablum *rell.* : solitum *Mayhoff.*
[2] *V.l.* portare : portavere *J. Müller.*
[3] et ut *Warmington* : et.

---

[a] Who was in Spain 221–219 B.C. preparing war against
Rome.

from the dust of Pozzuoli, not to be used for an embankment against the sea and to act as a break-water against waves, but for the purpose of subduing men's bodies for the exercises of the wrestling school. At all events it used to be imported from there for Patrobius, a freedman of the emperor Nero, and moreover I also find that this sand was carried with other military commodities for Alexander the Great's generals Craterus, Leonnatus and Meleager, though I shall not say more about this part of the subject any more than, by heaven, I shall mention the use of earth in making ointments, employed by our young men while ruining their vigour of mind by exercising their muscles. XLVIII. Moreover, are there not in Africa and Spain walls made of earth that are called framed walls, because they are made by packing in a frame enclosed between two boards, one on each side, and so are stuffed in rather than built, and do they not last for ages, undamaged by rain, wind and fire, and stronger than any quarry-stone? Spain still sees the watchtowers of Hannibal [a] and turrets of earth placed on the mountain ridges. From the same source is also obtained the substantial sods of earth suitable for the fortifications of our camps and for embankments against the violent flooding of rivers. At all events everybody knows that party-walls can be made by coating hurdles with clay, and are thus built up as if with raw bricks.

XLIX. Bricks should not be made from a sandy *Bricks.* or gravelly soil and far less from a stony one, but from a marly and white soil or else from a red earth; or even with the aid of sand, at all events if coarse male sand is used. The best time for making bricks

385

sabulo, masculo certe. finguntur optime vere, nam
solstitio rimosi fiunt. aedificiis non nisi bimos
probant; quin [1] et intritam ipsam eorum, priusquam
fingantur, macerari oportet.

171　　Genera eorum fiunt [2] tria: didoron,[3] quo nos [4]
utimur, longum sesquipedem, latum pedem, alterum
tetradoron, tertium pentadoron. Graeci enim antiqui δῶρον palmum vocabant et ideo δῶρα munera,
quia manu darentur; ergo a quattuor et quinque
palmis, prout sunt, nominantur. eadem est et lati-
tudo. minore privatis operibus, maiore in publicis
utuntur in Graecia. Pitanae in Asia et in ulteriore
Hispania civitatibus Maxilua et Callet [5] fiunt lateres,
qui siccati non merguntur in aqua. sunt enim e
terra pumicosa, cum subigi potest, utilissima.

172 Graeci, praeterquam ubi e silice fieri poterat struc-
tura, latericios parietes praetulere. sunt enim
aeterni, si ad perpendiculum fiant. ideo et publica
opera et regias domos sic struxere: murum Athenis,
qui ad montem Hymettum spectat, Patris aedes
Iovis et Herculis, quamvis lapideas columnas et
epistylia circumdarent, domum Trallibus regiam
Attali, item Sardibus Croesi, quam gerusian fecere,

---

[1] quia *Mayhoff.*
[2] fiunt *Mayhoff coll. Vitruv.* II. 3 : qui *B* : que *rell.*
[3] didoron *Hermolaus Barbarus* : lydion (lidron *cd. Par.* 6801).
[4] nos *add. Mayhoff coll. Vitruv.* : volgo *J. Müller.*
[5] Callet *Urlichs, Detlefsen coll.* III. 12 : callent *B* : canlent *aut* canlento *aut* calento *rell.*

---

[a] As a measure this could be 4 inches or 9, but here it is 9.
[b] *Silex* is in particular the strong concrete made by the Romans from a lava mixed with lime and *pozzolana* (for which see § 166, note).
[c] Attalus I of Pergamum, 241–197 B.C.

is in spring, as at midsummer they tend to crack.
For buildings, only bricks two years old are recommended; moreover the material for them when it
has been pounded should be well soaked before they
are moulded.

Three kinds of bricks are made: the 'didoron,'
the one employed by us, eighteen inches long and
a foot wide, second the 'tetradoron' and third the
'pentadoron,' *doron* being an old Greek word meaning
the palm of the hand [a]—from which comes *doron*,
meaning a gift, because a gift was given by the
hand. Consequently the bricks get their names
from four or five palms' length as the case may be.
Their breadth is in all cases the same. In Greece
the smaller kind is used for private structures and
the larger in public buildings. At Pitana in Asia
Minor as also in the city states of Maxilua and
Callet in Further Spain bricks are made which when
dried will not sink in water, being made of pumice-
like earth, which is an extremely useful material
when it is capable of being worked. The Greeks
preferred brick walls except in places permitting
of a stone [b] structure, as brick walls last for ever
if built exactly perpendicular. Consequently that
was how they built both public works and kings'
palaces—the wall at Athens that faces towards
Mount Hymettus, at Patrae the Shrines of Zeus
and of Heracles (although the columns and archi-
traves with which they surrounded these were of
stone), and the royal palace of Attalus [c] at Tralles
and likewise the palace of Croesus at Sardis, which 560-546 B.C.
they converted into a house of elders,[d] and that of

---

[a] Pliny's source Vitruvius II. 8. 10 takes γερουσία here as a
home for the aged, but it must mean council-house.

Halicarnasi Mausoli, quae etiam nunc durant.
173 Lacedaemone quidem latericiis parietibus excisum
opus tectorium propter excellentiam picturae ligneis
formis inclusum Romam deportavere in aedilitate
ad comitium exornandum Murena et Varro. cum
opus per se mirum esset, tralatum tamen magis
mirabantur. in Italia quoque latericius murus
Arreti et Mevaniae est. Romae non fiunt talia
aedificia, quia sesquipedalis paries non plus quam
unam contignationem tolerat, cautumque est, ne
communis crassior fiat, nec intergerivorum ratio
patitur.

174   L. Haec sint dicta de lateribus. in terrae autem
reliquis generibus vel maxime mira natura est
sulpuris, quo plurima domantur. nascitur in insulis
Aeoliis inter Siciliam et Italiam, quas ardere diximus,
sed nobilissimum in Melo insula. in Italia quoque
invenitur in Neapolitano Campanoque agro collibus,
qui vocantur Leucogaei. ibi e cuniculis effossum
175 perficitur igni. genera IIII: vivum, quod Graeci
apyron vocant, nascitur solidum[1] solum[2]—cetera[3]
enim liquore constant et conficiuntur oleo incocta—;
vivum effoditur tralucetque et viret. solo ex
omnibus generibus medici utuntur.[4] alterum genus
appellant glaebam, fullonum tantum officinis fa-

---

[1] solidum hoc est glaeba *cdd.* : *del.* h.e.g. *Urlichs, Detlefsen.*
[2] solum *del. Urlichs, Detlefsen.*
[3] cetera *cd. Leid. Voss. m.*2 : ex omnibus generibus (*om.* B) medici utuntur. alterum genus cetera *B, cd. Leid. Voss. m.*1, *cd. Flor. Ricc.* : *vide infra* : ex omnibus . . . utuntur cetera *rell.*
[4] solo . . . utuntur *om. cd. Par. Lat.* 6797, *cd. Par.* 6801, *cd. Tolet.*

Mausolus[a] at Halicarnassus, buildings still standing. 377–353 B.C.
Murena and Varro in their aedileship had some
plaster work on brick walls at Sparta cut away,
and because of the excellence of its painting had
it enclosed in wooden frames and brought to Rome
to decorate the Assembly-place. It was in itself a
wonderful piece of work, yet its transfer caused even
more admiration. In Italy also there is a brick wall
at Arezzo and at Mevania. Structures of this sort
are not erected in Rome, because an eighteen-inch
wall will only carry a single storey, and there is a
regulation forbidding any partition exceeding that
thickness: nor does the system used for party-walls
permit of it.

L. Let this be what we say about bricks. Among *Sulphur.*
the other kinds of earth the one with the most remarkable properties is sulphur, which exercises a great
power over a great many other substances. Sulphur
occurs in the Aeolian Islands between Sicily and
Italy, which we have said are volcanic, but the most III, 92 ff.
famous is on the island of Melos. It is also found in
Italy, in the territory of Naples and Campania, on
the hills called the Leucogaei. It is there dug out of
mine-shafts and dressed with fire. There are four
kinds: live sulphur, the Greek name for which
means 'untouched by fire,' which alone forms as a
solid mass—for all the other sorts consist of liquid
and are prepared by boiling in oil; live sulphur is
dug up, and it is translucent and of a green colour;
it is the only one of all the kinds that is employed
by doctors. The second kind is called 'clod-
sulphur,' and is commonly found only in fullers'

---

[a] The remains of his monument the Mausoleum were brought
to England in 1859.

miliare. tertio quoque generi unus tantum est usus ad lanas suffiendas,[1] quoniam candorem mollitiamque confert. egula vocatur hoc genus, quartum autem[2] ad ellychnia maxime conficienda; cetero tantum[3] vis est ut morbos comitiales deprehendat nidore inpositum igni. lusit et Anaxilaus eo, addens in calicem vini prunaque subdita circumferens, exardescentis repercussu[4] pallorem dirum velut
176 defunctorum effundente in[5] conviviis.[6] natura eius excalfacit, concoquit, sed et discutit collectiones corporum, ob hoc talibus[7] emplastris malagmatisque miscetur. renibus quoque et lumbis in dolore cum adipe mire prodest inpositum. aufert et lichenas faciei cum terebinthi resina et lepras; harpax ita vocatur a celeritate praebendi,[8] avelli enim subinde
177 debet. prodest et suspiriosis linctu,[9] purulenta quoque extussientibus et contra scorpionum ictus. vitiligines vivum nitro mixtum atque ex aceto tritum et inlitum tollit, item lendes, et in palpebris aceto sandaracato admixtum. habet et in religionibus locum ad expiandas suffitu domos. sentitur vis eius et in aquis ferventibus, neque alia res facilius accen-

---

[1] suffiendas *coll. Isid. Gelen* : sufficiendas.
[2] autem *cd. Par.* 6801 : caute *rell.* (cate *cd. Leid. Voss.*: aptum *Isid.*) : *coni.* καντήρ *Mayhoff.*
[3] tanta *cd. Par.* 6801 *ex Isid.* : tamen *vel* vis tantum *coni. Mayhoff.*
[4] repercussu *edd. vett. ex Isid.* : percussu *B* : supercussu *aut* se percussu *rell.*
[5] effundente in *Mayhoff* : effundentem *B* : effundente *rell.*
[6] convivis *B.*
[7] albis *Fröhner.*
[8] praebendi *B* : uellendi *cd. Par.* 6801 : praeuelli *rell.* avellendi *edd. vett.* : prendendi *Ian.*
[9] linctu *Sillig* : linctum *aut* linctus *aut* lictus *cdd.* (unctu *B*).

workshops. The third kind also is only employed for one purpose, for smoking woollens from beneath, as it bestows whiteness and softness; this sort is called egula. The fourth kind is specially used for making lamp-wicks. For the rest, sulphur is so potent that when put on the fire it detects epilepsy by its smell. Anaxilaus even made a sport with it by putting some in a cup of wine and placing a hot coal underneath and handing it round at dinner-parties, when by its reflection it threw on their faces a dreadful pallor as though they were dead. Its property is calorific and concoctive,[a] but it also disperses abscesses on the body, and consequently is used as an ingredient in plasters and poultices for such cases. It is also remarkably beneficial for the kidneys and loins if in cases of pain it is applied to them with grease. In combination with turpentine it also removes lichenous growths on the face and leprosy; so it is called harpax,[b] owing to the speed with which it has to be applied, which is caused by the need for immediate removal. Used as an electuary it is good for cases of asthma, and also purulent expectoration after coughing and as a remedy for the sting of scorpions. Live sulphur mixed with soda and pounded in vinegar and used as a liniment removes cutaneous eruptions, and also eggs of lice, and in combination with vinegar mixed with realgar it is useful on the eyelids. Sulphur also has a place in religious ceremonies, for the purpose of purifying houses by fumigation. Its potency is also perceptible in hot springs of water, and no other substance is more easily ignited,

---

[a] I.e. brings boils, etc., to a head.
[b] ἅρπαξ ' rapacious,' from ἁρπάζω, ' seize,' ' snatch.'

ditur, quo apparet ignium vim magnam ei inesse.
fulmina, fulgura quoque sulpuris odorem habent,
ac lux ipsa eorum sulpurea est.

178   LI. Et bituminis vicina natura est. aliubi limus,
aliubi terra est, limus e Iudaeae lacu, ut diximus,
emergens, terra in Syria circa Sidonem oppidum
maritimum. spissantur haec utraque et in densi-
tatem coeunt. est vero liquidum bitumen, sicut
Zacynthium et quod a Babylone invehitur; ibi
quidem et candidum gignitur. liquidum est et
Apolloniaticum, quae omnia Graeci pissasphalton
179 appellant ex argumento picis ac bituminis. gignitur
et pingue oleique liquoris in Sicilia Agragantino
fonte, inficiens rivum. incolae id harundinum pani-
culis colligunt, citissime sic adhaerescens, utunturque
eo ad lucernarum lumina olei vice, item ad scabiem
iumentorum. sunt qui et naphtham, de qua in
secundo diximus volume, bituminis generibus adscribant, verum eius ardens natura et ignium cognata
180 procul ab omni usu abest. bituminis probatio ut
quam maxime splendeat sitque ponderosum, graveolens;[1] atrum[2] modice, quoniam adulteratur pice.
vis quae sulpuri: sistit, discutit, contrahit, glutinat.
serpentes accensum nidore fugat. ad suffusiones

---

[1] graveolens *Mayhoff coll. Diosc.* : grave leve (lene *cd. Par.*
6801) : graveolens, leve *Külb.*

[2] atrum *Mayhoff coll. Diosc.* : autem.

---

[a] This occurs as a liquid (petroleum), as a liquid solid
(mineral pitch and tar) and as a solid (asphalt).

[b] The Dead Sea.

showing that it contains a powerful abundance of fire. Thunderbolts and lightning also have a smell of sulphur, and their actual light has a sulphurous quality.

LI. Near to the nature of sulphur is also that of bitumen.[a] In some places it is a slime and others an earth, the slime being emitted, as we have said, from the lake[b] of Judaea and the earth being found in the neighbourhood of the seaside town of Sidon in Syria. Both of these varieties get thickened and solidify into a dense consistency. But there is also a liquid sort of bitumen, for instance that of Zacynthus and the kind imported from Babylon; at the latter place indeed it also occurs with a white colour. The bitumen from Apollonia also is liquid, and all of these varieties are called by the Greeks *pissasphalt*, from its likeness to vegetable-pitch and bitumen. There is also an unctuous bitumen, of the consistency of oil, found in Sicily, in a spring at Girgenti, the stream from which is tainted by it. The inhabitants collect it on tufts of reeds, as it very quickly adheres to them, and they use it instead of oil for burning in lamps, and also as a cure for scab in beasts of burden. Some authorities also include among the varieties of bitumen naphtha about which we spoke in Book II, but its burning property and liability to ignition is far removed from any practical use. The test of bitumen is that it should be extremely brilliant, and that it should be massive, with an oppressive smell; when quite black, its brilliance is moderate, as it is commonly adulterated with vegetable pitch. Its medical effect is that of sulphur, as it is astringent, dispersive, contractive, and agglutinating. Ignited it drives away snakes

Bitumen.

V, 72.

II, 235.

393

o 2

oculorum et albugines Babylonium efficax traditur,
item ad lepras, lichenas pruritusque corporum.
inlinitur et podagris. omnia autem eius genera
incommodos oculorum pilos replicant, dentium
181 doloribus medentur simul nitro intrito. lenit[1]
tussim veterem et anhelitus cum vino potum; dysintericis etiam datur eodem modo sistitque alvum.
cum aceto vero potum discutit concretum sanguinem
ac detrahit. mitigat lumborum dolores, item articu-
lorum, cum farina hordeacia inpositum emplastrum
peculiare facit suo nomine. sanguinem sistit, volnera colligit, glutinat nervos. utuntur etiam ad
quartanas bituminis drachma et hedyosmi pari
182 pondere cum murrae obolo subacti. comitiales
morbos ustum deprendit. volvarum strangulationes
olfactu discutit cum vino et castoreo, procidentes
suffitu reprimit, purgationes feminarum in vino
potum elicit. in reliquo usu aeramentis inlinitur
firmatque ea contra ignes. diximus et tingui solitum
aes eo statuasque inlini. calcis quoque usum prae-
buit ita feruminatis Babylonis muris. placet in
ferrariis fabrorum officinis tinguendo ferro clavorum
capitibus et multis aliis usibus.
183   LII. Nec minor est aut adeo dissimilis aluminis
opera, quod intellegitur salsugo terrae. plura et

---

[1] intrito. lenit *Mayhoff coll. Diosc.* : inlitum (*Sillig*) lenit
Detlefsen : inlitus *B* : illitum *aut* illini *aut* inlini *rell.*

---

[a] Several astringent substances were included in the word
alumen, especially, it seems, aluminium sulphates, sulphate of
iron, and common potash-alum; also kalinite, and perhaps
also certain halotrichites (K. C. Bailey, *The Elder Pliny's
Chapters on Chemical Subjects*, II, p. 233).

by its smell. Babylonian bitumen is said to be
serviceable for cataract and film in the eye, and also
for leprosy lichen and itch. It is also used as a
liniment for gout; while all varieties of it are used
to fold back eyelashes that get in the way of sight,
and also to cure toothache, when smeared on with
soda. Taken as a draught with wine it alleviates
an inveterate cough and shortness of breath; and it
is also given in the same way in cases of dysentery,
and arrests diarrhoea. Drunk however with vinegar
it dissolves and brings away coagulated blood.
It reduces pains in the loins and also in the joints,
and applied with barley-meal it makes a special
kind of plaster that bears its name. It stops a flow of
blood, closes up wounds, and unites severed muscles.
It is employed also for quartan fevers, the dose
being a dram of bitumen and an equal weight of
wild mint pounded up with a sixth of a dram of
myrrh. Burnt bitumen detects cases of epilepsy,
and mixed with wine and beaver-oil its scent dissipates suffocations of the womb; its smoke when
applied from beneath relieves prolapsus of the
womb; and drunk in wine it hastens menstruation.
Among other uses of it, it is applied as a coating to
copper and bronze vessels to make them fireproof.
We have stated that it also used to be the practice XXXIV, 15.
to employ it for staining copper and bronze and
coating statues. It has also been used as a substitute
for lime, the walls of Babylon being cemented with it.
In smithies also it is in favour for varnishing iron and
the heads of nails and many other uses.

LII. Not less important or very different is the *Alum.*
use made of alum,[a] by which is meant a salt exudation from the earth. There are several varieties of

eius genera. in Cypro candidum et nigrius, exigua
coloris [1] differentia, cum sit usus magna,[2] quoniam
inficiendis claro colore lanis candidum liquidumque
utilissimum est contraque fuscis aut obscuris nigrum.
184 et aurum nigro purgatur. fit autem omne ex aqua
limoque, hoc est terrae exudantis natura. con-
rivatum hieme aestivis solibus maturatur. quod
fuit ex eo praecox, candidius fit. gignitur autem in
Hispania, Aegypto, Armenia, Macedonia, Ponto,
Africa, insulis Sardinia, Melo, Lipara, Strongyle,
laudatissimum in Aegypto, proximum in Melo.
huius quoque duae species, liquidum spissumque.
liquidi probatio ut sit limpidum lacteumque, sine
offensis fricandi, cum quodam igniculo coloris.[3] hoc
phorimon [4] vocant. an sit adulteratum, deprehenditur suco Punici mali; sincerum enim mixtura ea
non nigrescit.[5] alterum genus est pallidi et scabri
et quod inficiatur et [6] galla, ideoque hoc vocant

[1] coloribusque *B* : coloris visusque *coni. Mayhoff.*
[2] magna *Gelen* : magni.
[3] caloris *edd. vett.*
[4] phorimon *edd. vett. coll. Galen.*, κατὰ τόπους, vi. 3 : porth-
mon *aut* portmon (*B*) *aut* pontinon *aut* posthonon.
[5] mixtura ea non nigrescit *K. C. Bailey* : mixtura ea
nigrescit *cd. Flor. Ricc. ut videtur* : mixturam fugit *cd. Par.*
6801 : mixtura *rell.* : mixtura inficitur *coni. Mayhoff.*
[6] et *Mayhoff* : a *B* : om. *rell.*

---

[a] Sulphate of aluminium would be useful for dyeing;
potash-alum and alunogen could provide the bright colour,
and alums containing metals the sombre colours (K. C. Bailey).
[b] Cf. XXXIII, 65; also for removing baleful influences of
gold held above the head, cf. XXXIII, 84.
[c] Where potash-alum is found.

it. In Cyprus there is a white alum and another sort of a darker colour, though the difference of colour is only slight; nevertheless the use made of them is very different, as the white and liquid kind is most useful for dying woollens a bright colour whereas the black kind is best for dark or sombre hues.[a] Black alum is also used in cleaning[b] gold. All alum is produced from water and slime, that is, a substance exuded by the earth; this collects naturally in a hollow in winter and its maturity by crystallisation is completed by the sunshine of summer; the part of it that separates earliest is whiter in colour. It occurs in Spain, Egypt, Armenia, Macedonia, Pontus, Africa, and the islands Sardinia, Melos, Lipari[c] and Stromboli; the most highly valued is in Egypt and the next best in Melos. The alum of Melos also is of two kinds, fluid[d] and dense. The test of the fluid kind is that it should be of a limpid, milky consistency, free from grit when rubbed between the fingers, and giving a slight glow of colour[e]; this kind is called in Greek 'phorimon' in the sense of 'abundant.' Its adulteration can be detected[f] by means of the juice of a pomegranate, as this mixed with it does not turn it black if it is pure. The other kind[g] is the pale rough alum which may be stained with oak-gall also, and consequently this is called 'paraphoron,'

[d] Apparently the solid kind (potash-alum especially) in solution.

[e] So MSS.; *caloris* ('heat') is a change based on what is probably a corruption in the text of Dioscorides.

[f] That is, an alum supposedly free from iron would, if it contained iron, turn juice of pomegranate black.

[g] Probably light yellow halotrichite (hydrated iron sulphate with aluminium) and green vitriol (ferrous sulphate).

185 paraphoron. liquidi aluminis vis adstringere, indurare, rodere. melle[1] admixto sanat oris ulcera, papulas pruritusque. haec curatio fit in balneis ii mellis partibus, tertia aluminis. virus alarum sudorisque sedat. sumitur pilulis contra lienis vitia pellendumque per urinam sanguinem. emendat et scabiem nitro ac melanthio admixtis.

186 Concreti aluminis unum genus σχιστὸν appellant Graeci, in capillamenta quaedam canescentia dehiscens, unde quidam trichitim potius appellavere. hoc fit e lapide, ex quo et aes—chalcitim vocant—, ut[2] sudor quidam eius lapidis in spumam coagulatus. hoc genus aluminis minus siccat minusque sistit umorem inutilem corporum, et auribus magnopere prodest infusum, vel inlitum et oris ulceribus dentibusque et si[3] saliva cum eo contineatur. et oculorum medicamentis inseritur apte verendisque utriusque sexus. coquitur in catinis,[4] donec liquari

187 desinat. inertioris est alterum generis, quod strongylen vocant. duae est eius species, fungosum atque omni umore dilui facile, quod in totum damnatur. melius pumicosum et foraminum fistulis spongeae simile rotundumque natura, candido propius, cum quadam pinguitudine, sine harenis, friabile, nec inficiens nigritia. hoc coquitur per se carbonibus

[1] melli B : melli admixtum coni. Mayhoff.
[2] vocant—, ut Mayhoff : vocant ut sit edd. vett. : vocamus B : vocatus cd. Leid. Lips. : vocatur rell.
[3] et si Mayhoff : et is B[1] : et his B[2] : si cd. Par. 6801 : et rell.
[4] catinis B : patinis rell.

---

[a] The following medical uses are like the modern uses of potash-alum.

[b] Including potash-alum, halotrichite, etc.

[c] Both potash-alum and aluminium sulphate, if heated, melt, swell, and solidify into ' burnt alum.'

' perverted ' or adulterated alum. Liquid alum [a] has an astringent, hardening and corrosive property. Mixed with honey it cures ulcers in the mouth, pimples and eruptions; this treatment is carried out in baths containing two parts of honey to one of alum. It reduces odour from the armpits and perspiration. It is taken in pills against disorders of the spleen and discharge of blood in the urine. Mixed with soda and chamomile it is also a remedy for scabies.

One kind [b] of solid alum which is called in Greek schiston, ' splittable,' splits into a sort of filament of a whitish colour, owing to which some people have preferred to give it in Greek the name of trichitis, ' hairy alum.' This is produced from the same ore as copper, known as copperstone, a sort of sweat from that mineral, coagulated into foam. This kind of alum has less drying effect and serves less to arrest the detrimental humours of the body, but it is extremely beneficial as an ear-wash, or as a liniment also for ulcers of the mouth and for the teeth, and if it is retained in the mouth with saliva; or it forms a suitable ingredient in medicines for the eyes and for the genital organs of either sex. It is roasted in crucibles until it has quite lost its liquidity.[c] There is another alum of a less active kind, called in Greek strongyle, ' round alum.' Of this also there are two varieties, the fungous which dissolves easily in any liquid and which is rejected as entirely worthless, and a better kind which is porous and pierced with small holes like a sponge and of a round formation, nearer white in colour, possessing a certain quality of unctuousness, free from grit, friable, and not apt to cause a black stain. This is roasted by itself on

188 puris, donec cinis fiat. Optimum ex omnibus quod
Melinum vocant ab insula, ut diximus. nulli vis
maior neque adstringendi neque denigrandi neque
indurandi, nullum spissius. oculorum scabritias
extenuat, combustum utilius epiphoris inhibendis,
sic et ad pruritus corporis. sanguinem quoque
sistit intus potum,[1] foris inlitum. evulsis pilis ex
aceto inlinitur renascentesque mollit in languinem.

189 summa [2] omnium generum vis in adstringendo, unde
nomen Graecis. ob id oculorum vitiis aptissima
sunt, sanguinis fluctiones inhibent. cum adipe
putrescentia ulcerum compescit [3]—sic et infantium
ulcera [4] et hydropicorum eruptiones siccat—et
aurium vitia cum suco Punici mali et unguium scabritias cicatricumque duritias et pterygia ac perniones, phagedaenas ulcerum ex aceto aut cum galla
pari pondere cremata, lepras cum suco olerum, cum
salis vero II partibus vitia, quae serpunt, lendes et

190 alia capillorum animalia aquae [5] permixtum. sic et
ambustis prodest et furfuribus corporum cum sero
picis. infunditur et dysintericis uvamque in ore
comprimit ac tonsillas. ad omnia, quae in ceteris

[1] potum *Sillig, Ian* : totum *aut* tutum.
[2] *V.l.* summam.
[3] *Post* compescit *del.* cum adipe *K. C. Bailey*.
[4] sic . . . ulcera *supra ante* putrescentia *cd. Par.* 6801.
[5] aquae *coni. Ian, Sillig* : aque *B²* : atque *B¹* : quae *cd. Leid. Lips.* : que *rell.*

---

[a] In § 184 Pliny implies that the best is the Egyptian.
[b] Στυπτηρία.

clean hot coals till it is reduced to ash. The best [a]
of all kinds is that called Melos alum, after the
island of that name, as we said; no other kind has a §184.
greater power of acting as an astringent, giving a
black stain and hardening, and none other has a closer
consistency. It removes granulations of the eyes,
and is still more efficacious in arresting defluxions
when calcined, and in that state also it is applied to
itchings on the body. Taken as a draft or applied
externally it also arrests haemorrhage. It is applied
in vinegar to parts from which the hair has been
removed and changes into soft down the hair that
grows in its place. The chief property of all kinds
of alum is their astringent effect, which gives it its
name [b] in Greek. This makes them extremely
suitable for eye troubles, and effective in arresting
haemorrhage. Mixed with lard it checks the spread
of putrid ulcers—so applied it also dries ulcers in
infants and eruptions in cases of dropsy—and, mixed
with pomegranate juice, it checks ear troubles and
malformations of the nails and hardening of scars,
and flesh growing over the nails, and chilblains.
Calcined with vinegar or gallnuts to an equal weight
it heals gangrenous ulcers, and, if mixed with cabbage
juice, pruritus, or if with twice the quantity of salt,
serpiginous eruptions, and if thoroughly mixed with
water, it kills eggs of lice and other insects that
infest the hair. Used in the same way it is also
good for burns, and mixed with watery fluid from
vegetable pitch for scurf on the body. It is also
used as an injection for dysentery, and taken in the
mouth it reduces swellings of the uvula and tonsils.
It must be understood that for all the purposes
which we have mentioned in the case of the other

generibus diximus, efficacius intellegatur ex Melo
advectum. Ad [1] reliquos usus vitae in coriis lanisque
perficiendis quanti sit momenti, significatum est.

191   LIII. Ab his per se ad medicinam pertinentia
terrae genera tractabimus. Samiae II sunt, quae
collyrium et quae aster appellantur. prioris laus
ut recens sit ac lenissima [2] linguaeque glutinosa,
altera glaebosior [3]; candida utraque. uritur, lavatur.
sunt qui praeferant priorem. prosunt sanguinem
expuentibus; emplastrisque, quae siccandi causa
componuntur, oculorum quoque medicamentis
miscentur.

192   LIV. Eretria totidem differentias habet, namque
est alba et cinerea, quae praefertur in medicina.
probatur mollitia et quod, si aere perducatur,
violacium reddit colorem. vis et ratio eius in
medendo dicta est inter pigmenta.

193   LV. Lavatur omnis terra—in hoc enim loco dicemus
—perfusa aqua siccataque solibus, iterum ex aqua
trita ac reposita, donec considat et digeri possit in
pastillos. coquitur in calicibus crebro concussis.

194   LVI. Est in medicaminibus et Chia terra candicans.
effectus eius idem [4] qui Samiae; usus ad mulierum
maxime cutem. idem et Selinusiae. lactei coloris

---

[1] advectum. Ad K. C. Bailey : advectum nam ad cdd.
pro nam coni. iam Bailey.
[2] lenissima cdd. (lenis cd. Par. 6801) : levissima Detlefsen,
Urlichs : levis Hermolaus Barbarus.
[3] glaebosior Sillig : glebosior aut globosior.
[4] eius idem Mayhoff : eiusdem.

---

[a] Kaolinite or china-clay, which is sometimes found in
fan-shaped (star-like) arrangements of plates, but generally
in white, greyish, or yellowish masses (K. C. Bailey). The
latter would be those used for eye-salves.

kinds the alum imported from Melos is more efficacious. It has been indicated how important §183. it is for the other requirements of life in giving a finish to hides and woollens.

LIII. Next to these we will deal with the various *Uses of various earths.* kinds of earth which are connected with medicine. There are two sorts of Samos earth,[a] called collyrium, *Samian.* ' eye-salve,' and star-earth. The recommendation of the former is that it must be fresh and very soft and sticky to the tongue; the second is more lumpy; both are white in colour. The process is to calcine them and then to wash them. Some people prefer the former kind. They are beneficial for people spitting blood, and for plasters made up for drying purposes, and they are also used as an ingredient in medicines for the eyes.

LIV. Earth of Eretria [b] has the same number of *Eretrian.* varieties, as one is white and one ash-coloured, the latter preferred in medicine. It is tested by its softness and by its leaving a violet tint if rubbed on copper. Its efficacy and the method of using it as a medicine have been spoken of among the pigments. §38.

LV. All these earths—we will mention it in this place—are washed by having water poured over them and dried in the sun, and then after being put in water again ground up and left to stand, till they settle down and can be divided into tablets. They are boiled in cups that are repeatedly well shaken.

LVI. White earth [c] of Chios is also among *Chian and other earths.* medicaments; its effect is the same as that of Samos earth. It is specially used as a cosmetic for the skin of women, and Selinunte earth is used in the same way. The latter is of the colour of milk, and it

[b] Cf. §§ 30, 38.　　[c] Some kind of china-clay.

haec et aqua dilui celerrima[1]; eadem lacte diluta
tectoriorum albaria interpolantur. pnigitis[2] Ere-
triae simillima est, grandioribus tantum glaebis
glutinosaque. effectus eius idem qui Cimoliae,
infirmior tantum. bitumini simillima est ampelitis.[3]
experimentum eius, si cerae modo accepto oleo
liquescat et si nigricans colos maneat tostae. usus
ad molliendum discutiendumque, et ad haec medicamentis additur, praecipue in calliblepharis et
inficiendis capillis.

195   LVII. Cretae plura genera. ex iis Cimoliae duo
ad medicos pertinentia, candidum et ad purpurissum
inclinans. vis utrique ad discutiendos tumores,
sistendas fluctiones aceto adsumpto. panos quoque
et parotidas cohibet et lienem inlita pusulasque, si
vero aphronitrum et cyprum[4] adiciatur et acetum,
pedum tumores ita, ut in sole curatio haec fiat et
196 post VI horas aqua salsa abluatur. testium tu-
moribus cypro et cera addita prodest. et refri-
gerandi quoque natura cretae est, sudoresque
immodicos sistit inlita atque ita papulas cohibet
ex vino adsumpta in balineis. laudatur maxime
Thessalica. nascitur et in Lycia circa Bubonem,
Est et alius Cimoliae usus in vestibus. nam Sarda

---

[1] celerrima *edd. vett.*: ceterrima *aut* ceterum (teterrima
cd. Par. Lat. 6797).

[2] pnigitis *Hermolaus Barbarus*: phinicis *aut* pnitis *aut sim.*

[3] ampelitis *Hermolaus Barbarus*: appellitis.

[4] cyprium *cd. Par.* 6801: cyprus *Brotier*: nitrum *Gelen.*

---

[a] The word means any fullers' earths, here particularly
calcium montmorillonite from the island Argentiera or Cimolo
in the Aegean.

[b] ἀφρόνιτρον, more properly ἀφρὸς νίτρον, ' foam of soda ';
probably pure soda or possibly partly causticised soda, whereas
ordinary *nitrum* was carbonate of soda.

[c] Obtained from the flowers of *Lawsonia alba.*

dissolves very quickly in water, and likewise dissolved in milk it is used for touching up the whitewash on plastered walls. Pnigitis, or 'suffocating' earth closely resembles that of Eretria, only it is in larger lumps and is sticky. It produces the same effect as Cimolian earth, although it is less powerful. Ampelitis or 'vine' earth is very like bitumen. The test for it is whether it dissolves when oil is put in it, like wax, and whether when roasted it retains a blackish colour. It is used for an emollient and dissipant, and is added to drugs for these purposes, especially in the case of eye-lash beautifiers and for hair dyes.

Asphaltic deposit

LVII. There are several sorts of white earth. Among them there are two sorts of Cimolian earth [a] that concern doctors, one bright white and one inclining to purple. Either is effective for dispelling tumours, and, with vinegar added, for stopping fluxes. They also check swellings and inflammation of the parotid glands, and applied as a liniment, troubles of the spleen and pimples; while if foam-soda [b] and oil of cypros [c] and vinegar are added, they cure swollen feet, provided the treatment is applied in the sun, and the application is washed off again with salt water six hours later. A mixture of this earth with oil of cypros and wax is good for swellings of the testicles. Cretaceous earth also possesses cooling properties, and applied in a liniment it stops immoderate sweating, and likewise taken in wine while in a bath it removes pimples. The kind from Thessaly is most esteemed, but it is also found in the neighbourhood of Bubo in Lycia. Another use also made of Cimolus earth is in regard to cloth. The kind called Sarda, which is brought

Cimolian earths.

quae adfertur e Sardinia, candidis tantum adsumitur,
inutilis versicoloribus, et[1] est vilissima omnium
Cimoliae generum; pretiosior Umbrica et quam
197 vocant saxum. proprietas saxi quod crescit in
macerando; itaque[2] pondere emitur, illa mensura.
Umbrica non nisi poliendis vestibus adsumitur.
neque enim pigebit hanc quoque partem adtingere,
cum lex Metilia extet fullonibus dicta, quam C.
Flaminius L. Aemilius censores dedere ad populum
198 ferendam. adeo omnia maioribus curae fuere.
ergo ordo hic est: primum abluitur vestis Sarda,
dein sulpure suffitur, mox desquamatur Cimolia
quae est coloris veri. fucatus enim deprehenditur
nigrescitque et funditur sulpure, veros autem et
pretiosos colores emollit Cimolia et quodam nitore
exhilarat contristatos sulpure. candidis vestibus
saxum utilius a sulpure, inimicum coloribus. Graecia
pro Cimolia Tymphaico[3] utitur gypso.

199 LVIII. Alia creta argentaria appellatur nitorem
argento reddens, set vilissima qua circum praeducere
ad victoriae notam pedesque venalium trans maria
advectorum denotare instituerunt maiores; talemque

---

[1] ea *coni. Mayhoff.*   [2] itaque *Mayhoff*: atque.
[3] Tymphaico *Hermolaus Barbarus coll. Theophr.* : tympaigo
B : tympauco *rell.*

---

[a] Sarda would be strong calcium montmorillonite; Umbrian
earth, some kaolinite; and *saxum*, bentonite. Cf. R. H. S.
Robertson, *Class. Rev.*, LXIII, 51–3. K. C. Bailey thinks
saxum is quicklime.   [b] Cf. § 44.

from Sardinia, is only used for white fabrics, and is of no use for cloths of various colours; it is the cheapest of all the Cimolus kinds; more valuable are the Umbrian and the one called ' rock.' [a] The peculiarity of the latter is that it increases in size when it is steeped in liquid; consequently it is sold by weight, whereas Umbrian is sold by measure. Umbrian earth is only employed for giving lustre to cloths. It will not be out of place to touch on this part of the subject also, as a Metilian law referring to fullers still stands, the law which Gaius Flaminius and Lucius Aemilius as censors put forward 220 B.C. to be carried in parliament: so careful about everything were our ancestors. The process then is this: the cloth is first washed with earth of Sardinia, and then it is fumigated with sulphur, and afterwards scoured with Cimolian earth provided that the dye is fast; if it is coloured with bad dye it is detected and turns black and its colour is spread by the action of the sulphur; whereas genuine and valuable colours are softened and brightened up with a sort of brilliance by Cimolian earth when they have been made sombre by the sulphur. The ' rock ' kind is more serviceable for white garments, after the application of sulphur, but it is very detrimental to colour. In Greece they use Tymphaea gypsum instead of Cimolian earth.

LVIII. There is another cretaceous earth [b] called *Silversmiths' earth.* silversmiths' powder as used for polishing silver; but the most inferior kind is the one which our ancestors made it the practice to use for tracing the line indicating victory in circus-races and for marking the feet of slaves on sale that had been imported from over-seas; instances of these being

Publilium Antiochium,[1] mimicae scaenae condi-
torem, et astrologiae consobrinum eius Manilium
Antiochum, item grammaticae Staberium Erotem
200 eadem nave advectos videre proavi. sed quid hos
referat aliquis, litterarum honore commendatos?
talem in catasta videre Chrysogonum Sullae, Am-
phionem Q. Catuli, Hectorem [2] L. Luculli, Demetrium Pompei, Augenque Demetri, quamquam et
ipsa Pompei credita est, Hipparchum M. Antoni,
Menam et Menecraten Sexti Pompei aliosque
deinceps, quos enumerare iam non est, sanguine
201 Quiritium et proscriptionum licentia ditatos. hoc
est insigne venaliciis gregibus obprobriumque inso-
lentis fortunae. quos et nos adeo potiri rerum
vidimus, ut praetoria quoque ornamenta decerni a
senatu iubente Agrippina Claudi Caesaris videremus
tantumque non cum laureatis fascibus remitti illo,
unde cretatis pedibus advenissent.

---

[1] Antiochium *O. Jahn* : lucilium *cd. Par.* 6801 : lochium
rell.
[2] Hectorem *Urlichs, Detlefsen* : interfectorem *cd. Par.
Lat.* 6797, *cd. Tolet.* : rectorem *rell.* : interfectorem, Heronem
edd. vett. : Heronem *ed. Basil.*

---

[a] This would be Publilius Syrus, fl. *c.* 45 B.C.
[b] Probably father or grandfather of Manilius who wrote the
extant *Astronomica.*
[c] Teacher of Brutus and Cassius.
[d] From the period 80–30 B.C.
[e] Demetrius of Gadara whose native city, destroyed by the
Jews, was rebuilt by Pompey at Demetrius' request.

Publilius of Antioch [a] the founder of our mimic stage and his cousin Manilius [b] Antiochus the originator of our astronomy, and likewise Staberius Eros [c] our first grammarian, all of whom our ancestors saw brought over in the same ship. But why need anybody mention these men, recommended to notice as they are by their literary honours? Other instances [d] that have been seen on the stand in the slave market are Chrysogonus freedman of Sulla, Amphion freedman of Quintus Catulus, Hector freedman of Lucius Lucullus, Demetrius [e] freedman of Pompey, and Auge freedwoman of Demetrius, although she herself also was believed to have belonged to Pompey; Hipparchus freedman of Mark Antony, Menas [f] and Menecrates freedmen of Sextus Pompeius, and a list of others whom this is not the occasion to enumerate, who [g] have enriched themselves by the bloodshed of Roman citizens and by the licence of the proscriptions.[h] Such is the mark set on these herds of slaves for sale, and the disgrace attached to us by capricious fortune!— persons whom even we have seen risen to such power that we actually beheld the honour of the praetorship awarded to them by decree of the Senate at the bidding of Claudius Caesar's wife Agrippina,[i] and all but sent back with the rods of office wreathed in laurels to the places from which they came to Rome with their feet whitened with white earth ![j]

[f] Admiral of Sextus Pompeius *c.* 40 B.C. He deserted twice to Octavian. Hipparchus likewise deserted to Octavian. Menecrates killed himself after ill success under Menas against Octavian's fleet, 38 B.C.

[g] Especially Chrysogonus and perhaps Hipparchus.

[h] By Sulla in 82 and by Antony, Octavian, Lepidus in 43 B.C.

[i] She married Claudius in A.D. 49.   [j] See § 199.

202   LIX. Praeterea sunt genera terrae proprietatis suae, de quibus iam diximus, sed et hoc loco reddenda natura : ex Galata insula et circa Clupeam Africae scorpiones necat, Baliaris et Ebusitana serpentes.

NOTE ON THE PAINTERS NAMED ARISTIDES.

It would appear that an elder Aristides (XXXV. 75, 108, 111, and 122 ?—the statuary of XXXIV. 50 and 72 may be the same) had as pupils his sons Nicomachus (XXXV. 108, 109), Niceros (111) and Ariston (110, 111), and two others (not sons), namely Euphranor (111, 128) and Antorides (111).

LIX. Moreover there are other kinds of earth with a special property of their own about which we have spoken already, but the nature of which must again be stated here: soil taken from the island of Galata and in the neighbourhood of Clupea in Africa kills scorpions, and that of the Balearic Islands and Iviza is fatal to snakes.

III, 78.
V, 42.

Note however that the reading *Aristidis* in XXXV. 108 is uncertain and that Nicomachus is not mentioned in 111. Nicomachus had a son and pupil the younger Aristides (A. of Thebes 98–100, 110) who was thus grandson of A. the elder. The younger is named also in XXXV. 24, and VII. 126. Pliny shows some confusion of the two.

INDEX OF ARTISTS

413

INDEX OF ARTISTS

INDEX OF ARTISTS

415

INDEX OF ARTISTS

MUSEOGRAPHIC INDEX

417

MUSEOGRAPHIC INDEX

INDEX OF MINERALS

419

INDEX OF MINERALS

INDEX OF MINERALS

PRINTED IN GREAT BRITAIN BY
RICHARD CLAY AND COMPANY, LTD.,
BUNGAY, SUFFOLK.

THE LOEB CLASSICAL LIBRARY

VOLUMES ALREADY PUBLISHED

Latin Authors

AMMIANUS MARCELLINUS. Translated by J. C. Rolfe. 3 Vols. (*2nd Imp. revised.*)

APULEIUS : THE GOLDEN ASS (METAMORPHOSES). W. Adlington (1566). Revised by S. Gaselee. (*7th Imp.*)

ST. AUGUSTINE, CONFESSIONS OF. W. Watts (1631). 2 Vols. (Vol. I. *7th Imp.*, Vol. II. *6th Imp.*)

ST. AUGUSTINE, SELECT LETTERS. J. H. Baxter.

AUSONIUS. H. G. Evelyn White. 2 Vols. (*2nd Imp.*)

BEDE. J. E. King. 2 Vols.

BOETHIUS : TRACTS and DE CONSOLATIONE PHILOSOPHIAE. Rev. H. F. Stewart and E. K. Rand. (*4th Imp.*)

CAESAR : CIVIL WARS. A. G. Peskett. (*5th Imp.*)

CAESAR : GALLIC WAR. H. J. Edwards. (*9th Imp.*)

CATO : DE RE RUSTICA ; and VARRO : DE RE RUSTICA. H. B. Ash and W. D. Hooper. (*2nd Imp.*)

CATULLUS. F. W. Cornish; TIBULLUS. J. B. Postgate; and PERVIGILIUM VENERIS. J. W. Mackail. (*12th Imp.*)

CELSUS : DE MEDICINA. W. G. Spencer. 3 Vols. (Vol. I. *3rd Imp. revised.*)

CICERO : BRUTUS, and ORATOR. G. L. Hendrickson and H. M. Hubbell. (*3rd Imp.*)

CICERO: DE FATO ; PARADOXA STOICORUM ; DE PARTITIONE ORATORIA. H. Rackham. (With De Oratore, Vol. II.) (*2nd Imp.*)

CICERO : DE FINIBUS. H. Rackham. (*4th Imp. revised.*)

CICERO : DE INVENTIONE, etc. H. M. Hubbell.

CICERO : DE NATURA DEORUM and ACADEMICA. H. Rackham. (*2nd Imp.*)

CICERO : DE OFFICIIS. Walter Miller. (*4th Imp.*)

CICERO : DE ORATORE. 2 Vols. E. W. Sutton and H. Rackham. (*2nd Imp.*)

CICERO : DE REPUBLICA and DE LEGIBUS. Clinton W. Keyes. (*3rd Imp.*)

CICERO : DE SENECTUTE, DE AMICITIA, DE DIVINATIONE. W. A. Falconer. (*5th Imp.*)

CICERO : IN CATILINAM, PRO FLACCO, PRO MURENA, PRO SULLA. Louis E. Lord. (*2nd Imp. revised.*)

CICERO : LETTERS TO ATTICUS. E. O. Winstedt. 3 Vols. (Vol. I. *6th Imp.*, Vols. II. and III. *3rd Imp.*)

CICERO : LETTERS TO HIS FRIENDS. W. Glynn Williams. 3 Vols. (Vols. I. and II. *3rd* Imp., Vol. III. *2nd Imp. revised.*)

CICERO : PHILIPPICS. W. C. A. Ker. (3rd Imp. revised.)
CICERO : PRO ARCHIA, POST REDITUM, DE DOMO, DE HARUS-
PICUM RESPONSIS, PRO PLANCIO. N. H. Watts. (2nd Imp.)
CICERO : PRO CAECINA, PRO LEGE MANILIA, PRO CLUENTIO,
PRO RABIRIO. H. Grose Hodge. (4th Imp.)
CICERO : PRO MILONE, IN PISONEM, PRO SCAURO, PRO FONTEIO,
PRO RABIRIO POSTUMO, PRO MARCELLO, PRO LIGARIO, PRO
REGE DEIOTARO. N. H. Watts. (2nd Imp.)
CICERO : PRO QUINCTIO, PRO ROSCIO AMERINO, PRO ROSCIO
COMOEDO, CONTRA RULLUM. J. H. Freese. (2nd Imp.)
CICERO : TUSCULAN DISPUTATIONS. J. E. King. (3rd Imp.)
CICERO : VERRINE ORATIONS. L. H. G. Greenwood. 2 Vols.
(Vol. I. 2nd Imp.)
CLAUDIAN. M. Platnauer. 2 Vols.
COLUMELLA : DE RE RUSTICA. H. B. Ash. 3 Vols. (Vol. I.
2nd Imp.)
CURTIUS, Q. : HISTORY OF ALEXANDER. J. C. Rolfe. 2 Vols.
FLORUS. E. S. Forster, and CORNELIUS NEPOS. J. C. Rolfe.
(2nd Imp.)
FRONTINUS : STRATAGEMS and AQUEDUCTS. C. E. Bennett and
M. B. McElwain. (2nd Imp.)
FRONTO : CORRESPONDENCE. C. R. Haines. 2 Vols.
GELLIUS. J. C. Rolfe. 3 Vols. (2nd Imp.)
HORACE : ODES and EPODES. C. E. Bennett. (13th Imp. revised.)
HORACE : SATIRES, EPISTLES, ARS POETICA. H. R. Fairclough.
(6th Imp. revised.)
JEROME : SELECTED LETTERS. F. A. Wright.
JUVENAL and PERSIUS. G. G. Ramsay. (7th Imp.)
LIVY. B. O. Foster, F. G. Moore, Evan T. Sage, and A. C.
Schlesinger. 14 Vols. Vols. I.–XII. (Vol. I. 3rd Imp.,
Vols. II.–VI., VII., IX.–XII., 2nd Imp. revised.)
LUCAN. J. D. Duff. (3rd Imp.)
LUCRETIUS. W. H. D. Rouse. (6th Imp. revised.)
MARTIAL, W. G. A. Ker. 2 Vols. (Vol. I. 5th Imp., Vol. II.
4th Imp. revised.)
MINOR LATIN POETS : from PUBLILIUS SYRUS to RUTILIUS
NAMATIANUS, including GRATTIUS, CALPURNIUS SICULUS,
NEMESIANUS, AVIANUS, and others with " Aetna " and the
" Phoenix." J. Wight Duff and Arnold M. Duff. (2nd Imp.)
OVID : THE ART OF LOVE AND OTHER POEMS. J. H. Mozley.
(3rd Imp.)
OVID : FASTI. Sir James G. Frazer. (2nd Imp.)
OVID : HEROIDES and AMORES. Grant Showerman. (4th Imp.)
OVID : METAMORPHOSES. F. J. Miller. 2 Vols. (Vol. I. 9th
Imp., Vol. II. 7th Imp.)
OVID : TRISTIA and EX PONTO. A. L. Wheeler. (2nd Imp.)
PERSIUS. Cf. JUVENAL.
PETRONIUS. M. Heseltine ; SENECA : APOCOLOCYNTOSIS.
W. H. D. Rouse. (8th Imp. revised.)
PLAUTUS. Paul Nixon. 5 Vols. (Vols. I. and II. 5th Imp., Vol.
III. 4th Imp., Vols. IV. and V. 2nd Imp.)

2

PLINY : LETTERS. Melmoth's Translation revised by W. M. L. Hutchinson. 2 Vols. (Vol. I. 6th Imp., Vol. II. 4th Imp.)

PLINY : NATURAL HISTORY. H. Rackham and W. H. S. Jones. 10 Vols. Vols. I.–V. H. Rackham. Vol. VI. W. H. S. Jones. (Vol. I. 3rd Imp., Vols. II. and III. 2nd Imp.)

PROPERTIUS. H. E. Butler. (5th Imp.)

PRUDENTIUS. H. J. Thomson. 2 Vols. Vol. I.

QUINTILIAN. H. E. Butler. 4 Vols. (2nd Imp.)

REMAINS OF OLD LATIN. E. H. Warmington. 4 Vols. Vol. I. (ENNIUS AND CAECILIUS.) Vol. II. (LIVIUS, NAEVIUS, PACUVIUS, ACCIUS.) Vol. III. (LUCILIUS and LAWS OF XII TABLES.) Vol. IV. (2nd Imp.) (ARCHAIC INSCRIPTIONS.)

SALLUST. J. C. Rolfe. (3rd Imp. revised.)

SCRIPTORES HISTORIAE AUGUSTAE. D. Magie. 3 Vols. (Vol. I. 2nd Imp. revised.)

SENECA : APOCOLOCYNTOSIS. Cf. PETRONIUS.

SENECA : EPISTULAE MORALES. R. M. Gummere. 3 Vols. (Vol. I. 3rd Imp., Vols. II. and III. 2nd Imp. revised.)

SENECA : MORAL ESSAYS. J. W. Basore. 3 Vols. (Vol. II. 3rd Imp., Vol. III. 2nd Imp. revised.)

SENECA : TRAGEDIES. F. J. Miller. 2 Vols. (Vol. I. 3rd Imp., Vol. II. 2nd Imp. revised.)

SIDONIUS : POEMS and LETTERS. W. B. Anderson. 2 Vols. Vol. I.

SILIUS ITALICUS. J. D. Duff. 2 Vols. (Vol. I. 2nd Imp., Vol. II. 3rd Imp.)

STATIUS. J. H. Mozley. 2 Vols.

SUETONIUS. J. C. Rolfe. 2 Vols. (Vol. I. 7th Imp., Vol. II. 6th Imp. revised.)

TACITUS : DIALOGUS. Sir Wm. Peterson. AGRICOLA and GERMANIA. Maurice Hutton. (6th Imp.)

TACITUS : HISTORIES and ANNALS. C. H. Moore and J. Jackson. 4 Vols. (Vols. I and II. 3rd Imp., Vols. III. and IV. 2nd Imp.)

TERENCE. John Sargeaunt. 2 Vols. (Vol. I. 6th Imp., Vol. II. 5th Imp.)

TERTULLIAN : APOLOGIA and DE SPECTACULIS. T. R. Glover. MINUCIUS FELIX. G. H. Rendall.

VALERIUS FLACCUS. J. H. Mozley. (2nd Imp. revised.)

VARRO : DE LINGUA LATINA. R. G. Kent. 2 Vols. (2nd Imp. revised.)

VELLEIUS PATERCULUS and RES GESTAE DIVI AUGUSTI. F. W. Shipley.

VIRGIL. H. R. Fairclough. 2 Vols. (Vol. I. 17th Imp., Vol. II. 13th Imp. revised.)

VITRUVIUS : DE ARCHITECTURA. F. Granger. 2 Vols. (Vol. I. 2nd Imp.)

Greek Authors

ACHILLES TATIUS. S. Gaselee. (*2nd Imp.*)

AENEAS TACTICUS, ASCLEPIODOTUS and ONASANDER. **The** Illinois Greek Club. (*2nd Imp.*)

AESCHINES. C. D. Adams. (*2nd Imp.*)

AESCHYLUS. H. Weir Smyth. 2 Vols. (Vol. I. *5th Imp.*, Vol. II. *4th Imp.*)

ALCIPHRON, AELIAN, PHILOSTRATUS LETTERS. A. R. Benner and F. H. Fobes.

ANDOCIDES, ANTIPHON. Cf. MINOR ATTIC ORATORS.

APOLLODORUS. Sir James G. Frazer. 2 Vols. (*2nd Imp.*)

APOLLONIUS RHODIUS. R. C. Seaton. (*4th Imp.*)

THE APOSTOLIC FATHERS. Kirsopp Lake. 2 Vols. (*7th Imp.*)

APPIAN: ROMAN HISTORY. Horace White. 4 Vols. (Vol. I. *3rd Imp.*, Vols. II., III. and IV. *2nd Imp.*)

ARATUS. Cf. CALLIMACHUS.

ARISTOPHANES. Benjamin Bickley Rogers. 3 Vols. Verse trans. (Vols. I. and II. *5th Imp.*, Vol. III. *4th Imp.*)

ARISTOTLE: ART OF RHETORIC. J. H. Freese. (*3rd Imp.*)

ARISTOTLE: ATHENIAN CONSTITUTION, EUDEMIAN ETHICS, VICES AND VIRTUES. H. Rackham. (*2nd Imp.*)

ARISTOTLE: GENERATION OF ANIMALS. A. L. Peck. (*2nd Imp.*)

ARISTOTLE: METAPHYSICS. H. Tredennick. 2 Vols. (*3rd Imp.*)

ARISTOTLE: MINOR WORKS. W. S. Hett. On Colours, On Things Heard, On Physiognomies, On Plants, On Marvellous Things Heard, Mechanical Problems, On Indivisible Lines, On Situations and Names of Winds, On Melissus, Xenophanes, and Gorgias.

ARISTOTLE: NICOMACHEAN ETHICS. H. Rackham. (*5th Imp. revised.*)

ARISTOTLE: OECONOMICA and MAGNA MORALIA. G. C. Armstrong; (with Metaphysics, Vol. II.). (*3rd Imp.*)

ARISTOTLE: ON THE HEAVENS. W. K. C. Guthrie. (*2nd Imp. revised.*)

ARISTOTLE: ON THE SOUL, PARVA NATURALIA, ON BREATH. W. S. Hett. (*2nd Imp. revised.*)

ARISTOTLE: ORGANON. H. P. Cooke and H. Tredennick. 3 Vols. (Vol. I. *2nd Imp.*)

ARISTOTLE: PARTS OF ANIMALS. A. L. Peck; MOTION AND PROGRESSION OF ANIMALS. E. S. Forster. (*2nd Imp. revised.*)

ARISTOTLE: PHYSICS. Rev. P. Wicksteed and F. M. Cornford. 2 Vols. (*2nd Imp.*)

ARISTOTLE: POETICS and LONGINUS. W. Hamilton Fyfe; DEMETRIUS ON STYLE. W. Rhys Roberts. (*4th Imp. revised.*)

ARISTOTLE: POLITICS. H. Rackham. (*4th Imp. revised.*)

ARISTOTLE: PROBLEMS. W. S. Hett. 2 Vols. (Vol. I. *2nd Imp. revised.*)

4

ARISTOTLE : RHETORICA AD ALEXANDRUM (with PROBLEMS, Vol. II.). H. Rackham.
ARRIAN : HISTORY OF ALEXANDER and INDICA. Rev. E. Iliffe Robson. 2 Vols. (2nd Imp.)
ATHENAEUS : DEIPNOSOPHISTAE. C. B. Gulick. 7 Vols. (Vols. I., V., and VI. 2nd Imp.)
ST. BASIL : LETTERS. R. J. Deferrari. 4 Vols. (Vols. I., II. and IV. 2nd Imp.)
CALLIMACHUS and LYCOPHRON. A. W. Mair; ARATUS. G. R. Mair. (2nd Imp.)
CLEMENT OF ALEXANDRIA. Rev. G. W. Butterworth. (2nd Imp.)
COLLUTHUS. Cf. OPPIAN.
DAPHNIS AND CHLOE. Thornley's Translation revised by J. M. Edmonds ; and PARTHENIUS. S. Gaselee. (3rd Imp.)
DEMOSTHENES I: OLYNTHIACS, PHILIPPICS and MINOR ORATIONS: I.–XVII. AND XX. J. H. Vince.
DEMOSTHENES II : DE CORONA and DE FALSA LEGATIONE. C. A. Vince and J. H. Vince. (2nd Imp. revised.)
DEMOSTHENES III : MEIDIAS, ANDROTION, ARISTOCRATES, TIMO-CRATES and ARISTOGEITON, I. AND II. J. H. Vince.
DEMOSTHENES IV–VI : PRIVATE ORATIONS and IN NEAERAM. A. T. Murray. (Vol. I. 2nd Imp.)
DEMONSTHENES VII: FUNERAL SPEECH, EROTIC ESSAY, EXORDIA and LETTERS. N. W. and N. J. DeWitt.
DIO CASSIUS : ROMAN HISTORY. E. Cary. 9 Vols. (Vols. I. and II. 2nd Imp.)
DIO CHRYSOSTOM. J. W. Cohoon and H. Lamar Crosby. 5 Vols. (Vols. I. and II. 2nd Imp.)
DIODORUS SICULUS. 12 Vols. Vols. I.–V. C. H. Oldfather. Vol. IX. R. M. Geer. (Vol. I. 2nd Imp.)
DIOGENES LAERTIUS. R. D. Hicks. 2 Vols. (Vol. I. 4th Imp., Vol. II. 3rd Imp.)
DIONYSIUS OF HALICARNASSUS : ROMAN ANTIQUITIES. Spelman's translation revised by E. Cary. 7 Vols. (Vols. I. and IV. 2nd Imp.)
EPICTETUS. W. A. Oldfather. 2 Vols. (Vols. I and II. 2nd Imp.)
EURIPIDES. A. S. Way. 4 Vols. (Vol. I. 7th Imp. and II., IV. 6th Imp., Vol. III. 5th Imp.) Verse trans.
EUSEBIUS : ECCLESIASTICAL HISTORY. Kirsopp Lake and J. E. L. Oulton. 2 Vols. (Vol. I. 2nd Imp., Vol. II. 3rd Imp.)
GALEN : ON THE NATURAL FACULTIES. A. J. Brock. (3rd Imp.)
THE GREEK ANTHOLOGY. W. R. Paton. 5 Vols. (Vols. I. and II. 4th Imp., Vols. III. and IV. 3rd Imp., Vol. V. 2nd Imp.)
GREEK ELEGY AND IAMBUS with the ANACREONTEA. J. M. Edmonds. 2 Vols. (Vol. I. 2nd Imp.)
THE GREEK BUCOLIC POETS (THEOCRITUS, BION, MOSCHUS). J. M. Edmonds. (7th Imp. revised.)
GREEK MATHEMATICAL WORKS. Ivor Thomas. 2 Vols. (2nd Imp.)

HERODES. Cf. THEOPHRASTUS : CHARACTERS.

HERODOTUS. A. D. Godley. 4 Vols. (Vols. I.–III. 4th Imp., Vol. IV. 3rd Imp.)

HESIOD and THE HOMERIC HYMNS. H. G. Evelyn White. (7th Imp. revised and enlarged.)

HIPPOCRATES and the FRAGMENTS OF HERACLEITUS. W. H. S. Jones and E. T. Withington. 4 Vols. (Vol. I. & II. 3rd Imp., Vols. III. & IV. 2nd Imp.)

HOMER : ILIAD. A. T. Murray. 2 Vols. (6th Imp.)

HOMER : ODYSSEY. A. T. Murray. 2 Vols. (7th Imp.)

ISAEUS. E. W. Forster. (2nd Imp.)

ISOCRATES. George Norlin and LaRue Van Hook. 3 Vols.

ST. JOHN DAMASCENE : BARLAAM AND IOASAPH. Rev. G. R. Woodward and Harold Mattingly. (2nd Imp. revised.)

JOSEPHUS. H. St. J. Thackeray and Ralph Marcus. 9 Vols. Vols. I.–VII. (Vol. V. 3rd Imp., Vol. VI. 2nd Imp.)

JULIAN. Wilmer Cave Wright. 3 Vols. (Vol. I. 2nd Imp., Vol. II. 3rd Imp.)

LUCIAN. A. M. Harmon. 8 Vols. Vols. I.–V. (Vols. I–III. 3rd Imp.)

LYCOPHRON. Cf. CALLIMACHUS.

LYRA GRAECA. J. M. Edmonds. 3 Vols. (Vol. I. 4th Imp., Vol. II. 2nd Ed. revised and enlarged, Vol. III. 4th Imp. revised.)

LYSIAS. W. R. M. Lamb. (2nd Imp.)

MANETHO. W. G. Waddell : PTOLEMY : TETRABIBLOS. F. E. Robbins. (2nd Imp.)

MARCUS AURELIUS. C. R. Haines. (3rd Imp. revised.)

MENANDER. F. G. Allinson. (2nd Imp. revised.)

MINOR ATTIC ORATORS (ANTIPHON, ANDOCIDES, DEMADES, DEINARCHUS, HYPEREIDES). K. J. Maidment and J. O. Burrt. 2 Vols. Vol. I. K. J. Maidment.

NONNOS : DIONYSIACA. W. H. D. Rouse. 3 Vols. (Vol. III. 2nd Imp.)

OPPIAN, COLLUTHUS, TRYPHIODORUS. A. W. Mair.

PAPYRI. NON-LITERARY SELECTIONS. A. S. Hunt and C. C. Edgar. 2 Vols. (Vol. I. 2nd Imp.) LITERARY SELECTIONS. Vol. I. (Poetry). D. L. Page. (3rd Imp.)

PARTHENIUS. Cf. DAPHNIS AND CHLOE.

PAUSANIAS : DESCRIPTION OF GREECE. W. H. S. Jones. 5 Vols. and Companion Vol. arranged by R. E. Wycherley. (Vols. I. and III. 2nd Imp.)

PHILO. 11 Vols. Vols. I.–V.; F. H. Colson and Rev. G. H. Whitaker. Vols. VI.–IX.; F. H. Colson. (Vols. I., II., V., VI. and VII. 2nd Imp., Vol. IV. 3rd Imp. revised.)

PHILOSTRATUS : THE LIFE OF APOLLONIUS OF TYANA. F. C. Conybeare. 2 Vols. (Vol. I. 4th Imp., Vol. II, 3rd Imp.)

PHILOSTRATUS : IMAGINES ; CALLISTRATUS : DESCRIPTIONS. A. Fairbanks.

PHILOSTRATUS and EUNAPIUS : LIVES OF THE SOPHISTS. Wilmer Cave Wright. (2nd Imp.)

PINDAR. Sir J. E. Sandys. (*7th Imp. revised.*)

PLATO : CHARMIDES, ALCIBIADES, HIPPARCHUS, THE LOVERS, THEAGES, MINOS and EPINOMIS. W. R. M. Lamb.

PLATO : CRATYLUS, PARMENIDES, GREATER HIPPIAS, LESSER HIPPIAS. H. N. Fowler. (*2nd Imp.*)

PLATO : EUTHYPHRO, APOLOGY, CRITO, PHAEDO, PHAEDRUS. H. N. Fowler. (*9th Imp.*)

PLATO : LACHES, PROTAGORAS, MENO, EUTHYDEMUS. W. R. M. Lamb. (*3rd Imp. revised.*)

PLATO : LAWS. Rev. R. G. Bury. 2 Vols. (*3rd Imp.*)

PLATO : LYSIS, SYMPOSIUM, GORGIAS. W. R. M. Lamb. (*4th Imp. revised.*)

PLATO : REPUBLIC. Paul Shorey. 2 Vols. (Vol. I. *4th Imp.*, Vol. II. *3rd Imp.*)

PLATO : STATESMAN, PHILEBUS. H. N. Fowler; ION. W. R. M. Lamb. (*4th Imp.*)

PLATO : THEAETETUS and SOPHIST. H. N. Fowler. (*4th Imp.*)

PLATO : TIMAEUS, CRITIAS, CLITOPHO, MENEXENUS, EPISTULAE. Rev. R. G. Bury. (*3rd Imp.*)

PLUTARCH : MORALIA. 14 Vols. Vols. I.–V. F. C. Babbitt; Vol. VI. W. C. Helmbold; Vol. X. H. N. Fowler. (Vols. I., III., and X. *2nd Imp.*)

PLUTARCH : THE PARALLEL LIVES. B. Perrin. 11 Vols. (Vols. I., II., and VII. *3rd Imp.*, Vols. III., IV., VI., and VIII.–XI. *2nd Imp.*)

POLYBIUS. W. R. Paton. 6 Vols.

PROCOPIUS : HISTORY OF THE WARS. H. B. Dewing. 7 Vols. (Vol. I. *2nd Imp.*)

PTOLEMY : TETRABIBLOS. Cf. MANETHO.

QUINTUS SMYRNAEUS. A. S. Way. Verse trans. (*2nd Imp.*)

SEXTUS EMPIRICUS. Rev. R. G. Bury. 4 Vols. (Vol. I. and III. *2nd Imp.*)

SOPHOCLES. F. Storr. 2 Vols. (Vol. I. *8th Imp.*, Vol. II. *6th Imp.*) Verse trans.

STRABO : GEOGRAPHY. Horace L. Jones. 8 Vols. (Vols. I. and VIII. *3rd Imp.*, Vols. II., V., and VI. *2nd Imp.*)

THEOPHRASTUS : CHARACTERS. J. M. Edmonds; HERODES, etc. A. D. Knox. (*2nd Imp.*)

THEOPHRASTUS : ENQUIRY INTO PLANTS. Sir Arthur Hort, Bart. 2 Vols. (*2nd Imp.*)

THUCYDIDES. C. F. Smith. 4 Vols. (Vol. I. *3rd Imp.*, Vols. II., III. and IV. *2nd Imp. revised.*)

TRYPHIODORUS. Cf. OPPIAN.

XENOPHON : CYROPAEDIA. Walter Miller. 2 Vols. (*3rd Imp.*)

XENOPHON : HELLENICA, ANABASIS, APOLOGY, and SYMPOSIUM. C. L. Brownson and O. J. Todd. 3 Vols. (Vols. I. and III. *3rd Imp.*, Vol. II. *4th Imp.*)

XENOPHON : MEMORABILIA and OECONOMICUS. E. C. Marchant. (*2nd Imp.*)

XENOPHON : SCRIPTA MINORA. E. C. Marchant. (*2nd Imp.*)

IN PREPARATION

Greek Authors

ARISTOTLE : DE MUNDO, ETC. D. Furley and E. M. Forster.
ARISTOTLE : HISTORY OF ANIMALS. A. L. Peck.
ARISTOTLE : METEOROLOGICA. H. D. P. Lee.
PLOTINUS : A. H. Armstrong.

Latin Authors

ST. AUGUSTINE : CITY OF GOD.
[CICERO] : AD HERENNIUM. H. Caplan.
CICERO : PRO SESTIO, IN VATINIUM, PRO CAELIO, DE PROVINCIIS
    CONSULARIBUS, PRO BALBO. J. H. Freese and R. Gardner.
PHAEDRUS. Ben E. Perry.

DESCRIPTIVE PROSPECTUS ON APPLICATION

London
Cambridge, Mass.

WILLIAM HEINEMANN LTD
HARVARD UNIVERSITY PRESS